The Manager's Guide to Statistics

Erol A. Peköz

2017 Edition

Published by ProbabilityBookstore.com, Boston, MA. No part of this publication may be reproduced, stored in a retrieval system, or transmitted in any form or by any means, electronic, mechanical, photocopying, recording, scanning, or otherwise, without the prior written permission of the Publisher. Requests to the Publisher for permission should be addressed to sales@probabilitybookstore.com.

Printed in the United States of America.

To order give your college bookstore the information below:

The Manager's Guide to Statistics
by Erol A. Peköz (2017)
Publisher: ProbabilityBookstore.com, Boston, MA

Download data sets at ProbabilityBookstore.com

4

Contents

Preface

This book is an introduction to statistics for someone who does not need to know all the details of statistical theory and would just like to know how statistics is commonly used in business. We cover some of the basic statistical tools, some pitfalls to watch out for when using them, and we give some intuitive explanations of how everything works. We try to explain everything in words rather than with Greek symbols or mathematical formulas. You'll also find the technical details and notation in technical notes so as a student you could be comfortable using other statistics books in the future.

It's difficult for students of statistics to learn, at the same time, the imposing mathematical notation and the subtle, fragile concepts behind the scenes. If a teacher tries to teach both at the same time, invariably attention paid by students to the concepts goes out the window. The student leaves the course with only a vague recollection of a jumble of Greek letters and of some frustrating time spent on the computer pointing and clicking and coping with error messages. This is why we make such an effort to teach the concepts first without the imposing notation and then include much of the notation and technical details as technical notes afterwards.

We recommend using Excel or R for the statistical calculations. Excel and R are ubiquitous in business and becoming more comfortable with them is of great value in itself. Though Excel is not the program of choice for professional statisticians, it is a program of choice for business people.

This book is designed to be the textbook for a one semester introductory course for undergraduate business students, MBA students, or other aspiring leaders and decision makers.

About the Author

Erol A. Peköz is a Professor in the Questrom School of Business at Boston University. He received his B.S. from Cornell University and Ph.D. in Operations Research from the University of California, Berkeley and has published a number of technical articles in applied probability and statistics. He also has taught at the University of California, Berkeley, the University of California, Los Angeles, and Harvard University. At Boston University he was awarded the 2001 Broderick Prize for Teaching, and is also a co-author of the textbook *A Second Course in Probability* (with Sheldon Ross).

Acknowledgements

I would like to thank Paul Berger, Andrew Gelman, Janelle Heineke, Barry Kadets, Mark Kean, Sheldon Ross, Michael Shwartz and Mustafa Yilmaz for their valuable contributions.

Chapter 1

Seeing the Real Story

Statistics are tools for seeing and telling a story from data. Computers can usually handle the number-crunching calculations we may need, but interpreting the output and seeing a story is up to us. Instead of starting off by teaching you how to do calculations, we will start with some advice on how to see the real story behind the numbers.

One theme throughout this book is that it is easy to be misled by statistics if you don't know what to watch out for. Even honestly gathered data may appear to be telling one story on the surface, when actually quite the opposite is the real story. But after this chapter you shouldn't just become skeptical of all of statistics. Being overly skeptical is just as bad as being overly gullible: both keep you from the real truth. Our goal for you is to become wise enough to know when to be skeptical and when to believe. In this chapter we will cover a few commonly encountered ways statistics tend to mislead people, so you will know what to watch out for.

1. Comparing Rates: Missing Denominators

Sometimes the ratio of two numbers tells more than either of the numbers do by themselves. Next are a few examples.

How many hospital beds does a city need? If your city has ten times more hospital beds than another city, does this mean your city probably has more beds than it needs? To start to answer this, we really need to know the number of people in each city. We should not just compare the **count** of beds, but should instead compare the ratio of the number of beds to the number of people in the city. We call this type of ratio a **rate**. For example, if a city of 200,000 people has 100 hospital beds, this corresponds to a rate of one hospital bed for every 2,000 people. Though assessing if a city has more beds than it needs is controversial and complex,[1] we should definitely start off by looking at rates instead of counts.

> Watch out if someone is comparing counts when they should instead be comparing rates.

Are you safer without a seat belt?   During the year 2002 in California, 1,524 people wearing seat belts were killed in car accidents but only 1,343 people without seat belts were killed in car accidents.[2] Do these numbers mean you are safer without a seat belt?

Solution.   No, you are not safer without a seat belt. Since most people wear seat belts, we would expect a large number of deaths among people wearing seat belts. To see the benefits of a seat belt you should compare the death rates for the people with and without seat belts. You could do this by dividing the number of deaths for each group by the total number of people (or the total number of accidents) for each group. This rate would turn out to be much higher for the group of people who weren't wearing seat belts. It's also true that many more people are killed riding in cars than riding motorcycles: only 318 motorcyclists were killed in 2002 in California. Even though motorcycles are more dangerous than cars (and the death rate turns out to be higher), there are fewer deaths because there are fewer motorcyclists.

Is New York City more dangerous than Iraq?   The death rate for Americans (that is, deaths per thousand Americans per year) in Iraq during the

---

[1]There is research that claims that the over-supply of hospital beds induces demand for them. A study by E. Peköz and M. Shwartz funded by the *Department of Health and Human Services, Agency for Healthcare Research and Quality* titled "Do More Hospital Beds in an Area Induce Excess Demand?" is investigating the evidence for this.

[2]See *2002 Annual Report of Fatal and Injury Motor Vehicle Traffic Collisions* at `http://www.chp.ca.gov/pdf/2002-sec4.pdf`, page 21.

war was actually lower than the death rate in New York City during the same time period.[3] Does this mean it was safer to be sent to Iraq than to New York City?

Solution.   No, Iraq was much more dangerous. It's true there were fewer American deaths per year in Iraq than in New York City during that time, and this is because there were more Americans in New York City than in Iraq. But that still doesn't explain why there were fewer deaths per thousand Americans in Iraq. The reason for this difference in rates is because Americans in Iraq had a different age range than people in New York City. In Iraq the Americans were primarily young healthy soldiers, while the New York City death rate included the sick and the elderly, groups that typically have high death rates. If you compared Americans in Iraq with people of the same age range in New York City you would find a much higher death rate in Iraq. This type of age difference is called a *confounding factor*; we will talk more about these types of factors in the next section.

Exercises

1. **Workplace safety.** Managers of a manufacturing plant keep track of on-the-job accidents. Last year there were 50 worker injuries that happened on the day shift and only five worker injuries that happened on the night shift. Does this mean the night shift was safer? Or are we comparing the wrong type of numbers?

2. **Automobile theft.** Insurance industry records show the Cadillac Escalade SUV has the highest theft rate of all cars in terms of the number of thefts per thousand vehicles. But in the same records if you just look at the total number of thefts, more Toyota Camrys are stolen each year than any other car—including Escalades.[4] The 1989 Camry, in particular, is the one most often stolen.

---

[3]The New York City mortality rate was around 700 per 100,000 people per year as measured in the year 2000 census. The article "Counting the Dead" by James Dunnigan posted on July 29, 2004 on strategypage.com reports the figure as 360 per 100,000 troops per year in Iraq. The yearly death rate in New York City for men 20-24 was 120 per 100,000 in 1999-2001 (it was only 40 per 100,000 for women, and for men over 85 it was 15,000 per 100,000). See `http://strategypage.com/dls/articles/200472922.asp`, and `www.nyc.gov/html/doh/pdf/vs/2002sum.pdf`.

[4]`http://www.auto-theft.info/Statistics.htm` and `http://money.cnn.com/2004/02/27/pf/autos/nicb_most_stolen/`

(a) How can you explain the difference here?

(b) If you are considering buying either a Camry or an Escalade, which car would be more likely to get stolen?

Which would a thief prefer?[5]

3. **Pharmacy errors.** A newspaper article reports about a dangerous surge in pharmacy prescription errors.[6] The article details how the vast majority of complaints to the Massachusetts Department of Public Health have been lodged against CVS Corporation's pharmacies. The article goes on to say this is particularly troublesome because CVS is one of the larger pharmacy chains in the state. Does this mean CVS pharmacies are having problems? Or are there different numbers we should be looking at?

2. Comparing Two Groups: Confounding Factors

It is typical for researchers to compare two groups of subjects they are studying and try to draw conclusions based on differences they see. To study the effectiveness of a drug, for example, medical researchers may compare outcomes for people who take the drug with people who don't take the drug. To test the effectiveness of an advertising campaign, market researchers may compare sales for two regions where two different advertising campaigns are used.

This may seem like a straightforward process, but it can be complicated by the presence of **confounding factors** (also called **confounders**). A confounding factor is an important difference between the two groups you are

---

[5]Photos from `http://www.geartekcorporation.com/dailyphoto/2005/toyotacamry.html`, and `http://www.cadillacforums.com/cadillac-models/cadillac-escalade.html`

[6]"Massachusetts pharmacist woes a prescription for peril," by Jessica Heslam, *The Boston Herald,* Thursday, July 14, 2005, page 2.

comparing, other than the one you're primarily interested in. Such a difference is important if it has a big impact on what you are measuring in the study.

In the example of the previous section where we compared Americans in New York City with Americans in Iraq, we were primarily interested in the locations of these two groups of people. The confounding factor—the other important difference—was the age difference between the two groups of people. Overlooking this confounding factor at first is what gave us the misleading conclusion that Iraq was safer than New York City. Of course the two groups also differed in that most Americans in Iraq were US government employees, but this difference is not as important as the age difference and probably would not be a confounding factor.

> A confounding factor is a characteristic that differs between two groups you are comparing, other than the one you're primarily interested in. This characteristic also should have a big impact on what you are measuring in the study.

How can you avoid confounding factors? One way is to recognize them and then try to arrange the groups you compare so you don't have confounding factors. For example suppose, as in the example of the previous section, the ages of people are very different in two groups you would like to compare and you believe that age will have an important impact on what you are primarily interested in. You can then divide the groups up into several smaller groups and then only compare groups of people with the same age. This process in this case is called **adjusting for** age or **controlling for** age.

Sometimes it can be difficult to adjust for confounding factors. For example, consider a study of the harmful health effects of smoking. Even though people who smoke tend to be less healthy than people who don't smoke, people who smoke as a whole tend to drink more than people who don't smoke. It's difficult to tell if it's smoking causing health problems, or if it's drinking. This type of study where researchers compare two naturally formed groups of people is called an **observational study**.

Another way to avoid confounding factors is to conduct a **randomized experimental study**. In a randomized experiment, researchers randomly divide subjects into a **treatment group** and a **control group** (a group that receives no treatment or an inactive **placebo** treatment such as a sugar pill). Then the researchers apply some form of treatment to the first group and compare the results of the two groups. Although there will always be some differences from person to person in any study, randomly dividing the subjects into two large groups usually fairly balances out any potential confounding factors so they don't cause problems.

Usually experimental studies are preferable to observational studies, but important decisions must very often be made on the basis of observational studies. Experimental studies can be too expensive or impractical to conduct. Human studies on smoking, for example, are inherently observational studies. In our country it is not possible to force people to smoke for twenty years just to compare their health to people who didn't smoke.

Is your company selling a dangerous product?   Shortly after a popular blood pressure drug entered the marketplace, a study of hospital records showed that people who had taken this drug subsequently had a higher rate of heart attacks than people who took other blood pressure drugs.[7] Did this mean the new drug was dangerous? Or was there a likely confounding factor to consider?

Discussion.   There was a confounding factor. This was a high blood pressure drug intended for the more serious blood-pressure cases so people taking this drug tended to have more heart problems even before they started taking the medication. The prior health history of the subjects in this study would therefore be an important confounding factor, since it was very different between the two groups. Randomized experiments showed that the drug in fact was safe.

Is the market for a drug secure?   During the 1990s Wyeth Pharmaceuticals was selling millions of hormone treatments to women annually. Large

---

[7]B. M. Psaty, S. R. Heckbert, T. D. Koepsell, D. S. Siscovick, T. E. Raghunathan, N. S. Weiss, F. R. Rosendaal, R. N. Lemaitre, N. L. Smith et. al., "The risk of myocardial infarction associated with antihypertensive drug therapies," *Journal of the American Medical Association*, Vol. 274 No. 8, August 23, 1995, and Robert M. Califf and Judith M. Kramer, "What Have We Learned From the Calcium Channel Blocker Controversy?" *Circulation*, 1998;97:1529–1531, and http://cumc.columbia.edu/news/journal/journal-o/archives/jour_v17n1_0009.html

observational studies over many years unquestionably showed that women who took these treatments after menopause had lower rates of heart disease than women of the same age who didn't take the treatments.[8] Pharmaceutical industry financial analysts believed the market for these hormones was secure, and the stock price of Wyeth was soaring. Then in 2002 a small randomized experimental study came out that surprisingly found that hormone therapy increased the risk of heart disease—the very problem it was supposed to prevent. Imagine you are an investor in Wyeth pharmaceuticals just learning this news. Should you trust the earlier large studies or should you rush to sell your stock?

Discussion. You should have rushed to sell your stock because the earlier studies had reached the wrong conclusion. Those studies were observational studies, which means that researchers just looked at the health records of people who had been prescribed hormone treatments by their doctors. These women weren't randomly selected for the treatment as they would be in an experimental study. Researchers now believe the type of women who sought out and stuck with hormone replacement therapy were the type of women who tended to get better health care overall, tended to lead more health-conscious lives and tended to be healthier to begin with than women who weren't taking the therapy or who didn't stick with it. Also, it is believed doctors were less likely to prescribe this therapy for women with risk factors for heart disease. The prior health of these women was a confounding factor, since it was very different between the two groups being compared. Although researchers at the time had tried, they weren't able to properly adjust for all these factors in the observational studies.

In the experimental study, researchers assigned volunteers randomly to different groups and so they avoided the confounding factors that existed in the observational studies. Researchers now see how the observational studies were very misleading and hormone replacement therapy is now not recommended for prevention of heart disease—though there are recent reports of benefits observed in some age groups that weren't well documented before.[9]

---

[8] "The WHI Estrogen Alone Trial—Do Things Look Any Better?," by S. Hully, and D. Grady, *Journal of the American Medical Association*, April 14th, 2004, volume 291, No. 14, Pages 1769–1771. See also Hersh AL, Stefanick ML, Stafford RS, "National use of postmenopausal hormone therapy: annual trends and response to recent evidence," *Journal of the American Medical Association*, 2004; 291: 47–53.

[9] "The Latest Wisdom on Hormone Therapy," by Katherine Hobson, 4/20/07, *US News and World Report*, http://www.usnews.com/usnews/health/articles/070420/

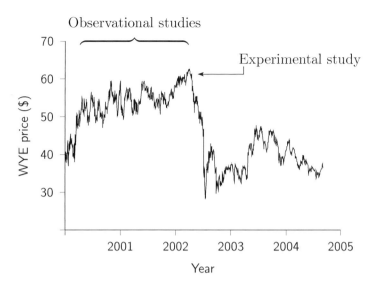

Figure 1.1. Wyeth pharmaceutical company's stock price plummets after an experimental study finds that hormone therapy – one of its most important products – increases the risk of heart disease.

This is an example of how a well-designed experimental study can overturn a conventional belief.

Incidentally, Wyeth's stock price (ticker symbol WYE) took a sharp nose dive shortly after the news of the experimental study came out (see Figure 1.1). This shows that as an investor or industry analyst, it pays to understand the difference between experimental studies and observational studies and the problems confounding factors can cause.

Can you have confounding factors in a randomized study? A study that looks at first as though it is experimental may actually be observational, and may be therefore susceptible to confounding factors. If any of the subjects in a study influence whether or not they are in the treatment group or the control group, the study is no longer experimental: it becomes an observational study. For example, researchers testing the effectiveness of a new drug with side effects may divide volunteers randomly into a treatment

20health.hormone.htm and "How NIH Misread Hormone Study in 2002," By Tara Parker-Pope, *The Wall Street Journal*, July 9, 2007, Page B1, http://online.wsj.com/article/SB118394176612760522.html

group and a control group. So far it looks as though the study is a randomized experiment, but almost always some of the volunteers drop out of the study in the middle. It may be tempting at the end of the study to ignore these dropouts, but this would introduce a confounding factor. If the drug has side effects, people who drop out are likely to be the ones who can't tolerate the side effects and so people who remain in the study are likely to be the strongest and healthiest members of the treatment group. If we allow subjects themselves to decide whether or not they drop out of the treatment group, the study becomes observational rather than experimental. In order to avoid this problem, the researchers should still consider people who drop out of the treatment group as part of the treatment group. The entire treatment group, including dropouts, should be compared with the entire control group, including dropouts. This way the study remains experimental and we will not have confounding factors.

When we detect an association between a treatment and a result it is tempting to conclude that the treatment causes the result. But in an observational study just because two things are associated with each other doesn't mean one causes the other. In fact, the causation may even go in the other direction. We will illustrate this next.

> Association is not always the same as causation.

Is it dangerous to quit a hazardous job?   Researchers studied high-voltage electrical utility workers and divided them into two groups: people currently on the job and people who had recently quit working.[10] It turned out that the first group was healthier than the second group. Should we conclude from this that, even though working near high-voltage wires may be hazardous, quitting such a job is even more hazardous?

Discussion.   We should first ask if the age or gender mix in the two groups was significantly different. Most likely the former workers were older than current workers, and this could be a confounding factor. To handle this difference, we can adjust for age by making a comparison for each age group separately. This means we compare young people in both groups, then com-

---

[10]Richardson et al., "Time Related Aspects of the Healthy Worker Survivor Effect," *Annals of Epidemiology,* 14:633-639, 2004.

pare middle-aged people in both groups, and so forth. Interestingly, the former workers were still less healthy even after adjusting for age.

In this study it turned out that many people quit their jobs because they were having health problems. This means poor health caused them to quit, and it was not quitting that caused the poor health. When it appears one thing may be causing another thing to happen, the causation may actually be running in reverse.

Incidentally, researchers also looked at the group of people who had quit their job more than five years ago. Surprisingly this group was healthier than both the other groups. Does this mean that high-voltage exposure could be good for you in the long run? No, unfortunately the sickest people who quit their job because of poor health did not survive five years. Those who survived had to have been extraordinarily healthy people to begin with. This was a way to effectively select out only the healthiest people, and is sometimes called **selection bias** or **survivor bias**. The next section gives more examples of this phenomenon.

Exercises

4. **Prison education programs.** Massachusetts is considering a new state-wide education program to better rehabilitate prisoners before their release with the hope of reducing the percentage who will be back in prison. This type of program has already been tried on a voluntary basis in one county where courses in anger management, addiction, and post-release planning were made available to inmates six months before their release. A newspaper article titled "Prison education programs cut rate of re-offending" details how inmates who participated in these courses were half as likely to re-offend as the ones who didn't participate.[11]

   (a) Is this study observational or experimental?

   (b) Do these results mean the counseling program is effective or is there an important confounding factor to consider here? Explain.

   (c) How could you conduct a better study where participation is still voluntary?

5. **Car alarms.** A newspaper article titled "Ban car alarms? They disrupt life & don't work" argues that, contrary to popular belief, car alarms

---

[11] "Prison education programs cut rate of reoffending," *The Boston Globe*, Tuesday, July 13, 2004, page A13, by Joseph McDonough.

do not effectively deter theft.[12] The article refers to a study that finds that car models that come with alarms installed are actually more likely to be stolen than models that don't come with alarms. Does this mean car alarms don't work or is there an important confounding factor to consider here?

6. **Corporate funded studies.** Many drug safety research studies are sponsored by pharmaceutical companies that would financially benefit if the results of the study are favorable.[13] Is this an example of a potential confounding factor?

7. **Antidepressants and suicide.** A high profile study finds that the use of antidepressants significantly increases the risk of suicidal behavior in adults under 25 while it decreases the risk of such behavior in older adults.[14] The study involved over 100,000 people and found that adults under 25 taking antidepressants were more than twice as likely as those taking a placebo to attempt, prepare for, or commit suicide. Someone criticizes the study saying that people who take antidepressants are generally more depressed to begin with compared with people who don't— and this could be a serious confounding factor. Is this valid criticism of the study?

8. **Study volunteers.** Since patients who volunteer to participate in a randomized experimental drug study tend to have health problems to begin with, they don't usually represent the general population. Is this an example of a confounding factor? Explain why or why not.

---

[12] "Ban car alarms? They disrupt life & don't work," by Aaron Naparstek, *New York Daily News*, Tuesday, July 8th, 2003.

[13] See "Financial ties to industry cloud depression study," Tuesday, July 11, 2006, by David Armstrong, *The Wall Street Journal* http://www.post-gazette.com/pg/06192/705022-114.stm and "JAMA says docs misled over industry ties," July 18, 2006, by Lindsey Tanner, The Associated Press. See http://www.abcnews4.com/news/stories/0706/345549.html

[14] "Study Finds Medication Raises Suicide Risks in Young Adults," by Benedict Carey, December 6, 2006, *The New York Times*, http://www.nytimes.com/2006/12/06/health/06drug.html?ex=1323061200&en=d727e889c3f935bc&ei=5090partner=rssuserland&emc=rss

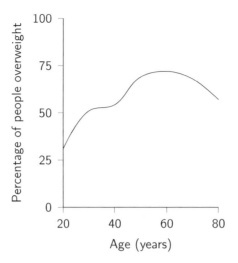

Figure 1.2. The percentage of people overweight versus age. Notice the decrease in the graph after age 60. Does this mean people are shaping up?

3. Selection Bias and Survivor Bias

Did you know that former astronauts have significantly fewer health problems than other people their age? Does this mean space travel is good for your health? Of course not, because one of the criteria for astronauts is that they should be in excellent health to begin with. This is an example of selection bias. Next are a few more examples.

Do people shape up as they age?   The percentage of people who were overweight was computed for different age groups in a large representative sample of people interviewed during a study in 1997.[15] As a graph of the results in Figure 1.2 shows, the percentage appears to rise until around age 60 and then starts to fall. Does this mean people tend to get in better shape after age 60?

Discussion.   People who are overweight generally do not live as long as people who are not overweight. This means the older people are less likely

---

[15]Data from Department of Health, Social Services and Public Safety, *1997 Northern Ireland Health and Social Wellbeing Survey.* See Table 5.5 in http://www.dhsspsni.gov. uk/publications/archived/2001/revised_health\%20_lifestyle_report.pdf

to be overweight. It may appear that people are getting in shape, but it's actually that the overweight people are dropping out of the sample. This form of selection bias is sometimes called **survivor bias**.

There is an important difference in the conclusions you can draw from **longitudinal** data, where subjects are followed over a long period of time, and **cross-sectional** data, where data are gathered at a single point in time. This study was cross-sectional. If the same subjects were instead followed over time, we would see that people do not tend to shape up after age 60. Cross-sectional data are usually much cheaper and faster to gather than longitudinal data, but one has to be cautious about drawing longitudinal-type conclusions from cross-sectional data. In some cases it can be reasonable to make longitudinal conclusions from cross-sectional data, but you should be very cautious when doing this.

> If a study claims things will happen over time, ask if the data are cross-sectional or longitudinal.

Mutual fund historical performance.   A mutual fund company shows the historical performance of all the funds they currently sell. Over the past five years every single one of these funds performed above the Standard & Poor's 500 index (S&P 500 index), a commonly used benchmark for investment performance. Does this mean that someone who bought a fund from this company five years ago would probably have beaten the S&P 500 index?

Discussion.   Mutual fund companies commonly eliminate, merge, or rename funds that don't perform well. The track records of these funds are erased from the databases that companies and the media generally use to measure performance. Funds that are currently advertised for sale (the "live" funds) are the ones that happened to have performed well. The unlucky people who invested in funds that were eliminated, merged, or renamed (the "dead" funds) very likely did not earn returns above the S&P 500. Eliminating these "dead" funds from the records creates a **survivor bias** and what people think of as the average mutual fund's return is greatly overstated. A recent *Wall Street Journal* article discusses how this bias can be upwards of several percentage points.[16]

---

[16] "Mutual Fund Math Puts a Sheen on Returns," by Ian McDonald, *The Wall Street Journal*, Friday Jul. 23rd, 2004, p. C1.

This survivor bias not only applies to people and mutual funds, but to publicly traded companies as well. When a company's stock performs very poorly, the company may itself go out of business. This means that the historical performance of currently traded stocks tends to look a lot better than for the average stock. Such survivor-biased historical data can also cause other more subtle problems when you use it in more complicated analyses.

> Survivor bias can appear in surprising places.

How difficult is it to find a job?   A university interested in knowing how difficult it is for recent graduates to find a job sends out a survey form to a random selection of graduates a few months after graduation. When they don't get many responses, they decide to keep sending out survey forms to more people until they get a reasonable number of responses. Is this a sensible approach for increasing the number of responses?

Solution.   No. Survey researchers have long realized that the type of people who take the time to fill out survey forms usually differ significantly from the type of people who don't, and only using their responses biases the survey. People who are poorly motivated will have difficulty finding a job and will also be less inclined to respond to a survey and, on the other hand, people who have recently started a new job may not have the time to fill out a survey form.[17] These are examples of what is called **non-response bias**. One way to reduce this bias is to contact the people who didn't initially respond either by telephone or repeated mailings or use other ways to induce them to respond. Sending out survey forms to new people simply recreates the same bias on a larger scale.

If someone tells you about the results of a survey they have conducted, don't just ask them how many responses they received. Also ask how many survey forms were sent out.

> A survey with a huge number of responses can still suffer from non-response bias.

---

[17]See Gerard J. van den Berg, Maarten Lindeboom & Peter J. Dolton, 2004, "Survey Non-response and Unemployment Duration," Tinbergen Institute Discussion Papers 04-094/3, Tinbergen Institute. `http://www.tinbergen.nl/discussionpapers/04094.pdf`

Initial public offering (IPO) brokerage firms.   A study of a developing country's stock exchange showed that companies employing prestigious American investment banking firms to handle their IPOs tended to be better respected by investors after the IPO and had stock prices that rose to higher levels. The researchers concluded from this that the prestige of these investment banks can create value even beyond the cost of their fee. Can you think of another explanation?

Discussion.   Their conclusion may be true, but this study doesn't effectively justify it. Companies that have money to hire a prestigious firm are certainly more likely to be in better financial shape than companies that don't have the money. Also, prestigious brokers are more likely to be willing to work with companies that are in better financial shape. Be cautious in concluding that prestigious investment banks can add value beyond the cost of their fee based on this study. An important confounding factor here is the financial strength of the company before the IPO; a strong company is more likely to be able to get a prestigious broker and is also more likely to be respected by investors on its own. It's also true that companies with expensive office furniture tend to be more successful than other companies, but it doesn't mean that the furniture itself is responsible. The successful companies are more likely to be able to afford such furniture.

Exercises

9. **Online banking.** A bank considering eliminating tellers and switching completely over to on-line banking services would like to know if its customers would welcome the idea. The bank conducts a survey on its web site and finds that almost everyone says they would welcome the idea. Does this mean the switch would be welcome or is there an important source of bias in this survey? Explain.

10. **Team-building and profitability.** A study shows that companies that send employees to expensive weekend team-building workshops in exotic locations tend to be the most profitable companies in their industries. Does this mean these programs work to help create a profitable company? Or is there a likely confounding factor to consider here? Explain.

11. **Hospital quality.** Massachusetts is considering a controversial proposal to publicize heart surgery survival rates for individual hospitals and

physicians so patients can be better informed consumers.[18] Insurance analysts profiling the performance of two hospitals find that the first hospital has a much higher survival rate for heart surgery than the second hospital, even though the second hospital has better equipment and more experienced and specialized surgeons. Does this mean the first hospital is doing a better job? Or is there a likely confounding factor to consider here? Explain.

4. Descriptive and Inferential Statistics

In this chapter we frequently made statements such as "a study shows that..." or "some data show that...," but we haven't really discussed yet how statistics can be used to draw such conclusions from data. The field of statistics can be divided into two broad categories: **descriptive** statistics and **inferential** statistics. Descriptive statistics covers methods for describing data you have gathered, and inferential statistics covers methods for using the data you have gathered to draw a conclusion about something unknown. For example, suppose we walk into a store and see only women customers. If we just reported we saw only women it could be classified as part of descriptive statistics. If we then used this to say we believe this store generally tends to have primarily women customers, it could be classified as part of inferential statistics. In this book we first study descriptive statistics and then we move on to inferential statistics.

5. Mathematical Summary

Mathematicians use special notation to explain statistical ideas and it's not always necessary to know the notation to understand the concepts. Sometimes, though, it can come in handy to know the notation when communicating technical details with other technical people. Here we summarize the concepts of this chapter using statistical notation.

Given a group of subjects and a factor A, let $X(i, A) = 1$ if subject number i has factor A and let $X(i, A) = 0$ otherwise. For example, three such factors

---

[18] "Heart surgery data may go public: State looking at patient death rates for doctors," By Liz Kowalczyk, *The Boston Globe*, July 4, 2006. http://www.boston.com/yourlife/health/diseases/articles/2006/07/04/heart_surgery_data_may_go_public/

for people could be smoking, drinking, and good health. Given factors A, B, and C let

$$P(A|B) = \frac{\sum_i X(i,A)X(i,B)}{\sum_i X(i,B)}$$

denote the fraction of subjects with factor B who also have factor A (read as "the probability of A given B"), and also let

$$P(A|BC) = \frac{\sum_i X(i,A)X(i,B)X(i,C)}{\sum_i X(i,B)X(i,C)}$$

denote the fraction of subjects with factor B and C who also have factor A (read as "the probability of A given B and C").

We say that C is a **confounding factor** for A and B if both $P(A|B) \neq P(A|BC)$ and $P(B|A) \neq P(B|AC)$ hold. This means the relation between A and B is different whether or not you take factor C into account. If C is a confounding factor, it's possible to see a situation where $P(A|B) < P(A) < P(A|BC)$, meaning the direction of the impact of B on A can be reversed by C.

6. Summary

Subtle sources of bias can leave an impressive-looking statistical analysis with major flaws. Being better able to recognize these situations will help you defend yourself from being misled. Some key concepts to have in mind are the following: comparing counts versus rates; distinguishing experimental and observational studies; identifying confounding factors; adjusting or controlling for a factor; recognizing association and causation, survivor bias, selection bias, and non-response bias.

7. Exercises

12. **On-the-job fatalities.** Government statistics for on-the-job fatalities in 2003 show there were 1,033 on-the-job deaths of construction workers as well as 1,388 transportation workers, 312 police and fire-fighters, 282 factory workers, and 630 management and executive-level workers.[19] Does

---

[19] http://www.bls.gov/news.release/cfoi.t03.htm

this mean management and executive jobs may be safer than construction jobs, but may actually be more dangerous than fighting fires or working in a factory? Or are we comparing the wrong numbers here?

13. **Dangerous occupations.** Below are some data from 2005 for on-the-job deaths in some dangerous jobs.[20] Which job seems the most dangerous? Which seems the least dangerous? Explain.

Occupation	Total deaths	Total employed
Driver/sales workers and truck drivers	993	3,412,370
Farmers and ranchers	341	829,680
Construction laborers	339	1,493,390
Miscellaneous agricultural workers	176	758,620
Aircraft pilots	81	121,070
Logging workers	80	86,110
Fishers and fishing workers	48	40,540
Electrical power line installers/repairers	36	110,090
Structural iron and steel workers	35	62,940
Refuse and recyclable material collectors	32	73,050

14. **Travel-related illness.** A newspaper article reports on a study of international travel-related illnesses, a topic of major interest in the tourism industry.[21] A diagram in the article, shown in Figure 1.3, gives the number of people who returned home with health problems after traveling to various regions of the developing world. Of the 15,704 cases of travel-related illnesses reported during the period of the study, 1,115 were from travel to the Caribbean, 1,326 from Central America, 1,675 from South America, 4,524 from sub-Saharan Africa, 2,403 from South Central Asia, 2,793 from Southeast Asia, and 3,517 from other regions.

(a) What story do the data tell about where the most hazardous regions are? Explain.

(b) A few days later a reader wrote in to criticize the diagram and the newspaper subsequently published a correction note. Can you guess what the reader's complaint was?

---

[20] "America's most dangerous jobs," by Les Christie, *CNNMoney.com*, August 16 2006: 6:19 PM EDT, http://money.cnn.com/2006/08/16/pf/2005_most_dangerous_jobs/index.htm?cnn=yes

[21] "The souvenir no tourist wants: dengue fever," February 6, 2006, Tina Cassidy, *The Boston Globe*, and also see the article "Spectrum of Disease and Relation to Place of Exposure among Ill Returned Travelers," Freedman D. O et al., *New England Journal of Medicine*, 2006; 354:119-130, Jan 12, 2006.

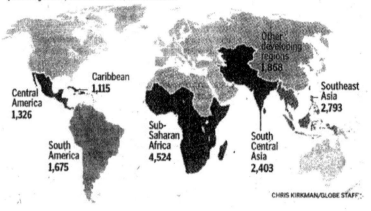

Figure 1.3. The number of people returning home with health problems after traveling to various regions of the developing world. What story does this graph tell about where the most hazardous regions are?

15. **Decision making.** A recent *Harvard Business Review* article titled "When to Sleep on It" discusses the usefulness of deliberating a business decision.[22] The article describes a study in which subjects were asked to make a number of decisions and each subject was given the option to decide immediately or deliberate while performing an unrelated task. The researchers found that subjects who answered immediately made the best decisions and that "the longer our participants thought about their answers, the more likely they were to include irrelevant information at the expense of relevant information." The article concludes that "conscious deliberation, however long and careful, can be a surprisingly crude and ineffective tool."

    (a) Is this study observational or experimental?

    (b) Can you think of a confounding factor here?

    (c) If you could re-design this study, how would you do it?

---

[22] "The HBR Breakthrough Ideas for 2007: When to Sleep on It," by Ap Dijksterhuis, February 2007, *Harvard Business Review* pp. 2–54.

16. **Online course materials.** A textbook publisher wants to market a new online system for teaching math. As a trial of the system, students in a large number of classrooms are given access to the online activities. In nearly every classroom they see that students who complete most of the on-line exercises later score significantly higher on standardized tests than the students who don't do the exercises. The publisher says "these results show this is a powerful tool for boosting student scores and abilities," but a school administrator says "the results look promising, but this same type of study needs to be replicated on a larger scale nationwide for more credibility."

    (a) Do you agree with the publisher? Do you agree with the administrator? Explain.

    (b) Was the fact this study was funded by the textbook publisher an example of a confounding factor? Explain.

17. **Asthma education.** A health insurer would like to reduce the cost of treating asthma through better education of patients. In one city a special course is offered to educate asthmatic patients on how to properly monitor their condition and how to administer medications. The goal is to help reduce mistakes that may require an expensive hospital visit. People who complete the course are followed for two years and insurance records show that on average they require significantly fewer emergency room visits per person compared with asthmatics of the same age and gender in the city who did not take the course.

    (a) Does this mean the education program is effective or is there an important confounding factor here?

    (b) How could you conduct an experimental study here?

    (c) Someone criticizes the study saying that the specific reasons for the emergency room visits need be taken into account to check if they are asthma-related or not. Is this an example of a confounding factor? Explain.

18. **Online polls.** Yahoo Travel's "best beaches" web page contains an online poll where 74% of the thousands of respondents say that a beach is the most likely place they would visit if taking a leisure trip.[23] Does this mean beaches are one of the most popular destinations for leisure travelers? Or is there an important source of bias in this poll?

---

[23]http://travel.yahoo.com/p-promo-3088819

19. **Selling music.** A record label would like to see if sending out posters of their artists can generate orders for their records. They choose a large sample of people and randomly send half of them a poster of their most popular artist along with a catalog, while the other half just gets a catalog. They observe that more orders come from the first group than the second group. They conclude the poster is generating the additional sales.

    (a) Is this study experimental or observational?

    (b) Someone criticizes the study saying "we can't tell if the poster actually motivated people to purchase more music or if those people were going to purchase more music anyway." Is this a potential confounding factor? Is it a reasonable criticism of the study?

20. **Music and performance.** Researchers would like to see if music can affect people's performance. They randomly divide a very large group of volunteers into two groups: one group gets to listen to loud rap music and the other group listens to soft classical music. The physical strength of each person is tested just after listening to the music. Interestingly, they find that the rap music group is much stronger. They conclude that rap music can make you stronger.

    (a) Is this study experimental or observational?

    (b) Someone criticizes the study saying "people who tend to listen to rap music tend to be much younger than people who listen to classical music, and this would be a confounding factor." Is this valid criticism of the study?

    (c) Someone else criticizes the study saying "you couldn't claim causation unless you also measured everyone's strength before they listened to the music." Is this valid criticism of the study?

More Challenging Exercises

21. **Voting felons.** Most convicted felons are prohibited by law from voting, but several states are reconsidering these laws. An editorial in *The New York Times* titled "Why Felons Deserve the Right to Vote" argues that the privilege of voting may help ex-offenders become responsible members of society once they are released from prison and may reduce the chance of re-offending.[24] The article refers to data showing that former

---

[24] "Why Felons Deserve the Right to Vote," *The New York Times*, Monday February 7, 2005, Page A26.

offenders who vote are less likely to return to jail and goes on to say "this lesson has long since been absorbed by democracies abroad, some valuing the franchise so much that they take the ballot boxes right to the prisons."

(a) Does this mean that voting might help reform prisoners or is there an important confounding factor to consider here?

(b) Someone criticizes this study saying social and economic factors play the most important role in whether or not an ex-offender re-offends. Are these examples of confounding factors? Explain?

22. **Arbitration effectiveness.** An online business-to-business auction firm mediates disputes between client businesses and wants to reduce costs by encouraging clients to settle disputes without a formal hearing. Currently 50% of disputes filed get settled without a formal hearing. They decide to run an experiment to see if they can increase this percentage by providing an optional counseling session before the formal hearing where both sides are encouraged to settle the dispute. They randomly select a large number of disputes filed and offer the clients involved the option to attend a counseling session. Half decide to go to counseling and 90% of their disputes get settled without a hearing. Of the other half that does not go to counseling, the percentage is 40%. Is there evidence here that counseling works?

23. **Airline arrivals.** An airline's reputation for being on-time is often considered in consumer ratings. Based on one year of data below (also in the file `airlines.xls`) for Alaska Airlines and America West, which of these airlines would you say is doing a better job overall in keeping its flights on time?[25] Justify your answer.

---

[25] Taken from Arnold Barnett, "How Numbers Can Trick You," *Technology Review*, October, 1994.

	Alaska Airlines		America West Airlines	
Arrival City	# flights arriving on time	# flights arriving late	# flights arriving on time	# flights arriving late
LA	496	63	694	117
Phoenix	220	13	4839	416
San Diego	212	20	383	65
San Francisco	502	103	320	129
Seattle	1841	305	200	62
Totals	3271	504	6436	789

24. **Soda and diabetes.** A recent newspaper article titled "Study links soda to diabetes in women" describes a high profile study of great concern to executives in the soft drink industry.[26] This study of 91,000 nurses finds "women drinking one or more sugar-sweetened soft drinks a day were twice as likely to develop diabetes as women who drank less than one a month." It goes on to say that pure fruit juice and diet soft drinks were not linked to diabetes, but fruit "punch" (fruit juice with sugar and water added) was linked.

    (a) The age of the women was a potential confounding factor, since younger women tend to prefer different drinks than older women. But the researchers properly took this factor into account so it didn't cause bias in the study results. Instead of just comparing all women who drank sugared sodas with all women who didn't drink sugared sodas, what type of comparison(s) would the researchers need to make so that age differences don't bias the study's conclusions?

    (b) A soft drink industry spokesman criticizes the study saying that there are some additional important confounding factors that were not completely taken into account. Give an important confounding factor, other than age, that would play a role here.

    (c) Explain why the fact that all the subjects in the study were nurses would not be a confounding factor.

    (d) Would genetics be a likely confounding factor? Explain why or why not.

25. **Cancer-causing drug.** The stock price of Teva Pharmaceutical Industries plummeted after results of a study of 892 women came out showing one

---

[26] "Study links soda to diabetes in women," by Tara Burghart, Associated Press, August 25, 2004

of the company's drugs may be linked to increased risk of breast cancer in women.[27] In defense of the drug, a company spokesman reported that the rate of breast cancer among the 60,000 women who have taken the drug since 1996 was actually lower than the breast cancer rate for women of the same age in the general population. Can you think of a possible confounding factor that would weaken the spokesman's defense? If so, explain why it is a confounder.

26. **CFA exam.** There are three levels of exams that must be passed in sequence over a number of years to be certified as a financial analyst by the CFA Institute.[28] Part of the prestige of the certification is the difficulty of the exams, which is regularly monitored. The percentage of exam takers who passed the Level I, Level II, and Level III exams in 2004 was 34%, 32%, and 64% respectively. Does this mean the Level III exam was easier than the other two exams or is there an important source of bias here?

27. **Best places to work.** Each year Fortune magazine ranks companies that are rated highly by employees as a good place to work.[29] The rankings are based in part on surveys sent out randomly to employees, not all of which are returned. Since employees' opinion of their company would affect the chances they return the survey form, is this likely to bias the rankings? Explain briefly.

28. **Insiders on fund boards.** In 2004 the Securities and Exchange Commission was considering legislation aimed at reducing corruption, which would require that three-quarters of a mutual-fund's board of directors come from outside the company that runs the fund. Critics worried that this requirement may hurt fund performance. A *Business Week* article described a relevant study finding that funds that have board chairmen from outside the company tend to perform worse than other funds that don't.[30] Some experts criticized the study saying that smaller funds are

---

[27] "Teva Shares Fall on MS Drug and Cancer Paper," Reuters (Chicago), Tue Jan 18, 2005 06:05 PM. See http://www.reuters.com/newsArticle.jhtml?type=topNews&storyID=7361089

[28] See http://www.cfainstitute.org/cfaprogram/cfaprofile/cfa_exam.html

[29] See http://www.fortune.com/fortune/bestcompanies

[30] "Who's Right, The SEC Or Ned Johnson?" by Amy Borrus with Paula Dwyer, *Business Week*, New York: Jun 28, 2004, p. 45. See http://www.businessweek.com/magazine/content/04_26/b3889048_mz011.htm

more likely to have chairmen from outside the company and are also more likely to be less profitable (due to overhead expenses) compared with larger funds. How could this confounding factor have been properly taken into account in the study to help better understand the link between having outside chairmen and performance?

29. **Perils of benchmarking.** A study of one industry finds companies that engage in management practices that are conventionally viewed as risky, such as mandating the use of inter-disciplinary teams, tend to perform better than companies that don't.[31] The study concludes that risky practices are good in that industry.

    (a) Do you think this study most likely was observational or experimental?

    (b) Describe how survivor bias could have been responsible for the findings.

30. **Cell phones and fertility.** A recent newspaper article describes a study that finds heavy use of mobile phones may damage men's fertility.[32] It describes how doctors believe damage could be caused by the electromagnetic radiation emitted by handsets or the heat they generate when used close to the body for long periods. The article goes on to say, "the findings suggest millions of men may encounter difficulties in fathering a child due to the widespread use of mobile phones and offers another possible explanation for plummeting fertility levels among British males." Later in the article Dr. Allan Pacey, a senior lecturer in andrology at the University of Sheffield, is quoted as saying "this is a good quality study but I don't think it tackles the issue," and he then mentions several likely confounding factors. Can you think of a few?

31. **Immigrants and longevity.** A recent *Wall Street Journal* article with the headline "Immigrants Outlive the U.S.-Born Population" describes how black men born in the U.S. live into their 60s on average, black men born in Africa but immigrate to the U.S. live into their 70s on average, but, surprisingly, black men born in Africa who don't immigrate to the

---

[31] "Selection bias and the perils of benchmarking," Jerker Denrell, *Harvard Business Review*, June 2005.

[32] "Men who use mobile phones face increased risk of infertility," by Jenny Hope, *Daily Mail*, 23rd October 2006, http://www.dailymail.co.uk/pages/live/articles/news/news.html?in_article_id=412179&in_page_id=1770. Also see http://news.bbc.co.uk/2/hi/health/6079782.stm

U.S. live only into their 50s on average.[33] Since the same type of trend is observed for many other ethnic groups as well, how does immigrating to the U.S. increase your expected lifespan even beyond the average native-born person? Or is there an important confounding factor to consider here? Explain.

32. **Quality and IT systems.** The health care industry is a large market for information technology (IT), which many believe can both reduce costs and improve quality. A number of hospitals implement new state-of-the-art IT systems and two years later researchers find that the hospitals with the new IT systems had lower cost per patient and higher levels of quality measures than those hospitals without the new IT systems.

    (a) Is this an observational or experimental study?

    (b) Does this mean the IT systems improve costs and quality or is there confounding factor to consider?

    (c) How could you improve the study design without using randomization or forcing hospitals to do things?

33. **Cell phones and cancer.** A news article describes a study that finds that cell phones do not raise cancer risk.[34] The study looked at all cases of cancer in the entire population of Denmark over a decade and found that people who frequently used cell phones in fact had a much lower incidence of cancer than people who didn't use cell phones. (a) Is this study observational or experimental? (b) Does the lower incidence of cancer mean that using a cell phone may help protect you from getting cancer? Explain.

34. **Gender wage gap.** The "Fair Pay Act of 2007" is proposed federal legislation to address the troubling fact that, according to the Bureau of Labor Statistics, in 2005 female full-time wage and salary workers earned only 81% of what men did. A newspaper article writes that such a law

---

[33] *The Wall Street Journal*, May 24, 2004

[34] "Cell phones don't raise cancer risk: study," Wed Dec 6, 2006, by Will Dunham, *Reuters,* http://today.reuters.co.uk/news/articlenews.aspx? type=healthNews&storyID=2006-12-06T152441Z_01_N05280092_RTRIDST_0_ HEALTH-CANCER-CELLPHONES-DC.XML&pageNumber=0&imageid=&cap=&sz=13&WTModLoc= NewsArt-C1-ArticlePage2 On the other hand, a recent news article discusses studies from neighboring Finland and Sweden that find significantly increased cancer risk: see http://news.independent.co.uk/uk/health_medical/article2472140.ece

"would bureaucratize most of the labor market" and gives some confounding factors, other than discrimination, that could alternatively explain the salary gap. What factors do you think the article mentions?[35]

35. **R&D and profitability.** A study of high-tech companies in one industry shows that companies that spend more on Research and Development (R&D) tend to have higher profits: companies devoting 4% of revenues to R&D earned 20% more profit on average than companies devoting 2% of revenues to R&D.

   (a) What could explain this type of relationship?

   (b) If a company in the industry currently devotes 2% of revenues to R&D, what type of increase in profit might be expected if it increased R&D spending to 4% of revenues?

36. **IQ and birth order.** A recent study based on the measured IQ scores of 240,000 Norwegian army conscripts finds that "firstborn children are smarter than their siblings—and the reason is not genetics but the way their parents treat them."[36] Researchers attempted to determine whether the IQ difference was caused by genetics or by social interactions within families. Do you have any ideas on how this could be done?

37. **Marijuana and psychosis.** A study described in a news article finds that "Using marijuana seems to increase the chance of becoming psychotic...even infrequent use could raise the small but real risk of this serious mental illness by 40 percent." To examine the effect of marijuana on mental health, researchers examined 35 studies that tracked tens of thousands of people for periods ranging from one year to 27 years. [37]

   (a) Is this study experimental or observational? Explain.

   (b) The article also mentions some possible confounding factors. Can you think of one? Justify your answer.

---

[35] "Obama flunks Econ 101," by Cait Murphy, June 5 2007: 7:26 AM EDT, http://money.cnn.com/2007/06/04/magazines/fortune/muphy_payact.fortune/index.htm?cnn=yes. See also "The Gender Gap in Wages, circa 2000," June O'Neill, *American Economic Review,* May 2003, Vol.93 No.2, p.309-314.

[36] "Study: Firstborn kids are smarter," by Denise Gellene, *Los Angeles Times,* Friday, June 22, 2007. http://archives.seattletimes.nwsource.com/cgi-bin/texis.cgi/web/vortex/display?slug=birthorder22&date=20070622

[37] "Marijuana may increase psychosis risk, analysis says," CNN.com, July 27,2007, http://www.cnn.com/2007/HEALTH/07/27/marijuana.psychosis.ap/index.html.

38. **Breast implants and suicide.** A news article writes "Women who get cosmetic breast implants are nearly three times as likely to commit suicide as other women, U.S. researchers reported on Wednesday."[38] Can you think of a possible confounding factor?

39. **Obesity networks.** A high profile study of 12,067 people followed for 32 years concludes that obesity can spread from person to person in a social network of friends, much like a virus.[39] One commentator writes that this study only shows that obese people tend to associate with other obese people and not that obesity can be spread through social networks. The researchers, however, had a method for distinguishing the effects of these two different theories. Do you have any ideas on how they did this?

40. **Working smokers.** A news article writes "a new study shows smokers have poorer-than-average work performance and productivity; they also tend to call in sick more. In a study of more than 14,000 Swedish workers, Petter Lundborg, Ph.D., an economist at the Free University of Amsterdam in the Netherlands, found smokers took an average of almost 11 more sick days than non-smokers."[40] The article goes on to give some potential confounding factors that were adjusted for in the study. Can you think of some?

41. **Heart attacks.** An observational study concludes that the use of a certain drug lowers the risk of death from heart attacks over a 5-year period. Half the people in the study used the drug and the other half did not. Researchers also asked people in the study about how much exercise they do, whether they smoke, their weight, and their diet. From these data, a single number was assigned to each person ranging from 1 (most healthy behaviors) to 5 (least healthy behaviors).

---

[38] "Breast implants linked with suicide in study," Wed Aug 8, 2007 7:10PM, by Maggie Fox, Reuters, `http://www.reuters.com/article/healthNews/idUSN0836919020070808?feedType=RSS&rpc=22&sp=true`.

[39] See "Study Says Obesity Can Be Contagious," by Gina Kolata, July 25, 2007, *The New York Times*, `http://www.nytimes.com/2007/07/25/health/25cnd-fat.html?ex=1187064000&en=6d11916887617c73&ei=5070`, and "The Spread of Obesity in a Large Social Network over 32 Years," Nicholas A. Christakis, and James H. Fowler, *New England Journal of Medicine*, 2007;357:370-9.

[40] "Smokers drag down a workplace, study says," by Rachel Zupek, CNN.com, August 1, 2007, `http://www.cnn.com/2007/LIVING/worklife/08/14/cb.smokers/index.html`.

(a) The researchers stated they adjusted for differences in healthy behaviors as measured by the scale between those who took the drug and those who didn't. Instead of just comparing people who took the drug and people who didn't take the drug, what type of comparison(s) did the investigators need to make?

(b) A critic claims that the investigators did not adjust for differences in age, and that age would be a serious confounding factor since the risk of heart attacks increases with age. What other information would you need to determine if age was truly a confounding factor?

(c) Later, a randomized experiment was conducted: one group was randomized to receive the drug and another to receive a placebo. After 5 years, people in the treatment group who said they took more than 80% of the pills prescribed had fewer heart attacks than people who took the placebo or who took less than 80% of the pills. An analyst says that since this is an experimental study there are no confounding factors and the conclusion that the drug works – if more than 80% of the pills are taken - is sound. Do you agree with the analyst here? Explain briefly.

42. **Television and violence.** A newspaper reports the following results of a 17-year study of more than 700 randomly selected families: "Adolescents who watch more than one hour of television a day are more likely to commit aggressive and violent acts as adults...'It's a very important study...it very niftily isolates television as a causal factor' said George Comstock."[41]

(a) Noting that the families were randomly selected, does this mean the study is observational or experimental? Explain.

(b) Someone suspects that gender might be a confounder. Before deciding to adjust for gender, describe how you could check if gender is a confounder.

(c) Someone else suspects that income might be a confounder. Before deciding to adjust for income, describe how you could check if income is a confounder.

43. **IT systems and profitability.** A study reports a strong positive relationship between a company's profit margin and the extent to which it had a well-implemented state-of-the-art information system. To what extent

---

[41] *The Los Angeles Times*, March 29, 2002.

is this evidence that a well implemented state-of-the-art information system can improve profit margins? Briefly discuss.

44. **Do confident kids do worse in math?** A news article titled "Confident Students Do Worse in Math; bad news for U.S." writes "Kids who are turned off by math often say they don't enjoy it, they aren't good at it and they see little point in it. Who knew that could be a formula for success? The nations with the best scores have the least happy, least confident math students, says a study by the Brookings Institution's Brown Center on Education Policy. Countries reporting higher levels of enjoyment and confidence among math students don't do as well in the subject, the study suggests. The results for the United States hover around the middle of the pack, both in terms of enjoyment and in test scores. In essence, happiness is overrated, says study author Tom Loveless."[42] Later in the article it says "Loveless is not suggesting it makes sense to undermine kids' confidence or make math revolting." Based on this study, to what extent does it makes sense to undermine kids' confidence or make math revolting?

45. **Autism rates.** An article writes "The alarming rise in autism rates in the U.S. and some other developed nations is one of the most anguishing mysteries of modern medicine - and the source of much desperate speculation by parents...Autism may be caused by watching too much television at a tender age...Waldman and colleagues found that reported autism cases within certain counties in California and Pennsylvania rose at rates that closely tracked cable subscriptions, rising fastest in counties with fastest-growing cable."[43] Can you think of any confounding factors?

46. **Air-traffic control errors.** A news article writes "Air traffic control errors like the one that almost caused two airliners to collide near Chicago this week remain extremely rare and staffing levels are adequate despite controllers' complaints of fatigue and overwork, a federal aviation official said Friday...Evidence points to most of such errors occurring when a

---

[42] "Confident Students Do Worse in Math; bad news for U.S." POSTED: 7:22 p.m. EDT, October 18, 2006, WASHINGTON (AP), `http://www.ams.org/mathmedia/archive/11-2006-media.html`.

[43] "Does Watching TV Cause Autism?" Friday, Oct. 20, 2006, by Claudia Wallis, *Time.com*, `http://www.time.com/time/health/article/0,8599,1548682,00.html?cnn=yes`.

controller is handling fewer numbers of flights, Cory said, indicating there is no reason to expect an increased likelihood of errors during a hectic air travel period such as Thanksgiving week. She also cited a 2003 FAA study that found 78 percent of errors happen when air traffic is light or average, with only 22 percent occurring when a controller is handling a higher workload of 11 to 15 flights."[44] To what extent do the findings mentioned from the 2003 study shed light on the risks associated with overworked air traffic controllers? Explain briefly.

47. **Corporate social performance.** A *Harvard Business Review* article under the headline "Do Well by Doing Good? Don't Count on It" writes "Is there, in fact, a link between corporate social performance and corporate financial performance? Not a strong one, according to an analysis of 167 such studies that were conducted over 35 years...we found only a very small correlation between corporate behavior and good financial results."[45] Then the article goes on to propose an alternative reason for the correlation. What do you think the article says?

48. **Cell phone radiation.** A newspaper article describes a study concluding that the radiation from using a cell phone at bedtime can lead to headaches, confusion and depression because it "causes people to take longer to reach the deeper stages of sleep and to spend less time in them, interfering with the body's ability to repair damage suffered during the day."[46] The treatment group of subjects in the study were exposed to radiation identical to what is received when using a cell phone, while members of the control group were placed in the same setting but received no radiation.

    (a) Is this study experimental or observational?

    (b) Suppose instead of using cell phone radiation the experimenters simply asked treatment group subjects to talk on a real cell phone near

---

[44]"FAA reassures travelers after near-miss," by Dave Carpenter, *Associated Press*, Fri, Nov 16, 2007.http://news.yahoo.com/s/ap/20071117/ap_on_re_us/faa_near_collision_1

[45]"Do Well by Doing Good? Don't Count on It," by J. Margolis and H.A. Elfenbein, *Harvard Business Review*, Jan 2008, Vol. 86, Issue 1, p19-20.

[46]"Mobile phone radiation wrecks your sleep: Phone makers' own scientists discover that bedtime use can lead to headaches, confusion and depression," by Geoffrey Lean, *The Independent*, January 20, 2008, http://news.independent.co.uk/sci_tech/article3353768.ece.

bedtime. Can you think of a likely confounding factor with this design?

(c) The article also mentions that another study "following 1,656 Belgian teenagers for a year, found most of them used their phones after going to bed. It concluded that those who did this once a week were more than three times - and those who used them more often more than five times - as likely to be "very tired." Can you think of a likely confounding factor with this study?

49. **Gender and unemployment.** A *Business Week* article in May, 2008 discusses data from the Bureau of Labor Statistics and writes "American men are in recession, and American women are not....The share of all men aged 20 and over with jobs has fallen since last November, when private-sector employment peaked, going from 72.9% to 72.2% in April. For women the ratio rose, from 58.1% to 58.3%."[47] What could explain this phenomenon?

50. **Charter schools.** A news article writes "Charter schools, which are privately run but publicly financed, have been faring well on standardized tests in recent years. But skeptics have discounted their success by accusing them of 'creaming' the best students, saying that the most motivated students and engaged parents are the ones who apply for the spots...Most of the city's 99 charter schools admit students by lottery." The article also describes a well designed study to address this issue. What do you think the design is?[48]

51. **Quitting smoking.** There is a great deal of interest in determining the effect of a doctor's advice to quit smoking on the behavior of patients who smoke. Researchers study a large representative group of people and compute the percentage of smokers who quit among those who had been advised by their doctor to quit, and then compare this to the percentage among those who had seen a doctor but had not been advised to quit. They see that the first percentage is larger than the second and conclude that a doctor's advice to quit works.

---

[47] "The Slump: It's a Guy Thing," by Peter Coy, Monday, May 12, 2008, provided by *Business Week,* http://finance.yahoo.com/career-work/article/105040/The-Slump:-It's-a-Guy-Thing

[48] Study Shows Better Scores for Charter School Students, By Jennifer Medina, *The New York Times,* September 22, 2009. http://www.nytimes.com/2009/09/22/education/22charters.html?_r=2&hpw

(a) The presence of smoking-related illnesses might be a confounding factor because these motivate people to quit regardless of their doctor's advice. What else would have to be true in order for it to actually be a confounding factor? Explain briefly.

(b) The length of time a patient has smoked could be a confounding factor that causes researchers to underestimate the benefit of the advice, because people who have been smoking for a long time find it more difficult to quit. What else would have to be true in order for it to actually be a confounding factor? Explain briefly.

(c) What could be done to reduce the effect of the confounding factor from Part (b)? Explain briefly.

(d) Suppose investigators divide patients into two groups: the first group contains patients of doctors who always advise their smoking patients to quit, and the second group contains patients of doctors who don't always advise their smoking patients to quit. And suppose investigators see that the percentage of smokers who quit was higher in the first group than it was in the second. What else would have to be true for this design to avoid the confounding factors from (a) and (b) above? Explain briefly.

52. **Unemployment rates.** The graph in Figure 1.4 is from a recent article in *The Wall Street Journal.* The caption says "Jobless rates for workers at each education level have passed peaks reached in recession of the early 1980s. But the combined unemployment rate hasn't eclipsed its prior peak." Though the figure only shows selected education levels, the article mentions that the picture is similar for other educational levels. Since the caption seems to contradict itself, is there an error here or is there another explanation? Explain briefly.[49]

53. **College majors.** Figure 1.5 from the *Wall Street Journal* shows the results of a survey asking working college graduates if they were satisfied with their career paths. Satisfaction was lowest among psychology majors, where only 26% were "satisfied" or "very satisfied" with their career paths. The article goes on to attribute the low percentage to the fact that "few professions recruit for psychology undergraduate degrees

---

[49] When Combined Data Reveal the Flaw of Averages, by Cari Tuna, *The Wall Street Journal*, December 2, 2009. http://online.wsj.com/article/SB125970744553071829.html

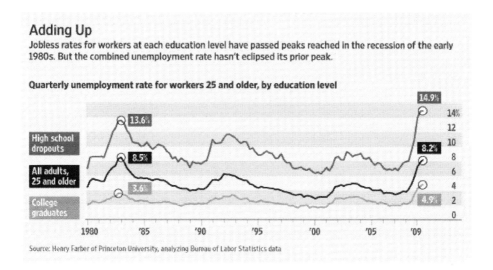

Adding Up
Jobless rates for workers at each education level have passed peaks reached in the recession of the early 1980s. But the combined unemployment rate hasn't eclipsed its prior peak.

Quarterly unemployment rate for workers 25 and older, by education level

Source: Henry Farber of Princeton University, analyzing Bureau of Labor Statistics data

Figure 1.4. Jobless rates for workers at selected education levels from 1980–2009.

specifically" and that "there's not a lot of appetite for their major without a graduate degree." Can you think of an alternate explanation?[50]

54. **Driver licensing and crashes.** In most US states, teenagers can obtain a driver's license when they turn 16 years old. In response to high accident rates among young drivers, many states have implemented graduated driver licensing (GDL) systems that place restrictions on 16-year-old drivers. For example, 16-year-olds must drive accompanied by an adult for the first 6 months after obtaining their license and then for the next 6-12 months cannot drive at night or with other passengers under age 18. To evaluate its GDL, researchers in one state looked at different age ranges and compared fatal crash rates (the number of accidents that resulted in deaths divided by the number of licensed drivers in the age range) for the 5 years prior to implementation of its GDL and for the 5 years after implementation of the program.[51]

---

[50]Psych Majors Aren't Happy With Options, by Joe Light, *The Wall Street Journal*, October 11, 2010. http://online.wsj.com/article/SB10001424052748704011904575538561813341020.html

[51]See "Graduated Driver Licensing and Fatal Crashes Involving 16- to 19-Year-Old Drivers" by S. Masten, R. Foss and S. Marshall, *Journal of the American Medical Association*, 2011;306(10):1098-110.

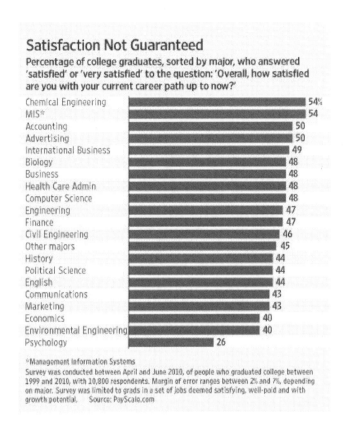

Figure 1.5. The percentage of graduates satisfied with their career for different college majors.

(a) Since the data show there were lower rates of fatal crashes for 16-year-old drivers after implementation of the program compared with before implementation, it looks as though GDL is effective for 16-year-olds. Give an example of a possible confounding factor here and explain in a couple sentences why it is likely to be a confounding factor.

(b) Crash rates for 16-year-olds from another state that had never implemented GDL are also available for comparison for each of the years covered in the study above, but an analyst says that, since rates have always been lower in the other state, the two states are not directly comparable. Could you use the rates from the other state to improve your conclusion about the impact of the GDL system? If yes, explain briefly how. If no, explain briefly why not.

(c) Someone criticizes the study saying that the responsibility level of drivers is a likely confounding factor: many 16-year-olds are responsible drivers and would have avoided accidents regardless of whether or not they were accompanied by an adult as part of the GDL requirements. On the other hand, being accompanied by an adult would probably not prevent an accident among the worst 16-year-old drivers. Is this a sensible criticism? Explain briefly.

(d) The results also show that for 18-year-olds there was a higher rate of fatal crashes after implementation of the program compared with before implementation. What do you think might explain this higher fatal accident rates for 18-year-olds? Explain briefly.

55. **Noise and heart disease.** A study shows that older people exposed to aircraft noise, especially at high levels, may face an increased risk of being hospitalized for heart disease. For seniors living near airports, the study found that every 10-decibel increase in noise from planes was tied to a 3.5 percent higher hospital admission rate for cardiovascular problems. The study also mentions that the researchers adjusted for socioeconomic status. It is also well known that better socioeconomic status tends to be associated with better health. Knowing this, explain why socioeconomic status is likely to be a confounding factor in the aircraft noise study.[52]

56. **Does Medicaid hurt or help?** A newspaper article discusses a study showing that Medicaid, the health insurance program for the poor, actually hurts its recipients. The evidence given is that Medicaid patients tend to be sicker than the uninsured, and slower to recover from surgery. The article goes on to give an important confounding factor. Can you guess what it is?[53]

57. **Obesity, breakfast and television.** A news article discusses studies finding that teenagers who eat breakfast regularly and watch less television have less weight gain and lower body mass index when contacted years

---

[52]Correia AW, Peters JL, Levy JI, Melly S, and Dominici F. Residential exposure to aircraft noise and hospital admissions for cardiovascular diseases: multi-airport retrospective study. *BMJ* 2013;347:f5561

[53]Lousy Medicaid Arguments, by Paul Krugman, *The New York Times*, October 20, 2013

later. The article concludes "encouraging children and teens to eat breakfast and cut back on TV time are important ways to combat obesity." Can you think of some important confounding factors?[54]

58. **Anti-anxiety drugs.** A news article discusses a study finding that people who regularly took anti-anxiety drugs had more than double the risk of death during the seven-year study period. It mentions that researchers controlled for age, smoking, alcohol use, socioeconomic status, sleep disorders, and anxiety disorders. However, the article says they were not able to control for the severity of the illnesses suffered by the study participants.[55]

    (a) Why was it important to control for all those factors? Explain briefly how they could impact the study.

    (b) Do you think not controlling for severity of illnesses could impact the study results? If so, explain how.

59. **Happiness and sun.** A news article titled "Study in Europe eclipses notion home in the sun equals happiness" discusses a study showing that people who migrate from northern Europe to live in warmer climates are marginally less happy than those who stay in northern Europe. The article explains this by saying that, while migrants tended to have higher incomes, "migration itself may be disruptive to other dimensions of people's lives social ties, sense of belonging possibly with consequences for their happiness." Can you think of a confounding factor that could be important here?[56]

---

[54]Eating Breakfast May Beat Teen Obesity, by Jennifer Warner, WebMD Health News, March 3, 2008, http://www.webmd.com/diet/news/20080303/eating-breakfast-may-beat-teen-obesity

[55]Anti-Anxiety Drugs Tied to Higher Mortality, by Nicholas Bakalar, *The New York Times*, March 27, 2014 3:57 PM, http://well.blogs.nytimes.com/2014/03/27/anti-anxiety-drugs-tied-to-higher-mortality/?_php=true&_type=blogs&_php=true&_type=blogs&_r=1

[56]Study in Europe eclipses notion home in the sun equals happiness, Reuters, April 22, 2014 8:09 PM

Chapter 2

Summarizing and Displaying Data

In this chapter we are going to introduce you to some ways of revealing a story which may be hidden in data. We will first show you how to display data so you can see some of its important features and then how a large amount of data can be numerically summarized to better reveal the big picture. By the chapter's end, you should have a better sense of how you can display and summarize data to both find and illustrate some of its important features.

We are going to look at a few important ways for displaying data. We will briefly consider different types of data, and then look at a familiar type of graphical display, the bar chart. We will then cover histograms, spending extra time on them because they are essential to understanding the rest of statistics. Finally, we will look at other types of graphs and figures used to display data.

1. What Are Data?

When people talk about having **data**, or having a **data set**, they could be talking about having just about any type of information.[1] **Quantitative**

---

[1] The word "data" is already plural so it's proper English to write "the data are interesting." Many dictionaries say it's also acceptable to write "the data is interesting," so

data are numerical, such as heights, weights, and salaries, and **qualitative** (or **categorical**) data are non-numerical, such as colors, geographical locations, or genders. We first focus on quantitative data, and then later in the book we introduce how qualitative data can be analyzed.

Often your data will be a **sample** of the possible data you could have gathered from some larger **population**. For example if you select random people in a city for a survey, your sample is the people you select and the population is all the people in the city you could have possibly selected.

Exercises

1. Give an example of four pieces of data you might consider gathering about a company before investing in it. Pick two that are quantitative and two that are qualitative.

2. Classify each as being quantitative or qualitative: year of birth, city of birth, mother's maiden name, social security number.

2. Bar Charts and Time Series Plots

A **bar chart** is one of the simplest ways to display data and it can reveal things you might not have noticed otherwise. To draw a bar chart, just make a bar for each data value with a height equal to the data value. For example, a bar chart for the numbers 3, 5, 4, would be the following:

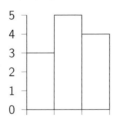

Since the first data value is 3, the first bar goes up to 3 on the scale at the left. The second and third data values are 5 and 4, so the second and third bars go up to 5 and 4 respectively.

Next is an example where the bar chart tells an interesting story you might not otherwise have seen as easily.

---

you can write and say it either way.

Table 2.1. Snickers Bars and Hershey's Kisses sales. Can you tell which is which?

Month	Product A	Product B
Jan	7816658	8465858
Feb	17865227	9307039
Mar	4128771	8945254
Apr	4672123	9056663
May	5877760	10390335
Jun	5324896	10944793
Jul	4998175	10794320
Aug	5354155	10361255
Sep	5502833	11306859
Oct	5318757	10972435
Nov	7212337	10294797
Dec	9074508	8227434
Jan	8401754	8586088
Feb	18990757	10026389
Mar	4391132	9616869
Apr	4617467	10024582
May	5440681	8927481
Jun	5753250	10115303
Jul	5523823	10135208
Aug	5782933	10674501
Sep	6113621	10831740
Oct	6016061	10013292
Nov	7983319	10380192
Dec	10192890	9133822

Revealing sales trends.   Table 2.1 shows monthly sales figures in dollars over a two-year period for two different chocolate candy products: Product A and Product B. One of these products is Snickers Bars and the other is Hershey's Kisses. Looking at the sales figures, can you tell which one is which?[2]

After looking at the table you probably have no idea which is which and you probably don't even know what you're supposed to be looking for in the numbers to figure it out. It's often difficult to make sense of a list of numbers by itself. We will draw a bar chart for each product to help reveal the story.

To draw the bar chart, we make a bar for each month that goes up to that

---

[2]Data courtesy of AC Nielsen.

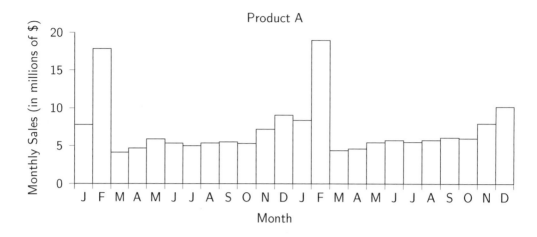

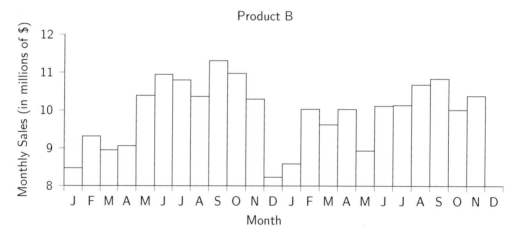

Figure 2.1. Monthly sales for two chocolate candies. Can you tell which one is for Snickers Bars and which one is for Hershey's Kisses?

month's sales figure on the vertical scale at the left. For example Product A in January had sales of about $7.8 million, so the first bar for Product A's chart goes up to around 7.8 on the scale on the left. The two bar charts are shown in Figure 2.1. Immediately we can see a pattern here. Sales for Product A are very high in February and are low through the summer. Sales for Product B are high in the summer, and relatively lower in the winter months of December through February. Can you guess which product is which now?

Product A is Hershey's Kisses, and it has the highest sales in February around Valentine's Day. Snickers Bars are not such a popular Valentine's Day gift, as the graph for Product B shows. Also, there seems to be a smaller bump in sales for Hershey's Kisses around Christmas time when Kisses are given as holiday gifts. Snickers Bars are more popular in the summer.

Sometimes people draw a **time series** plot for this type of seasonal data, and for this you draw a dot at the top of each bar and connect the dots. You would get the graphs in Figure 2.2, which essentially tell the same story but are preferable because they use less "ink."

3. Histograms

Histograms are special graphs that can help you more easily see where your data are concentrated and where they are spread out. This can also reveal things about your data that you can't easily see by looking at the numbers. We'll next give an example of the results of a salary survey and show you how the histogram can help tell the story.

How to Draw a Histogram

MBA expected salaries. Twenty-three first-year Master of Business Administration (MBA) students at Boston University were confidentially surveyed about the type of annual income they anticipated after graduating. The answers (in thousands of dollars) are given in Table 2.2 below. Do you see anything interesting in these numbers?

Table 2.2. Expected Salaries of selected MBAs

110	75	120	60
120	120	50	90
140	75	110	80
70	115	80	70
125	60	60	110
100	110	70	

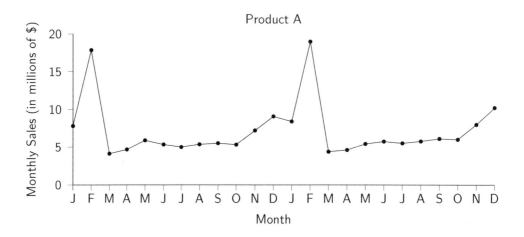

Figure 2.2. Two time series plots.

Most likely all you see here is a jumble of numbers, and you don't really notice much else. And if we draw a bar chart, shown in Figure 2.3, there doesn't appear to be anything striking in the picture.

A big list of numbers is hard for the mind to process, and a bar chart doesn't always reveal a story. A histogram can be used to more easily see a story.

Drawing the histogram.   A histogram shows you how many data values fall in various ranges of your data. Here are the steps to follow in creating a

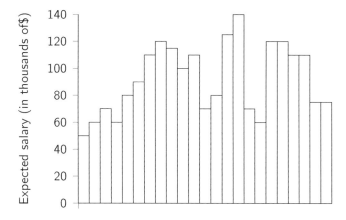

Figure 2.3. A bar chart for expected MBA salaries

histogram:

(1) Divide the range of the data into intervals of equal width.

(2) Count how many data values fall into each interval.

(3) Make a bar chart where the height of a bar is equal to this count.

We draw the histogram for this salary data by following this procedure:

Step 1. The data range from $50k to $140k ($50,000 to $140,000), so let's divide the data into nine income ranges that each are $10k wide: $50k–$59k, $60k–$69k, and so on. Each range must have the same width in terms of dollars, otherwise the graph can be misleading.

Step 2. We count the number of data values in each of the ranges (you can sort the data to make this task easier) and get the data arranged in Table 2.3 below.

Step 3. We then make a bar chart for the counts. This gives the histogram shown in Figure 2.4.

When you look at histogram in Figure 2.4, you should also be able to see the basic curve shape shown in Figure 2.5 and not get confused by the details of each bar's exact height.

Table 2.3. Counts of salary values in different ranges

Salary range ($1,000s)	Number of people
50 – 59	1
60 – 69	3
70 – 79	5
80 – 89	2
90 – 99	1
100 – 109	1
110 – 119	5
120 – 129	4
130 – 139	0
140 – 149	1

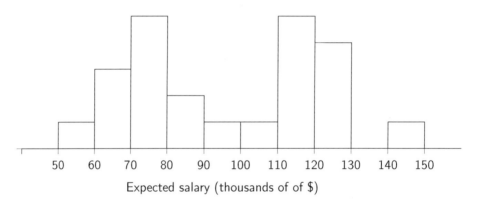

Expected salary (thousands of of $)

Figure 2.4. Histogram of expected post-MBA salaries.

Properties of a Histogram

On a histogram, the percentage of the area under the graph over some range represents the percentage of data values in that range. This also means the total area represents 100 percent of the data. In other words, two income ranges that have about the same area under the curve will have about the same number of data values in them. For example, if you look carefully at the histogram in Figure 2.6 you can see that the area under the histogram between 80 and 85 is about the same as it is between 90 and 110. This means there are about the same number of data values in each of these two ranges.

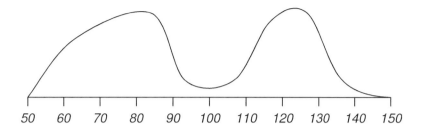

Figure 2.5. A sketch of the histogram of expected post-MBA salaries.

Since you can easily compare relative areas without even referring to the vertical axis, in many cases it is not even important to label the vertical axis of a histogram. In fact, you can often completely leave off the vertical axis. You can learn a lot by just looking at the relative area under the graph in various ranges.

> The percentage of the area under the histogram in some range represents the percentage of data values in that range.

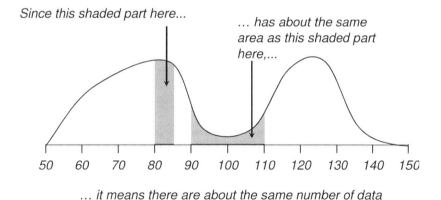

... it means there are about the same number of data values in the range 80-85 as there are in the range 90-110.

Figure 2.6. A histogram for expected MBA salaries. The area under a histogram in some range represents the percent of data values in that range.

The height of the histogram over a range represents how crowded the values are in that range. In the histogram in Figure 2.6, values are more crowded in

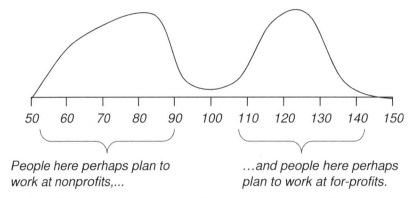

Figure 2.7. A histogram for expected MBA salaries in a class. Notice the bimodal shape.

the first shaded range than in the second shaded range—even though there is approximately the same number of values in each range. Being able to see areas of relative crowding in the data can reveal a story which initially was hidden.

In this case the two bumps in the histogram's shape represent two distinct regions of crowding, and there is a noticeable gap between them. What explains the fact there are two separate regions? One possible explanation is that the MBA class consisted of two types of people: people planning for a career in the nonprofit sector and people planning for a career in the for-profit sector. This is illustrated in Figure 2.7. The value below the highest peak of the histogram is called the **mode** and a shape with two distinct peaks is called a **bimodal** shape.

When another section of MBA students at Boston University was asked about their expected annual salaries, the histogram came out looking like Figure 2.8. What explains the different shape? Notice that the scale on this graph starts at zero.

In this group, it turned out that there were no students interested in the nonprofit sector. The small bump on the left represented the small handful of international students who planned to return to a country where $20,000 a year is a high salary.

The difference between a histogram and a bar chart.   There is a big dif-

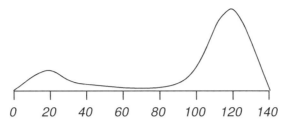

Figure 2.8. Expected MBA salaries for a different class. What explains the small bump at the left?

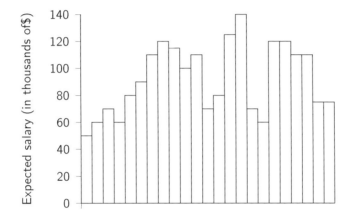

Figure 2.9. A bar chart for expected MBA salaries does not reveal very much.

ference between a histogram and the graph produced by clicking on "bar chart" in Excel even though they may look similar at first glance. In an Excel "bar chart" the vertical axis is in thousands of dollars and there will be a separate bar for each person as shown in Figure 2.9.

The bar chart in Figure 2.9 is not a histogram. In a histogram, the units on the horizontal axis are the same as the units of the data (thousands of dollars in this case) and the vertical axis measures the frequency with which values appear in a given range. The connection between bar charts and histograms is that a histogram is a bar chart for the count of the values in each range.

> A histogram is not the same as a bar chart for the data.

```
 5 │ 0
 6 │ 000
 7 │ 00055
 8 │ 00 ┴──── This corresponds to the number "75,"...
 9 │ 0
10 │ 0
11 │ 00005
12 │ 0005
13 │   ┴────── ...and this corresponds to the number "120."
14 │ 0
```

Figure 2.10. Stem-and-leaf plot for expected MBA salaries.

Stem-and-Leaf Plot

A stem-and-leaf plot is another commonly drawn graph similar to a histogram. Here instead of drawing bars, digits are plotted. For the example of the expected MBA salaries at the beginning of the chapter, below is the stem-and-leaf plot. To read a value from the graph, just look at the left-hand column to find the first digit(s) and go across to find the last digit. For example, the third row represents the five numbers 70, 70, 70, 75, 75: rather than repeating the ten's digit "7" each time, we put it in the left column of the row and then just put the five one's digits "00055" in the second column of the row. The third digit in the row starting with "12" is a zero, so this corresponds to the number 120. Figure 2.10 shows the stem-and-leaf plot.

Exercises

3. **Telling a story from sales data.** Figure 2.11 shows the annual sales (in millions of dollars) for all the companies in one particular industry.

    (a) Draw a histogram or a stem-and-leaf plot for this data by hand.

    (b) How would you describe the shape?

    (c) What story does this tell about company sizes in this industry?

4. **Histograms and bar charts.** Below on the left is a list of account balances (in thousands of dollars). Is the graph on the below right the corresponding histogram or is it a bar chart for the data? Explain your answer without doing any calculations.

33	14	12	50	15
47	8	46	8	41
21	13	42	14	3
11	12	18	12	35
11	29	4	14	7

Figure 2.11. Annual sales (in millions of dollars) for companies in an industry.

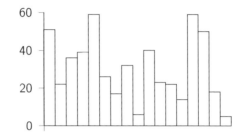

51	26	23	18
22	17	22	5
36	32	14	
39	6	59	
59	40	50	

5. **High wage earners.** Below is a histogram for hourly wages earned by employees at a company. Is the percentage of people who earn above $50 per hour closest to 1%, 5%, or 20%? Choose one and explain.

Hourly wage ($)

4. Interpreting Histogram Shapes

The shape of a histogram can reveal interesting features of data. High points in the histogram tell you where your data are heavily concentrated and unusual features in the histograms shape usually have a story behind them.

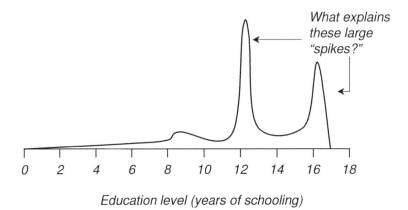

Education level (years of schooling)

Figure 2.12. A histogram for years of education completed for adult Americans.

Here we will give a few examples of unusually shaped histograms and discuss what they reveal about the data.

"Spikes" in a histogram.   Figure 2.12 is a sketch of the histogram for the distribution of education level for adults in the United States. The first thing we notice is two large "spikes." What does it mean to have a spike in the data here and what could explain the presence of these two large spikes?

When a histogram has large spikes, it means that there is a high concentration of data values in the range just below the spike. The first spike, right around the 12 years of schooling point, represents people who completed high school, and the second spike, right around 16 years of schooling, represents people who completed college. Since the histogram trails off to the left of eight years of education, we say that the histogram has a **left-hand tail**, or is **skewed to the left**. If a histogram has a large tail it can cause certain statistics to be misleading; we will discuss this in a later section. This graph tells us usually people end their education at around either 12 years or 16 years and if people are close to getting either their high school or college degree, they tend to finish it. This phenomenon causes the large spikes in the histogram.

How many employees do companies tend to have?   A histogram for the number of people employed by each of the nearly 6 million U.S. companies

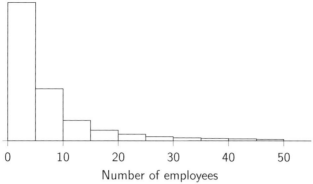

Figure 2.13. A histogram for the number of employees at U.S. companies in 2003.

that had a payroll in 2003 is shown in Figure 2.13.[3] Where is Microsoft on this graph? What percent of companies have fewer than 5 employees?

Microsoft has thousands of employees, so it would be way off the right edge of the graph. The graph does not show any companies with more than 50 employees because, although there are thousands of such companies, they are greatly outnumbered by companies with fewer than 50 employees—so the graph essentially goes down to nearly zero above 50. In fact, it appears that around half the total area under the graph is in the first bar (highlighted in Figure 2.14), meaning that around half the companies have fewer than five employees. Even though it's usually the big companies that make the news, they are outnumbered by the small companies.

Are profiles accurate online?   Within less than two years of being launched, the social networking web site Myspace.com was the fifth most viewed site on the internet; now Facebook has completely surpassed it. Users who set up profiles were asked to input their age, height and other statistics. To investigate the reliability of user profiles, we gathered data on people's heights as posted in the profiles. We searched for people living near Boston having different heights and counted the number of people. In Figure 2.15 the histogram for men's heights appears and below it is the histogram for women's heights. There appears to be quite a large spike in the men's histogram and a sudden sharp drop in the women's histogram. What could explain these

---

[3]See http://www.sba.gov/advo/research/data_uspdf.xls

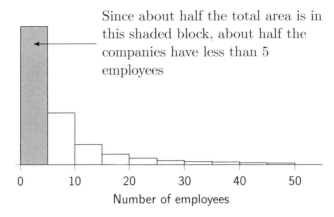

Since about half the total area is in this shaded block, about half the companies have less than 5 employees

Figure 2.14. Histogram for number of employees at U.S. companies in 2003. Notice that about half the companies have fewer than five employees.

two peculiarities?

Discussion. These histograms probably reveal how the two genders differently exaggerate their heights. In the men's histogram there is a sharp drop-off at 71 inches, which is 5 feet 11 inches, and then a large spike at 72 inches, which corresponds to 6 feet. This could be explained by 5'11" men exaggerating their heights up to 6'. In other words, many men who say they are 6' are actually only 5'11". For women, the exaggeration appears to go in the other direction; many of the women in the bar above 67 inches, or 5'7", actually probably should belong in the bar above 68 inches, or 5'8". It seems as though many men who are 5'11" wish they were 6 feet tall, and many women who are 5'8" wish they were only 5'7". The histogram in Figure 2.16 reveals this tendency that you could not see otherwise. We illustrate this in Figure 2.16.

How old are college students?   Figure 2.17 is a graph showing the age distribution of the 12,177 students enrolled at Southern Connecticut State University in 2004.[4] Notice that the graph drops off sharply after age 24 and rises slowly from age 25 to age 44. What explains this shape?

Discussion. You may at first think this graph tells a story about how most people leave college in their early 20s and then in increasingly larger numbers

---

[4]Source: http://www.southernct.edu/departments/research/all_university_age_distribution.html

Men

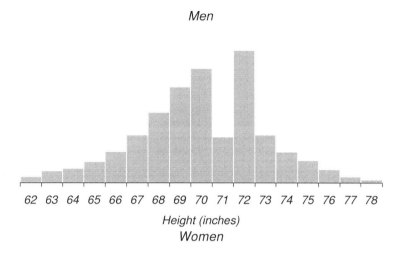

Height (inches)

Women

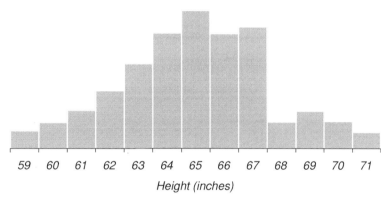

Height (inches)

Figure 2.15. Histograms for reported men's and women's heights.

return for more education throughout their 30s and 40s as they see the value of higher education in the workplace. But this graph is not a histogram—each of the categories does not have the same width in terms of years. The reason for the increase in the graph from age 25 to age 44 is simply that the ranges get wider and wider. One range covers age 23 to age 24—that's two years. Another range goes from age 31 to age 44—that's 14 years. If we draw the real histogram, we see it drops off smoothly as you would expect as people get older. Figure 2.18 is a correct histogram for the data, where

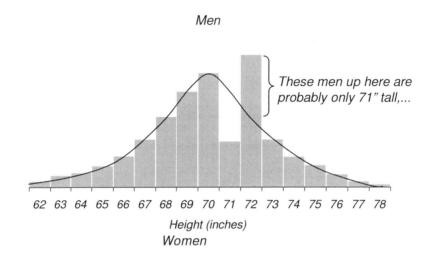

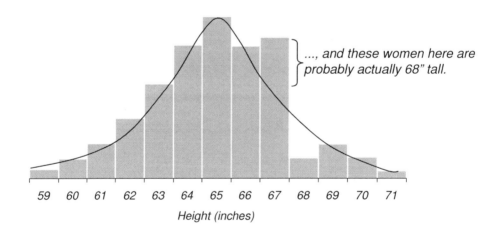

Figure 2.16. Two histogram showing how men and women probably exaggerate their heights differently.

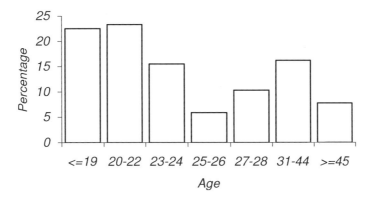

Figure 2.17. The distribution of the ages of students at Southern Connecticut State University in 2004.

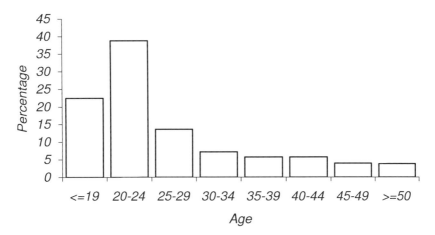

Figure 2.18. The correct histogram for the distribution of the ages of students at Southern Connecticut State University in 2004.

each bar (other than the first and last) represents a five-year range.

Exercises

6. **Age distribution in the United States.** Below is a sketch of the histogram of the ages of people living in the United States.[5]

---

[5]Source: US Census Bureau. See http://www.census.gov/ipc/www/idbpyr.html

(a) What could explain the bump in the middle?

(b) What explains the gradual decline on the right side?

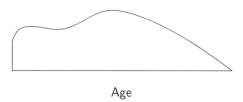

Age

7. **Age distribution in China.** China is an expanding market for products and it has also seen an increasing birth rate in recent years. Would the age histogram for people there look more like (i) or (ii) below? Explain briefly.

(i)   (ii)

8. **The future of online retailing.** How do online retailers compare with traditional "bricks and mortar" physical stores? A typical Barnes and Noble book store carries around 130,000 titles, but less than half the sales at the online retailer Amazon are of its top-selling 130,000 titles.[6] A similar phenomenon occurs in the music industry: a typical music store carries around 10,000 songs, but less than half the songs accessed from the online retailer Rhapsody are for its most popular 10,000 songs. If we record the book ranking each time a book is sold at Amazon (the book ranked number one has the highest number of sales, the book ranked number two has the second-highest, and so forth) and drew a histogram for these ranks, what do you think the shape would look like? Why? What does this say about the future of online retailing?

9. **Salary distributions.** CNN.com published a report on salaries in many common careers.[7]

---

[6]http://www.wired.com/wired/archive/12.10/tail_pr.html and http://www.thelongtail.com/

[7]http://money.cnn.com/magazines/moneymag/bestjobs/snapshots/2.html

(a) The average salary for accountants in 2006 was \$62,575, 75% earned more than \$53,925, 50% earned more than \$61,241, 25% earned more than \$70,234, and the top 5% earned more than \$119,943. Draw a rough sketch of the corresponding salary histogram.

(b) The average salary for college professors in 2006 was \$81,491, 75% earned more than \$61,780, 50% earned more than \$76,766, 25% earned more than \$105,928, and the top 5% earned more than \$567,753 (the highest earners are usually medical school professors). Draw a rough sketch of the corresponding salary histogram.

10. **Recognizing a histogram.** Could the graph below possibly be a histogram for some list of numbers somewhere? Explain why or why not.

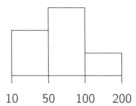

5. Measuring the Center: The Mean, Median, and Mode

In the previous section we saw how taking the data and displaying it in a graph can reveal a story. In this section we are going to discuss how you can summarize data numerically. Why go on to do this at all? Our displays of data were useful. Is there anything wrong with them?

The simple answer is that very often there is too much data to take the time to look at a graph carefully, and a single number may more quickly help you find what you are looking for. For example, suppose you have weekly stock price data for all stocks in the New York Stock Exchange during the year 2005. And suppose you are interested in identifying the stock that had the least price fluctuation and was the most consistently increasing in price over the year. You could spend hours looking at hundreds of different graphs or you could use a method to numerically summarize an entire year's worth of data for each stock as a single number and sort these summary numbers to quickly find the stock you're looking for. Computing the *standard deviation* is a common approach to numerically summarize variation in data, and we will discuss this later. We'll see later how in finance the *standard deviation* is a commonly used measure of risk.

For example, Figure 2.19 shows a graph of the weekly stock price for 14 different stocks over one year. Each line represents a different stock. Which stock had the least price fluctuation and was most consistently increasing over the year? One of the lines roughly second from the bottom looks like

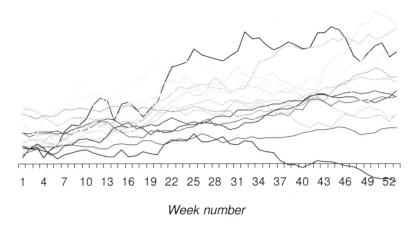

Week number

Figure 2.19. The weekly stock price for 14 different stocks. Which stock had the least price fluctuation and was most consistently increasing over the year?

it might be the one with the least fluctuation, because it seems a bit less crooked than most of the others. But it is difficult to say for sure in this tangled mess and this line stands out perhaps only because it is a bit lower than the others.

It is very difficult to answer this question, because there is so much data here. If we looked at data for the hundreds of stocks in the New York Stock Exchange, it would be almost impossible to do this visually. It is therefore useful to have a way to summarize an entire data set with a single number. First we discuss measures of the location of the data and later a measure of variability.

There are three most commonly used measures for summarizing the location of a data set: the mean, median, and mode. For a list of numbers, the **mean** or **average** is the sum of the numbers divided by the number of numbers. For example, if the list of numbers is 1, 3, 4, 2, 2 then the average equals

$(1 + 3 + 4 + 2 + 2)/5 = 2.4$.

The **median**, on the other hand, is the middle number when the list of numbers is arranged from lowest to highest. For the list 1, 3, 4, 2, 2 the middle number would be 2, since the numbers in order are 1, 2, 2, 3, 4. If there are two numbers in the middle, the tradition is to say that the median is halfway between these two numbers. For example, if 2 and 3 are the middle numbers, the tradition is to say that the median is 2.5.

The **mode** is the value that occurs most frequently and in this case it would equal 2. Although there are other ways to summarize the center of a list of data, these are the most common ones.

Technical notation The commonly used mathematical notation for the mean or average of a population is the Greek letter "mu," written as μ. This is the Greek letter corresponding to our letter "m," and you can remember it by imagining the "m" stands for "mean" (there doesn't seem to be any one commonly used Greek letter for the median). If we represent the first data value as X_1, the second as X_2 and the ith as X_i, we can write the formula for the average as

$$\mu = \frac{1}{n} \sum_{i=1}^{n} X_i = \frac{1}{n}(X_1 + X_2 + \cdots + X_n)$$

where the symbol $\sum_{i=1}^{n}$ is read as "the summation as i goes from 1 to n." This symbol represents adding up all the values of X_i from X_1 up to X_n.

If your data set does not represent the entire population but only represents a small sample of it, the convention is to refer to the mean as a "sample mean," and it is notated as "X-bar" or $\overline{X}$. This mathematical notation is not essential to understanding statistics or even for reading the rest of this book, but it can be important when reading reports or communicating with statisticians.

Which occupations pay the highest salaries? To answer this question we could look at a separate histogram for salaries in each of the hundreds of different occupations, but this would not be very efficient. The Bureau of Labor Statistics nicely summarizes salaries by computing the average annual salary for each occupation in 2004.[8] After sorting the list, we can easily come

---

[8] `ftp://ftp.bls.gov/pub/special.requests/oes/oesm05nat.zip`

Table 2.4. Salaries for the top ten highest paying occupations.

Rank	Occupation	Average Salary
1	Surgeons	$177,690
2	Anesthesiologists	$174,240
3	Obstetricians and gynecologists	$171,810
4	Orthodontists	$163,410
5	Oral and maxillofacial surgeons	$160,660
6	Internists, general	$156,550
7	Psychiatrists	$146,150
8	Prosthodontists	$146,080
9	Family and general practitioners	$140,370
10	Chief executives	$139,810

up with the occupations with the highest average salary – this is summarized in Table 2.4. This shows how numerically summarizing data can help you quickly answer questions. It's interesting to note that medical professionals take the top nine places and chief executives come in at number ten.

The chief executives we hear about in the news are usually the ones earning at least seven-figure salaries, but the average chief executive earns a relatively modest six-figure salary. If we just look at management occupations, Table 2.5 shows us that chief executives, engineering managers and information system managers are the highest-paid on average.

What do the different measures of center tell us? You can learn something about the relationship between the mean, median, and mode just by looking at the shape of the histogram. The median divides the histogram into two equal-sized areas. Half of the area under the histogram lies to the left of the median, and the other half lies to the right. This is illustrated in Figure 2.20.

The average is the "balancing point" of the histogram if we imagine that each bar of the graph is made out of a material with weight. The mode on a

Table 2.5. Salaries for the highest paying management occupations.

Rank	Occupation	Average Salary
10	Chief executives	$139,810
19	Engineering managers	$105,470
20	Computer and information systems managers	$102,360
21	Marketing managers	$101,990
23	Natural sciences managers	$99,140
24	Sales managers	$98,510
26	Financial managers	$96,620
29	General and operations managers	$95,470
36	Human resources managers, all other	$89,950

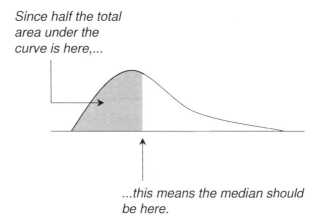

Figure 2.20. Finding the median on a histogram.

histogram is the value corresponding to the highest point of the graph and represents the value in the data that occurs most frequently.

In Figure 2.21 we look at a very simple histogram with three bars and we want to see the relationship between the average and the median. Notice the way it's drawn so the first bar is exactly twice the height of the other two bars, and therefore it contains the same area as the other two bars combined. Since half the area overall is in the first bar, the median would be equal to 75.

To see where the average would be in relation to the median, we can imagine we are balancing this shape by putting our finger right under the median

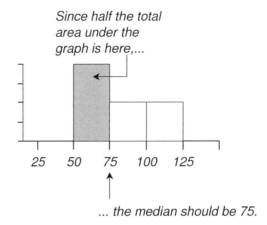

Since half the total area under the graph is here,...

... the median should be 75.

Figure 2.21. Finding the median on a simple histogram.

and visualizing which way it would tip over. This is illustrated in figure 2.22.

Sometimes the average and the median seem to tell different stories about the data if the histogram is not symmetric. In general, the average gets pulled away from the median in the direction of any "tail" the histogram shape may have (the histogram is said to be **skewed** in the direction of the "tail"). Extreme values (outliers) can easily influence the average, whereas the median tends to be more stable.

For example the list of numbers 1, 2, 3 has a median equal to 2. This is the same as the median for the list 1, 2, 100, even though the latter list has a much higher average. This explains why median housing prices are often reported instead of average prices: the listing of a few extremely expensive luxury homes could suddenly inflate the average but not the median.

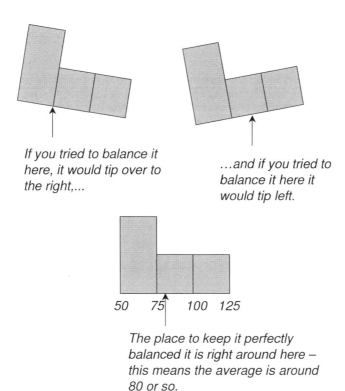

If you tried to balance it here, it would tip over to the right,...

...and if you tried to balance it here it would tip left.

The place to keep it perfectly balanced it is right around here – this means the average is around 80 or so.

If you instead had this histogram with only two bars,...

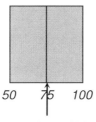

... it would balance at exactly 75 – so the average would equal 75 – and so would the median.

Figure 2.22. The relationship between the median and average.

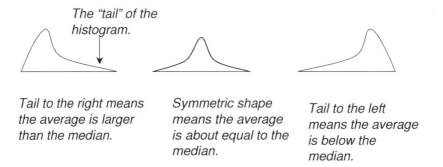

The "tail" of the histogram.

Tail to the right means the average is larger than the median.

Symmetric shape means the average is about equal to the median.

Tail to the left means the average is below the median.

Let us try a few examples to strengthen your intuition about the relationship between the average and median.

Example. Which list has a higher average? Which has a higher median?

(a) $1, 2, 3, 4, 100$

(b) $1, 2, 3, 4, 5$

Discussion. The first list has a higher average and both lists have the same median of 3. The average tends to get pulled in the direction of extreme values.

Example. The income histogram for people in the United States has a long tail to the right because most people earn at the low end of the spectrum and there are small numbers of people with extremely high incomes. Does this mean there are more people above average or below average?

Discussion. Obviously half the people are above the median income and the long tail to the right means the average is larger than the median—this is illustrated in Figure 2.23. This means that fewer than half the people are above average. Therefore, there are more people earning below average than above average.

The Five Number Summary and Inter-Quartile Range

In addition to the median, statisticians often compute **quartiles** for data. If you look at the data values below the median and take the median of these

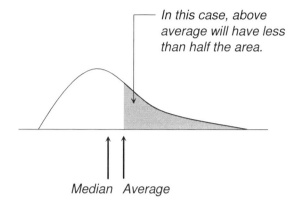

Figure 2.23. The relationship between the median and the average for a histogram.

data values, you get what is called the **first quartile**—because it tells you where the bottom quarter of the data values lie. If you look at the data values above the median and take the median of these data values, you get the **third quartile**. The **second quartile** equals the median. When statisticians report a **five-number summary** for data, it means reporting the minimum, the first quartile, the second quartile, the third quartile, and the maximum of the data values. The **inter-quartile range**, abbreviated IQR, equals the third quartile minus the first quartile.

Exercises

11. **Calculating mean and median.** Calculate the mean and median for the following list of numbers: 2, 4, 7, 1, 9.

12. **Estimating average and median.** Below is a histogram for some data.
    (a) Is the average closest to 10, 15, or 20?
    (b) Is the median closest to 10, 15, or 20?

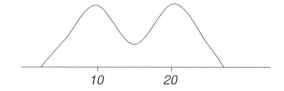

13. **Comparing mean and median.** Below is a histogram for some data.
    (a) Is the average closest to 10, 15, or 20?
    (b) Is the median closest to 10, 15, or 20?
    (c) Is the average higher or lower than the median?

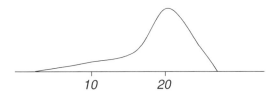

14. **Firm sizes.** In the United States in 2003 there were a total of 5.8 million companies employing a total of 114 million people, which is an average of about 20 people per company. The median company size, however, was only about five people. What does this tell you about the shape of the histogram for company sizes?[9]

6. Measuring Variability: Standard Deviation

We have looked at measures of the location of a data set. We now want to look at how to measure variability in data. This, for example, would be useful in identifying stocks to invest in that have the least amount of fluctuation in their prices.

The **standard deviation** is a number that tells you how spread out data tend to be. Usually abbreviated **SD**, it is a measure of how far numbers tend to be from their average. For many commonly encountered types of data, the majority of values (very often around two-thirds of the values) tend to be within one standard deviation away from the average. Very few values will be more than two or three standard deviations away. The standard deviation and average are commonly used together to summarize a large amount of data using just two numbers.

> The standard deviation (SD) is a measure of how far numbers tend to be from their average.

---

[9]http://www.sba.gov/advo/research/us_tot.pdf

Before we give you the mathematical formula for computing the standard deviation and have you actually compute it, we will give you some exercises that will strengthen your intuition for what standard deviation measures. Having an intuitive feel for what standard deviation measures will better help you understand how to interpret the number you calculate.

Example.   Below are five lists of numbers that each average out to 20. For each list would you guess the standard deviation (SD) is closest to 2, 4, or 8?

(a) 18, 22, 18, 22, 18, 22

(b) 16, 24, 16, 24, 16, 24

(c) 24, 20, 13, 24, 17, 22

(d) 16, 22, 19, 20, 22, 21

(e) 9, 25, 15, 14, 24, 33

Solutions.

(a) Guess 2 for the SD. The list of numbers 18, 22, 18, 22, 18, 22 has an average equal to 20. Here each number is exactly two points away from the average: The first number 18 is 2 points below average, and the second number 22 is 2 points above average. Guessing 2 for the SD would be best, since the SD measures how far numbers tend to be from average.

(b) Guess 4 for the SD. Again the list averages out to 20, but here each number is exactly four points away from the average: 16 is four points below average and 24 is four points above average.

(c) Again guess 4 for the SD, but this list is a bit more difficult. The first number 24 is 4 above average. Continuing similarly, the amounts away from average for all the numbers are 4, 0, 7, 4, 3, 2. The best choice out of 2, 4, or 8 as a standard deviation would probably be 4, since 4 seems to be in the middle of all these amounts away from average.

(d) Guess 2 for the SD. The amounts off from average are 4, 2, 1, 0, 2, 1 so 2 is probably the best choice for the SD.

(e) Here the SD is probably closest to 8.

Example.   Which set of numbers below has the largest SD and which has the smallest SD? Each has an average of 20.

(a) 0, 10, 20, 30, 40

(b) 0, 19, 20, 21, 40

(c) 0, 1, 20, 39, 40

Solution.   Although the numbers in list (a) are more spread out from one another, the numbers in list (c) are more spread out from the average. This means that list (c) should have the largest standard deviation. For the same reason, list (b) should have the smallest standard deviation.

Example.   Which of the following lists has the larger SD? They both have an average of 20.

(a) 5, 10, 20, 30, 35

(b) 5, 10, 20, 20, 20, 20, 30, 35

Discussion.   List (b) has three additional numbers that are concentrated right on the average. This means that, overall, the numbers tend to be closer to the average in (b) than in (a). So (a) should have the larger standard deviation.

The standard deviation for data can be visually estimated from a histogram, and knowing how to do this can help strengthen your intuition for what the standard deviation measures. After you learn the mathematical formula, you will also be able to determine if the number you get from it is in the right ballpark or not. We will use the following general rule of thumb mentioned previously:

> For many commonly encountered types of data, usually around two-thirds of the values are within one standard deviation from average. Few values will be more than two or three standard deviations away.

This rule of thumb is not a mathematical fact, but it is just something that statisticians have observed empirically over the years. For this reason it is often called **the empirical rule**.

Example.   Below is a histogram for hourly wages (in dollars) for employees at a company. The average wage is $35. Do you think the SD is closest to $10, $20 or $50?

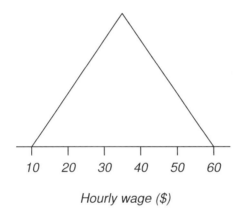

Hourly wage ($)

Solution.   Suppose we guess $20 for the SD. From the rule of thumb above this guess means we expect around two-thirds of the area to be within $20 from the average of $35 or, in other words, in the range from $15 to $55. When we shade this area in Figure 2.24, you can see that this range covers much more than two-thirds of the total area (it is almost all the area), and hence much more than two-thirds of the values fall there. This means that the SD should be less than $20.

Suppose instead we guess $10 for the SD. If we shade the area within $10 from average, we see in Figure 2.25 it is much closer to two-thirds of the total area. A standard deviation of $10 would be the best guess among the choices given.

Note that while the data cover a range that is $50 wide (from $10 to $60), the standard deviation is only $10. The standard deviation does not measure the full range of the data—it measures the typical distance of data values to the average.

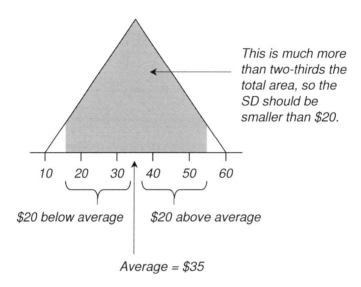

Figure 2.24. A histogram where the whole region within $20 of the average is shaded.

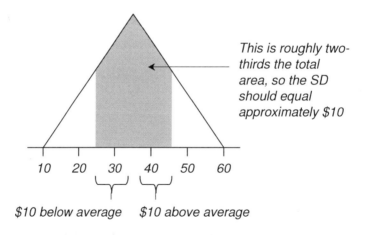

Figure 2.25. A histogram where the region within $10 of the average is shaded. The area is about two-thirds the total area.

Technical note: Chebyshev's Inequality

A famous statistical rule called **Chebyshev's Inequality** says that the fraction of data values more than k standard deviations away from the average is always less than $1/k^2$ for any $k > 1$. Using $k = 3$, the rule says that less than $1/3^2 = 1/9$ of the data values can be more than three standard deviations away from the average. Another interesting (but not as well-known) statistical rule is that the median can never be more than one standard deviation away from the average. In practice you'll usually find very few data values more than three standard deviations away from average and somewhere around two-thirds of the values within one standard deviation from average.

Exercises

15. Which of the following two lists of numbers would have a larger standard deviation? Explain briefly. (i) 101, 102, 103 (ii) 1, 3, 5

16. Would the standard deviation for the data in a histogram below be closest to 10, 20, or 30? Explain.

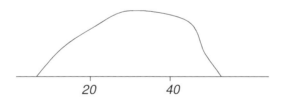

17. (a) For which histogram (i) or (ii) below is the standard deviation larger? Explain. (b) For which histogram is the mean larger? Explain. (c) Explain why for both histograms the median should be roughly the same.

7. Computing the Standard Deviation

In practice you will not need to plug numbers manually into a mathematical formula to compute the standard deviation, because software such as Excel has built-in functions to do this for you. Yet, actually walking through the calculations should help you better interpret the number that the computer comes up with and will also help you better understand the ways in which the standard deviation can be misleading.

To compute the standard deviation from a list of numbers, compute the distance from the average for each number in the list (called the "deviations from the average"), square each of these distances, average them, and then take the square root of this average. That's it.

Why do we square the distances from the average? This is a tradition that dates back to the days before computers, and it simplified other calculations that the standard deviation was used for (the more detailed reason has to do with the link between standard deviation and what is called the margin of error. We will cover this later when we delve into the topic of the margin of error.) And since it is the tradition now, it's important to have a good intuition for it.

Technical notation

The standard deviation for a population is represented by the Greek letter "sigma," written as σ, and we can write the formula in mathematical notation as

$$SD = \sigma = \sqrt{\frac{1}{n} \sum_{i=1}^{n}(X_i - \mu)^2}$$

where the ith data value in the list is X_i, and $\mu = \dfrac{1}{n}\displaystyle\sum_{i=1}^{n} X_i$.

In less technical language, we can summarize this procedure in the following formula:

$$SD = \sqrt{\text{Average of (deviations from average)}^2}$$

In step-by-step format, we mean the following:

1. Compute the average.

2. Find out how far from average each number is by subtracting the average from each number (these are called the "deviations from the average").

3. Square each of these deviations from Step (2).

4. Average these squared deviations and then take the square root.

The following example shows how to use this formula.

Example.   Compute the SD for the list 20, 25, 30, 25.

Solution.

1. The average equals 25 since $(20 + 25 + 30 + 25)/4 = 25$

2. To find out how far from average each number is, we can subtract the average from each number. We get the following list of deviations from average: -5, 0, 5, 0.

3. Squaring each one gives us 25, 0, 25, 0.

4. Averaging the deviations equals $(25 + 0 + 25 + 0)/4 = 12.5$, and taking the square root of 12.5 equals approximately 3.5

Together in one formula, we can write the calculations in these steps as

$$SD = \sqrt{\frac{(20 - 25)^2 + (25 - 25)^2 + (30 - 25)^2 + (25 - 25)^2}{4}} \approx 3.5.$$

Variance.

The term **variance** is also used to describe spread and is defined as the square of the standard deviation: Variance $=$ (Standard Deviation)^2.

Coefficient of variation.

The term **coefficient of variation** (abbreviated CV) equals the standard deviation divided by the average (CV $=$ SD/Average) and is used to measure variation relative to the average. It is useful when comparing the variability of two different distributions when their averages are very different and variation as a percentage of the mean is the important thing to be comparing. For example, suppose one stock has an average price around $30 with a standard deviation of $3 and a second stock has an average price around $3 with the same standard deviation of $3. Even though both stock prices have the same standard deviation, the second stock has a lot more variability relative to the average price, and would be a riskier investment. The CV for the second stock is 10 times the CV for the first stock, so the coefficient of variation picks up this variability difference.

Technical Note: the sample standard deviation.

If you have a very small sample of data, the formula above for the SD tends to give a slight underestimate of the true variability of the population from which the sample came. This is because small samples tend to appear less spread out from the sample average than from the true population average. To compensate for this, statisticians have mathematically justified[10] that you should "inflate" the SD you get from the formula by multiplying it by the factor

$$\sqrt{\frac{\text{sample size}}{\text{sample size} - 1}}.$$

For example, if you have a sample of three values which gives you an SD of 5 using the above SD formula, a better estimate of the SD of the population would be $5 \times \sqrt{3/(3-1)} \approx 6.1$. In almost all practical business situations your sample will be reasonably large and this factor will come out to be very close to 1 (for example with a sample of 25 we get $\sqrt{25/(25-1)} \approx 1.02$), so the factor can generally be safely ignored.

This "inflated" SD is called the **sample standard deviation**, and is usually denoted by the letter s. And then typically the sample average is represented by $\overline{X}$. This gives the formulas

$$\overline{X} = \frac{1}{n}\sum_{i=1}^{n} X_i \quad \text{and} \quad s = \sqrt{\frac{1}{n-1}\sum_{i=1}^{n}(X_i - \overline{X})^2}.$$

The regular "non-inflated" SD is called the **population standard deviation** and is denoted by the Greek letter σ.

---

[10]Though the sample variance is an unbiased estimator of the population variance, the sample standard deviation is still a biased estimator of the population standard deviation; see http://www.sec.gov/spotlight/regnms/volatility040605.pdf.

The Excel formula "=Stdev(...)" actually gives this "inflated" sample standard deviation, whereas the Excel formula "=Stdevp(...)" gives the regular population standard deviation.

Exercises

18. Compute the standard deviation for the following list of numbers: 10, 20, 30, 40.

19. Compute the standard deviation for the following list of numbers: 20, 20, 20, 20.

20. Compute the standard deviation for the following list of numbers: -10, $+10$, -5, $+5$.

8. Combining and Transforming Data

When you combine or transform sets of data the average and standard deviation can change in surprising ways. Below we will give you a couple of examples of what you might expect when you do this and what to watch out for.

Combining data sets.   Houses in the north half of a town have an average value of \$170k with a standard deviation of \$30k and houses in the south half have an average value of \$130k also with a standard deviation of \$30k. If we combine both data sets together, will the resulting standard deviation be larger than \$30k, smaller than \$30k, or equal to \$30k?

Discussion.   The standard deviation would probably become larger than \$30k. We would have a group of expensive homes combined with a group of less expensive homes, so overall the data would appear more spread out from the average—which would be somewhere in between the averages of the two groups. This is illustrated in Figure 2.26.

> The standard deviation may change in surprising ways when you combine two data sets.

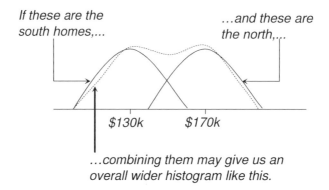

Figure 2.26. Combining two groups of homes.

Transforming data. In your company you decide to give everyone a $1,000 raise. How does the average salary change? How does the median salary change? How about the standard deviation? What would happen if instead you decide to give everybody a 10 percent raise? What if you just decide to give only the top executives a 10 percent raise?

Discussion. If you give everybody a $1,000 raise, it just shifts the histogram up by $1,000. This means that the average and median both increase by $1,000 and the standard deviation stays the same. If you give everybody a 10 percent raise, the sum of all the numbers will increase by 10 percent—and this means that the average and the median will increase by 10 percent.

To see what happens to the standard deviation, consider someone who is making $10,000 versus someone who is making $100,000. They are $90,000 apart currently. After you give each of them a 10 percent raise, they respectively make $11,000 and $110,000. Now they are $99,000 apart—a 10 percent increase. You can see that increasing everything by 10 percent increases the spread between any two data values by 10 percent and this increases the standard deviation by 10 percent.

If only top executives get a raise, it would probably raise the average, the standard deviation, but not the median (assuming, of course, less than half the employees are top executives). All this is summarized in the table below:

	Everyone gets a $1,000 raise	Everyone gets a 10% raise	Top executives only get a raise
Average	Increase by $1,000	Increase by 10%	Increase
SD	No change	Increase by 10%	Increase
Median	Increase by $1,000	Increase by 10%	Probably no change

> Adding a constant amount to all data values does not change the standard deviation, but increases the average by that constant amount.
>
> Multiplying all data values by the same positive constant amount multiplies the standard deviation and average by that same amount.

Technical justification

To understand why multiplying all the numbers by a positive number c multiplies the SD and the average by that same number, you can use algebra as follows:

$$\text{Average} = \frac{1}{n}\sum_{i=1}^{n} cX_i = c\frac{1}{n}\sum_{i=1}^{n} X_i$$

and

$$\text{SD} = \sqrt{\frac{1}{n}\sum_{i=1}^{n}(cX_i - c\mu)^2} = |c|\sqrt{\frac{1}{n}\sum_{i=1}^{n}(X_i - \mu)^2}.$$

Exercises

21. **Currency conversion.** An international company had monthly sales during July 2005 that averaged around $10 million per month, with a standard deviation of around $1 million. If the sales figures were instead reported in Euros, what would you get for the average and standard deviation in Euros (1 Dollar was equivalent to about 0.8 Euros in July 2005)? Explain.

22. **Long-term contracts.** A manufacturing company sells on average 1,000 units per month with a standard deviation of 300 units, but this includes 200 units per month sold to the government as part of a long-term

Figure 2.27. The histogram for a uniform distribution is completely flat.

contract. If the units sold to the government were to be excluded from
the sales figures, how would the average and standard deviation change?
What about the coefficient of variation? Explain.

23. **Home prices.** The average sales price in a city for one-bedroom apartments is $200k with a standard deviation of $100k, and for two-bedroom
apartments it's $300k with an SD of $100k. If we look at one- and two-
bedroom apartments combined
    (a) would the average sales price be larger or smaller than $300k?
    (b) would the SD be larger or smaller than $100k? Explain.

9. Uniform, Exponential, And Poisson Distributions

A number of commonly occurring histogram shapes are given names by
statisticians. In this section we will discuss three of them: the uniform, exponential, and Poisson. To describe these we first define the **cumulative
distribution function** $F(x)$ for a histogram to equal the fraction of the total
area under the histogram at or to the left of x.

When a histogram for data looks flat like Figure 2.27 and data values are
spread evenly throughout, the data are said to follow a **uniform distribution**.
A uniform distribution ranging between a and b has cumulative distribution
function

$$F(x) = \frac{x-a}{b-a}.$$

If the histogram looks like Figure 2.28 then the data are said to follow an
exponential distribution. An exponential distribution having mean m has
cumulative distribution function

$$F(x) = 1 - e^{-x/m}.$$

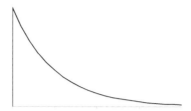

Figure 2.28. Histogram for an exponential distribution.

It can be shown using calculus that the standard deviation equals $(b-a)/\sqrt{12}$ for the uniform distribution and m for the exponential distribution.

Exponential distributions commonly arise with inter-event times, when events are the result of a large number of independently operating but rare possible events that could occur. For example if we record the time between the arrival of phone calls at a customer call center, the histogram might be expected to follow an exponential distribution. This type of shape also arises if we record the time between large stock market corrections. Below is a histogram of the number of weeks elapsing between one-day drops of more than 2% in the Standard & Poor's 500 stock market index between the years 1950 in 2007.[11] You can see the shape looks somewhat similar to an exponential distribution:

---

[11] Data from `finance.yahoo.com`.

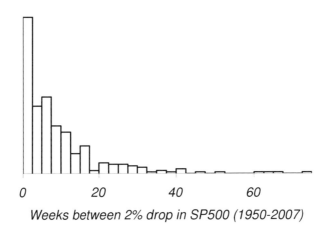

Weeks between 2% drop in SP500 (1950-2007)

A **Poisson distribution** with mean m and standard deviation $\sqrt{m}$ has cumulative distribution function equal to

$$F(x) = \sum_{k=0}^{x} \frac{e^{-m}m^k}{k!},$$

for $x = 0, 1, 2, \ldots$. For example, a mean of 2.6 corresponds to the histogram in Figure 2.29.

This shape often arises when counting the number of occurrences of rare events. For example, if we count the number of days in each month between 1950 and 2007 that the Standard & Poor's 500 stock market index changed by more than 3/4 of 1%, we get an average of 2.6 days per month and the histogram shape in Figure 2.30.[12]

This shape looks somewhat like the Poisson distribution above, though it's not exact since the height of the first bar and the standard deviation look somewhat different in each picture.

Exercises

24. Earthquake magnitudes. The United States Geological Survey reports that worldwide since 1900 there has been an average of 19.4 earthquakes

---

[12]Data from `finance.yahoo.com`.

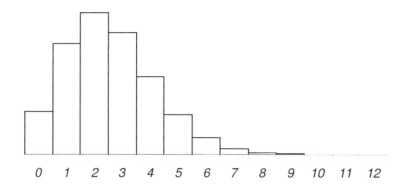

Figure 2.29. Histogram for a Poisson distribution with mean 2.6.

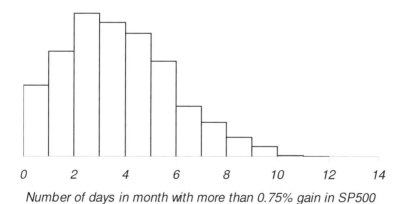

Number of days in month with more than 0.75% gain in SP500

Figure 2.30. Histogram for the number of days in each month where the stock market moved more than 0.75%.

Table 2.6. An excerpt from the file `earthquake.xls`.

Year	# earthquakes above 7.0 magnitude
1900	13
1901	14
1902	8
1903	10
1904	16
1905	26
1906	32
1907	27
1908	18
1909	32

of magnitude greater than 7.0 per year.[13] The data are given in the file `earthquake.xls` (an excerpt is in Table 2.6). Using the Poisson distribution as an approximation, how often will there be a year with more than 25 earthquakes of magnitude greater than 7.0? What do the data say?

25. **Order times.** The file `exponential.xls` contains the times of a large number of orders a company receives.

   (a) Compute the average number of days between orders.

   (b) Does the histogram for the number of days between orders look like it follows an exponential distribution?

   (c) Compute $F(5)$ for the cumulative distribution function of an exponential distribution with mean equal to what you computed in part (a). (d) What fraction of orders arrive within five days of the previous order? How does this compare with your answer in (c)?

26. **Order times.** The file `uniform.xls` contains the times of a large number of orders a company receives.

   (a) Does the histogram for the number of days between orders look like it follows a uniform distribution?

   (b) Compute $F(5)$ for the cumulative distribution function of the corresponding uniform distribution.

   (c) What fraction of orders arrive within five days of the previous order? How does this compare with your answer in (b)?

---

[13]See `http://neic.usgs.gov/neis/eqlists/7up.html`.

10. Using Excel

Drawing Histograms Using the Histogram Worksheet

Suppose we want to draw a histogram for the following salary numbers (in thousands of dollars):

110	75	120	60
120	120	50	90
140	75	110	80
70	115	80	70
125	60	60	110
100	110	70	

First type them into a column in Excel with the title "Salary" at the top (the data gets cut off at the bottom in the picture here):

	A	B	C
1	Salary		
2	110		
3	120		
4	140		
5	70		
6	125		
7	100		
8	75		
9	120		
10	75		
11	115		
12	60		
13	110		
14	120		
15	50		
16	110		

Then open the file `histogram.xls`. Copy the column of data from above into this worksheet by first clicking on the column letter "A" above the word "Salary" at the top of Column A, then pressing "Control-C" to copy it, then clicking on the column letter "A" in the file `histogram.xls` and finally pressing "Control-V" to paste it. Since we have such a small amount of data, enter "8" as the approximate number of bars in the gray box, and you should see the following histogram:

You may notice that there are nine bars in this histogram, instead of the eight bars we specified in that dialog box. Here Excel tries to get at least eight bars, but also tries to make sure the bars begin and end on "round" numbers. Usually you will see more bars than you specify, so you may have to adjust the number of bars to get the picture you want.

Drawing Histograms Using Standard Excel

Drawing histograms with the menu options that come standard in Excel is inconvenient and it's easier to use the add-in. But we'll show you how to do it anyway, because you'll probably find yourself someday without an add-in to help.

Suppose again we want to draw a histogram for the following salary numbers (in thousands of dollars):

110	75	120	60
120	120	50	90
140	75	110	80
70	115	80	70
125	60	60	110
100	110	70	

First type them into a column in Excel with the title "Salary" at the top (the data gets cut off at the bottom in the picture here):

	A	B	C
1	Salary		
2	110		
3	120		
4	140		
5	70		
6	125		
7	100		
8	75		
9	120		
10	75		
11	115		
12	60		
13	110		
14	120		
15	50		
16	110		

Then decide on the equal-sized ranges you would like to divide your data into. For example suppose we want to divide the data into income ranges that each are each $10k wide: $50k–$60k, $60k–$70k, and so on up to $140k–$150k. To do this, we type 50, 60, 70, and so on up to 150 in a new row of the spreadsheet off to the side of the data:

	A	B	C	D	E	F	G	H	I	J	K	L	M	N
1	Salary													
2	110		50	60	70	80	90	100	110	120	130	140	150	
3	120													
4	140													
5	70													

Then choose "Data analysis" from from the "Data" ribbon and then click on "Histogram," (you may need to install the "Analysis Toolpak" first by going to the "File" ribbon and clicking "Options/Add ins/Go" and checking off "Analysis Toolpak") and fill in the blanks as follows:

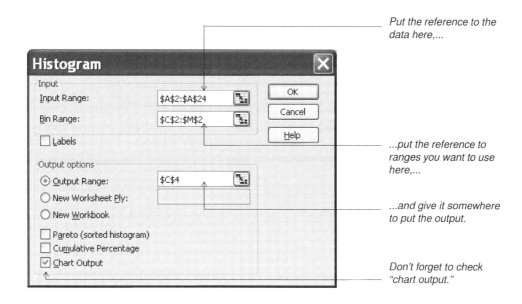

After you click "OK" and do a little formatting, you should get a picture like the following:

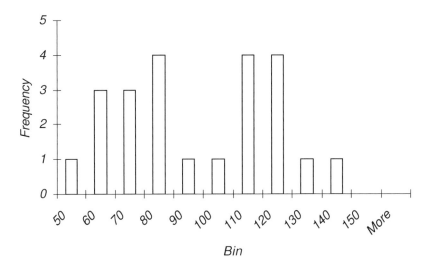

You can eliminate the spaces between the bars by double-clicking on one of the bars and then clicking on "options" and reducing the size of the "gap" down to zero. This is illustrated below:

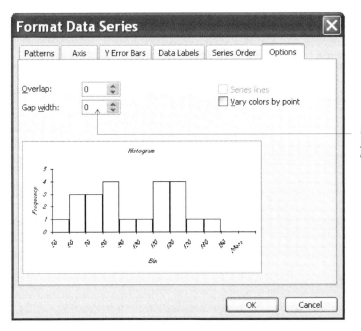

You may have noticed this histogram looks slightly different from the one we drew with the Excel add-in software. There the common convention is used that a histogram bar includes values at the left edge of the bar's range but not at the right edge (for example if a bar covers the range 10-20 it includes 10 but not 20). Standard Excel uses the slightly less common convention that a bar includes values at the right edge but not the left edge. Either of the conventions is okay, and you usually won't notice the difference when you have a large amount of data. When you have a very small amount of data, as we do here, you may really notice the difference, so you should make sure you know which convention is being followed.

Drawing histograms and computing descriptive statistics using R

R is a statistical programming language that has many powerful features Excel lacks. To create a histogram with R, you should first install R using the link

$$\text{https://cran.rstudio.com/}$$

and then install the editor R-studio using

$$\text{https://www.rstudio.com/products/rstudio/download/}$$

Also you should download the data sets for this textbook using

$$\text{http://smgpublish.bu.edu/pekoz/bookdata.zip}$$

Then start R-Studio (it should now be a newly installed program on your computer) and click on "File/Import dataset/From Excel" and navigate to the data file `salaryexample.xlsx` and click "Import". You should then see the list of salaries. Then click on "File/New File/R Script" and type the following:

```
hist(salaryexample$salary)
```

and then hold down the "Control" key while simultaneously pressing "Enter" (on the Mac you must press "Command" key while simultaneously pressing

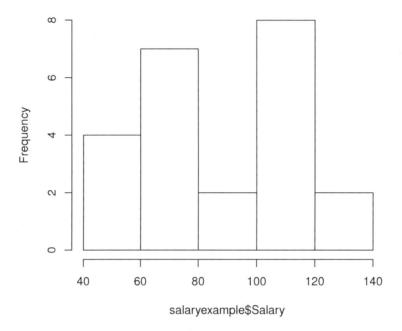

Figure 2.31. Salary histogram using R.

"Enter"). You should see the histogram in Figure 2.31. In R you must press control-enter (or command-enter on the Mac) after each line in order to execute it. You can go back to the end of a previous line and re-execute it if you want.

To compute descriptive statistics in R, it is first helpful to rename the data set to something shorter than "salaryexample" so we can refer to it more easily. The "data.frame" command we will use below has the added benefit of renaming any columns so they don't have any spaces in the names; R does not work well when column names have spaces in them. Type the following, pressing control-enter (or command-enter on the Mac) afterwards to execute it:

```
d=data.frame(salaryexample)
```

This renames the data set so it is now called "d." For the salary we can compute the average, standard deviation, median, mode and histogram using the following commands:

```
mean(d$salary)
sd(d$salary)
median(d$salary)
mode(d$salary)
hist(d$salary)
```

Computing Descriptive Statistics in Excel

In Excel it is very easy to compute the average, standard deviation, median, and mode using the following commands:

```
=Average(...)
=Stdevp(...)
=Median(...)
 =Mode(...),
```

where by the three dots "..." we mean you should put in the range where your data are. For example if you want to take the average of the numbers in cells A1 through A100, click on a blank cell, type the following command, and then press enter:

```
=Average(A1:A100)
```

You can also use the "descriptive statistics" option in Excel 2003 under the tools/data analysis menu (and under the "data" ribbon in Excel 2007).

Exercises

27. **Stock price return and variability.** The file sp500returns.xls (an excerpt is in Table 2.7) contains the daily percentage return for a large number of different stocks over the period of one year.

    (a) Which stock showed the most variability?

    (b) Which stock had the highest average return?

Table 2.7. An excerpt from the file `sp500returns.xls`.

Day #	A	AA	AAPL	ABC
1	-0.97%	0.23%	-1.97%	-2.17%
2	-1.59%	2.34%	-3.27%	-1.04%
3	-0.26%	0.85%	-2.26%	-8.07%
4	-0.79%	-0.54%	-0.67%	-9.11%
5	-0.38%	-4.53%	1.74%	-9.13%
6	-1.28%	-2.48%	3.78%	-9.84%
7	-0.77%	-2.31%	2.87%	-8.80%
8	-0.26%	-3.11%	5.34%	-8.65%
9	-0.31%	-3.25%	4.44%	-8.65%
10	-0.13%	-1.42%	5.38%	-8.69%

Mathematical Summary

Here we provide a quick summary of the main concepts of the chapter using standard statistical notation.

Definition 1. *Given a sample of data values $x_1, x_2, \ldots, x_n$ chosen from a larger population of values $y_1, \ldots, y_N$, we define the following.*

1. *The* **population mean** *is defined as*

$$\mu = \frac{1}{N} \sum_{i=1}^{N} y_i$$

2. *The* **sample mean** *is defined as*

$$\bar{x} = \frac{1}{n} \sum_{i=1}^{n} x_i$$

3. *For a sample of values the* **median** *is defined as*

$$m = \begin{cases} x_{((n+1)/2)} & \text{if } n \text{ is odd} \\ (x_{(n/2)} + x_{(1+n/2)})/2 & \text{if } n \text{ is even,} \end{cases}$$

where $x_{(i)}$ denotes the $\lfloor i \rfloor$th smallest data value, and $\lfloor i \rfloor$ denotes the integer portion of i. For a population, the median is defined similarly using the values y_i.

4. *The* **population standard deviation** *is defined as*

$$\sigma = \sqrt{\frac{1}{n}\sum_{i=1}^{n}(y_i - \mu)^2}$$

5. *The* **sample standard deviation** *is defined as*

$$s = \sqrt{\frac{1}{n-1}\sum_{i=1}^{n}(x_i - \bar{x})^2}.$$

6. *For a population the* **coefficient of variation** *is defined as*

$$cv = \sigma/\mu$$

and for a sample is defined as

$$cv = s/\bar{x}.$$

We state without proof the following proposition, which states that the sample standard deviation is a good way to estimate the population standard deviation.

Proposition 2. *Given a sample $x_1, \ldots, x_n$ selected randomly from a population $y_1, \ldots, y_N$, then s^2 is an unbiased estimator of σ^2 in the sense that*

$$\sigma^2 = \frac{1}{N^n}\sum_{z_1,\ldots,z_n}\frac{1}{n-1}\sum_{i=1}^{n}\left(z_i - \left(\frac{1}{n}\sum_{j=1}^{n}z_j\right)\right)^2,$$

where the outside summation is over all N^n possible samples $z_1, \ldots, z_n$ you can take with replacement from the population $y_1, \ldots, y_N$.

We will not prove the previous proposition, but we will illustrate it in a simple example. Suppose you have a population of two different possible data values: 2 and 4. This means $N = 2$, $y_1 = 2$, and $y_2 = 4$. Clearly the population variance (the square of the standard deviation) equals 1, and the average equals 3. Now suppose you take a sample of size 2 with replacement from the population and you would like to estimate the population variance. There are four possible samples (with replacement) you could get: 2, 2 or 2, 4 or $4, 2$ or $4, 4$. The first and last samples each have a variance of 0 and the second and third samples each have a variance of 1. Averaging together the four variances $0, 1, 1, 0$ gives 0.5, which is an underestimate of the population variance. Since we have a sample size of $n = 2$, the above proposition tells us we can instead inflate each variance by the a factor $n/(n-1) = 2/(2-1) = 2$ to get the four inflated variances $0, 2, 2, 0$ having an average of 1—the correct population variance!

We also state without proof a result stating that the mean and median respectively minimize the average squared distance and the average absolute distance to all the data points. This gives some justification for squaring the data values in the formula for standard deviation. We summarize this as follows.

Proposition 3. *Let μ and m respectively be the mean and median for the list of data values $x_1, \ldots, x_n$. Then for any c, we have*

1. $\dfrac{1}{n} \displaystyle\sum_{i=1}^{n} (x_i - \mu)^2 \leq \dfrac{1}{n} \sum_{i=1}^{n} (x_i - c)^2$

2. $\sqrt{\dfrac{1}{n} \displaystyle\sum_{i=1}^{n} |x_i - m|} \leq \sqrt{\dfrac{1}{n} \sum_{i=1}^{n} |x_i - c|}.$

Some useful properties about the average and standard deviation are the following, which state that a linear transformation of the variables affect the average by the same linear transformation, and scales the standard deviation by the absolute value of the slope of the transformation.

Proposition 4. *Given a sample $x_1, x_2, \ldots, x_n$, the following hold:*

1. $\dfrac{1}{n} \displaystyle\sum_{i=1}^{n} (a + bx_i) = a + b\dfrac{1}{n} \sum_{i=1}^{n} x_i$

2. $\sqrt{\dfrac{1}{n-1} \displaystyle\sum_{i=1}^{n} (a + bx_i - (a + b\bar{x}))^2} = |b| \sqrt{\dfrac{1}{n-1} \sum_{i=1}^{n} (x_i - \bar{x})^2}$

Finally, Chebyshev's inequality can be stated as follows.

Proposition 5. *Given data values* $y_1, y_2, \ldots, y_n$ *with standard deviation* σ, *mean* μ, *and function* $F(x)$ *which equals the fraction of data values less than or equal to* x, *we have*

$$F(\mu - k\sigma) + 1 - F(\mu + k\sigma) \leq 1/k^2$$

for any $k > 0$.

11. R lab: MBA graduate salaries

First install R using the link

$$\texttt{https://cran.rstudio.com/}$$

and then install the editor R-studio using

$$\texttt{https://www.rstudio.com/products/rstudio/download/}$$

If you want an introductory tutorial on either of these programs, you can find good ones at

$$\texttt{https://www.datacamp.com/}$$

Also you should have downloaded the data sets for this textbook using

$$\texttt{http://smgpublish.bu.edu/pekoz/bookdata.zip}$$

Then start R-Studio and click on "File/Import dataset/From Excel" and navigate to the data file "`BU MBA salalaries.xlsx`" and click "Import".

You should then see the data set. It contains data from four recent years for the MBA graduates at Boston University. The data set includes their incoming GMAT score, their undergraduate GPA, the number of years of work experience they had, the MBA program they applied to, their graduating GPA, a column called "OA at graduation" indicating whether or not they had accepted a job offer at graduation (1=Yes, 0=No), a column called "OA at 90 days" indicating whether or not they had accepted a job offered by 90 days after graduation (1=Yes, 0=No), a column indicating their graduating year, and a column with the salary of the job they took after graduation if they reported it.

Let's answer a few questions about the salaries using R. In R-Studio click on "File/New File/R Script" and type the commands below. After each command hold down the "Control" key while simultaneously pressing "Enter" (on the Mac you must press "Command" key while simultaneously pressing "Enter") to execute the command.

Our first command will rename the data set into something simpler and thus more convenient, and will also replace any spaces in any of the column names with periods (R does not work well with spaces in column names). The following creates a copy of the data set which we will be named "d":

```
d=data.frame(BU_MBA_salaries)
```

Make sure you hold down the "Control" key while simultaneously pressing "Enter" (on the Mac you must press "Command" key while simultaneously pressing "Enter") after each command. To see the list of transformed names you can use the command

```
names(d)
```

and to see the entire data set use

```
View(d)
```

Note that the "View" command is capitalized whereas the "names" command is not; since R is an open-source language developed by many people and influenced by many programming languages, there are many places where the syntax may seem strange. Next, let's take a look at the average salary. Type

```
mean(d$salary)
```

and you should see the answer "53102.52." You may notice after just typing a few of the letters that the column name "salary" appears in a pop-up menu; you can press the tab key or click on it to speed up the typing process. An average salary of $53k seems a little low for MBA graduates. Let's take a look at the histogram. Type

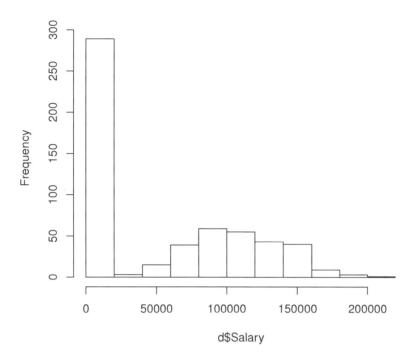

Figure 2.32. Salary histogram using R.

```
hist(d$salary)
```

and you should see the histogram in Figure 2.32 below. We see that there is a huge bar right above zero dollars, meaning that many of the salaries shown are for zero dollars. Looking back at the data set we see that not everybody has reported a salary, and apparently the data set has a zero whenever this happens. Let's draw a histogram for just the people who have reported a nonzero salary. In order to do this in R, we create a new condition inside square brackets as follows:

```
hist(d$Salary[d$Salary>0])
```

This gives us the histogram in Figure 2.33 below for just the people who

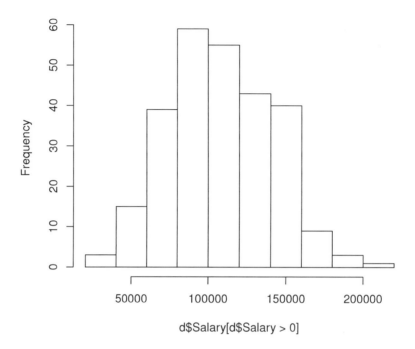

Figure 2.33. Salary histogram for just people reporting nonzero salaries using R.

have a salary number greater than zero. Let's find the average of this subset of people who report nonzero salaries:

```
mean(d$Salary[d$Salary>0])
```

and we see a much healthier salary average of "110580.5." Now let's take a look at how salaries differ depending on whether or not someone takes a job right after graduation or 90 days after graduation. First let's draw a histogram for the people who accepted an offer at graduation and have a nonzero salary listed. In order to do this we put two conditions inside square brackets separated by an "&" sign as follows:

```
hist(d$Salary[d$Salary>0 & d$OA.At.Graduation==1])
```

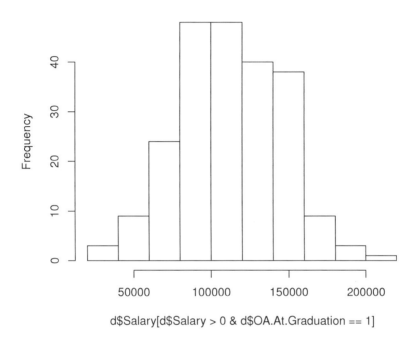

Figure 2.34. Salary histogram for people reporting a salary from a job accepted at graduation.

Note that we need to use a double "==" here even though it seems like a single "=" would make more sense to an Excel user. We then should see the histogram in Figure 2.34 below. Let's now draw a histogram for the people who did not accept a job at graduation but accepted one at 90 days:

```
hist(d$Salary[d$Salary>0 & d$OA.At.Graduation==0 & d$OA.At.90.Days==1])
```

We then should see the histogram in Figure 2.35 below. Looking at the histograms it appears that people who took jobs at graduation got higher salaries then the ones who had to wait 90 days. We can reconfirm this by computing the corresponding averages. Typing

```
mean(d$Salary[d$Salary>0 & d$OA.At.Graduation==1])
```

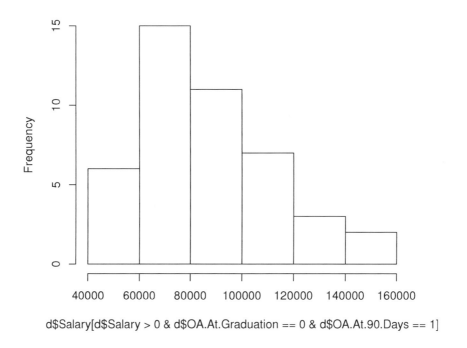

Figure 2.35. Salary histogram for people reporting a salary from a job accepted at graduation.

and

```
mean(d$Salary[d$Salary>0 & d$OA.At.Graduation==0 & d$OA.At.90.Days==1])
```

gives the respective averages "114663.7" and "89886.36," thus reconfirming that salaries tended to be generally higher for the people who took jobs at graduation.

Let's look at average salaries broken down by the program people applied to using the "by" command. Type

```
by(d$Salary[d$Salary>0],d$App.Type[d$Salary>0],mean)
```

and you should see the averages for each program. You can also use

```
by(d$Salary[d$Salary>0],d$App.Type[d$Salary>0],length)
```

to see the number of students in each program, or

```
by(d$Salary[d$Salary>0],d$App.Type[d$Salary>0],summary)
```

to look at more summary statistics for each program. From looking at these it looks like health sector management graduates are highly paid!

Summary

Summarizing data can reveal interesting stories that were hidden before, but can also be misleading unless you know what to watch out for. If you know what can go wrong, it will help make sure you get the right story. Some key concepts to have in mind while doing the problems are the following: histograms versus bar charts, bimodal and skewed histograms, average, median, mode, standard deviation, the empirical rule, and the effects of transforming or combining data.

12. Exercises

28. **Computing standard deviation.** Compute by hand the mean and standard deviation of the following list of numbers: 20, 30, 35, 40, 50.

29. **A funny thing happened to household income.** A recent *Harvard Business Review* article claims "a funny thing happened to U.S. household income while marketers weren't looking: Its distribution curve changed dramatically."[14] It then goes on to say "...a vast new space has opened up for offerings that were once not economical enough to pursue." The file incomes.xls (an excerpt is in Table 2.8) contains data for a representative sample of household incomes in 1970 and 2000 (adjusted for inflation). Using histograms, averages, medians, and standard deviations, discuss and illustrate how the distribution of household income has changed over this time period, and describe this "vast new space"

---

[14] "Selling to the Moneyed Masses," by Paul Nunes, Brian Johnson, and Timothy Breene, *Harvard Business Review*, July/August 2004 pp. 95–104.

Table 2.8. An excerpt from the file `incomes.xls`.

Incomes 1970 ($1,000s)	Incomes 2000 ($1,000s)
45	190
15	50
150	50
65	65
20	170
50	65
75	30
30	50
45	25
50	90
80	95
55	80
20	60

that has opened up (hint: it isn't the richest households). Can you think of some current products targeted to this space?

30. **Choosing suppliers.** A retailer has to decide which of two overseas suppliers (Supplier A and Supplier B) will manufacture its product. The retailer would like delivery times as short as possible in order to respond more efficiently to customer demand. The retailer has collected delivery time data on a large number of similar-sized orders placed with each of these suppliers in the past and sees that Supplier A has on average a 44 day delivery time, and Supplier B has on average a 51 day delivery time. Because 44 is smaller than 51, an analyst recommends using Supplier A. Looking at the file `retailer.xls` (an excerpt is in Table 2.9), what do you recommend? Explain your reasoning and any important features of the data you discover that are relevant.

31. **Working hours.** The file `hours.xls` (an excerpt is in Table 2.10) shows the reported hours worked at a job per week for a representative sample of 1000 Americans age 15 and over, as measured by the US Census Bureau.[15] How would you summarize this data and what story does it tell?

---

[15] http://www.bls.census.gov/cps/ads/1995/freq/frq_a_hrs1.htm

Table 2.9. An excerpt from the file `retailer.xls`.

Supplier A delivery times (in days)	Supplier B delivery times (in days)
31	37
81	50
30	34
37	47
24	46
20	56
82	42
25	34
17	52
23	54

Table 2.10. An excerpt from the file `hours.xls`.

Person #	# hours worked last week
1	0
2	0
3	0
4	0
5	0
6	0
7	0
8	40
9	0
10	56

32. **Gender differences in income.** The file `M&F incomes.xls` (an excerpt is in Table 2.11) gives the income distribution for Americans aged 15 and over in 2005 broken down by by gender. How would you characterize income differences between men and women? Explain briefly, using diagrams if necessary.

33. **Stock prices.** The file `stockprices.xls` (an excerpt is in Table 2.12) contains the daily closing prices for 100 different stocks over a thirty day period. Which stock showed the most price variability? Which stock had the highest average price?

34. **Do women reach the top sooner than men?** A recent *Harvard Busi-*

Table 2.11. An excerpt from the file M&F incomes.xls.

Income	Males	Females
Under $2,500	13.3%	21.2%
$2,500 to $4,999	2.5%	4.4%
$5,000 to $7,499	3.3%	6.9%
$7,500 to $9,999	3.2%	6.2%
$10,000 to $12,499	4.4%	6.4%
$12,500 to $14,999	3.7%	4.7%
$15,000 to $17,499	4.4%	5.2%
$17,500 to $19,999	3.4%	3.7%
$20,000 to $22,499	4.7%	4.6%
$22,500 to $24,999	2.9%	3.1%

Table 2.12. An excerpt from the file stockprices.xls.

Date	BTX	LLS	MFE	SEKO
3/2/2007	299.65	81.65	287.28	278.08
3/1/2007	302.56	81.02	291.05	278.96
2/28/2007	300.13	81.82	288.47	281.84
2/27/2007	302.52	82.44	281.69	281.26
2/26/2007	303.94	83.96	283.18	282.77
2/23/2007	303.78	85.68	288.52	283.37
2/22/2007	304.62	86.34	292.59	284.41
2/21/2007	301.08	85.06	293.19	284
2/20/2007	299.1	85.3	294.68	283.14
2/16/2007	308.05	83.11	301.33	284.8

ness Review article discusses differences in the ages of male and female executives.[16] The file execages.xls (an excerpt is in Table 2.13) is a large representative sample of ages of male and female executives holding similar jobs. The article goes on to say "Women may be scarce in senior management, but here's an intriguing finding: those who do make it into the executive ranks get there faster than men." Do the data tell this story? Explain briefly.

---

[16] "Younger Women at the Top," *Harvard Business Review*, 00178012, Apr2007, Vol. 85, Issue 4, and C.E. Helfat, D. Harris, and P.J. Wolfson, "The Pipeline to the Top: Women and Men in the Top Executive Ranks of U.S. Corporations," *The Academy of Management Perspectives*, November 2006.

Table 2.13. An excerpt from the file `execages.xls`.

Women	Men
52	50
46	57
48	55
41	57
43	55
47	45
53	46
34	52
44	40
53	44

Table 2.14. An excerpt from the file `ordersizes.xls`.

Order #	Amount ($)
1	311
2	282
3	812
4	74
5	49
6	588
7	836
8	94
9	680
10	155

35. **Order sizes.** The file `ordersizes.xls` (an excerpt is in Table 2.14) contains the dollar amounts of a large number of orders made at a company. Describe the histogram for this data.

36. **Household income.** The graph below shows the distribution of household income in 2003 in the United States.[17]

   (a) Assuming households are spread evenly within each bar, are the households earning in the range $10,000 to $25,000 spread fairly evenly over that range?

---

[17]Source: US Census Bureau. See `http://www.census.gov/prod/2004pubs/p60-226.pdf`

(b) Assuming households are spread evenly within each bar, are there more households earning in the range $60,000 to $70,000 or in the range $40,000 to $50,000?

(c) Is this graph a histogram? Explain briefly.

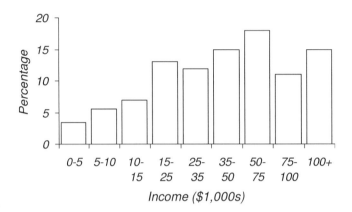

37. **Employee churn.** Among warehouse retailers, Costco, with 67,600 full-time employees in the United States, is number one in sales. Sam's Club, with 110,200 fulltime employees in the United States, is number two. A *Harvard Business Review* article details how when a Costco employee quits, the cost of replacing him or her on average is $21,216.[18] And due to the lower cost of labor, when a Sam's Club employee leaves the cost on average is $12,617. The article then says "at first glance, it may seem that the low-wage approach at Sam's Club would result in lower turnover costs." But it's believed Sam's Club's turnover rate per year is more than twice as high as Costco's: 44% versus 17% of employees quit every year. What is the total turnover cost at each company?

38. **Income distributions.** Figure 2.36 contains income histograms (in $1,000s) for three different groups of people.

(a) For each histogram, is the average closest to 90, 100, or 110?

(b) For which histogram is the median less than the average?

(c) Is the standard deviation (SD) in histogram (i) around 5, 15, or 50?

(d) Is the SD for histogram (i) a lot different than that for histogram (iii)?

---

[18] "The High Cost of Low Wages," *Harvard Business Review*, Dec 2006, p. 23, Wayne F. Cascio

Figure 2.36. Income histograms (in $1,000s) for three different groups of people.

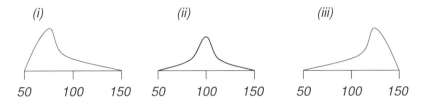

Table 2.15. An excerpt from the file `funds2005.xls`.

Mutual Fund Name	% Return 2004	% Return 2005
ABN AMRO Chic Capital Bal N CHTAX	5.02	-5.14
ABN AMRO/Veredus Aggr Grth VERDX	20.69	2.51
Acadian Emerging Markets AEMGX	33.53	25.14
Aegis Value AVALX	13.42	-6.41
Alpine Realty Income & Growth AIGYX	30.95	5.74
Alpine U.S. Real Estate Equity Y EUEYX	39.32	5.95
Amer Cent Global Growth TWGGX	15.11	17.05
Amer Cent Intl Bond BEGBX	10.1	-12.14
Amer Cent Real Estate REACX	28.85	2.62
Amer Cent Small Cap Quant ASQIX	28.26	-4.12

39. **Sales variation for seasonal products.** The average monthly demand for a product during the six month low-volume season is 200 units, and the standard deviation is 50 units. The average monthly demand rises to 500 units per month during the six month high-volume season, with a standard deviation of 70 units.

   (a) Can we say that the average monthly demand over the entire year will be somewhere around 350 units? If not, do you think it will be above 500 or below 200 units?

   (b) Can we say that over the entire year the standard deviation of demand per month will be somewhere around 60 units? If not, do you think it will be above 70 or below 50 units?

40. **Mutual fund returns.** The file `funds2005.xls` (an excerpt is in Table 2.15) contains the annual percentage return given by all mutual funds for sale in 2004 and 2005. Using summary statistics and histograms, illustrate how the distribution changed over this time period.

Table 2.16. An excerpt from the file `orders.xls`.

Order # (2000)	# of units	Order # (2005)	# of units
1	465	1	966
2	484	2	856
3	376	3	193
4	561	4	688
5	556	5	10
6	572	6	596
7	484	7	403
8	444	8	295
9	550	9	360
10	460	10	856

41. **IRS statistics.** The Internal Revenue Service (IRS) released data showing that the total income reported by all Americans combined in 2004 was down 1.4% from 2000.[19] The IRS data also showed that the average income reported per person by Americans in 2004 was down 3% from 2000, but the median income was down only 1%. Based on these figures, is it possible to determine if the number of Americans reporting income increased or decreased between 2000 and 2004? Explain briefly.

42. **Order sizes.** A company has noticed that the size of orders placed by its customers has changed over the years. The file `orders.xls` (an excerpt is in Table 2.16) contains the order size of all orders customers placed in the year 2000, as well as in the year 2005. Describe how the distribution has changed.

43. **Television watching.** As part of a marketing survey a large group of people are asked how many hours of television they watch per week. The average number of hours was about 5 and the SD was about 5. The histogram for the data looked like one of the three shapes (i), (ii) or (iii) below. Which shape do you think it was? Choose one and justify your choice.

---

[19] *Time* magazine, December 11, 2006, Notebook, page 34

Table 2.17. An excerpt from the file `option.xls`.

Simulated scenario	Profit (billions of $)
1	0.548
2	0.202
3	-0.331
4	-0.380
5	1.600
6	0.783
7	-2.189
8	0.727
9	0.807
10	-0.157

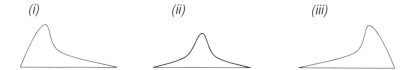

(i)   *(ii)*   *(iii)*

44. **Skewed histogram.** The average account balance of a large selection of customer savings accounts is $20,000 and the SD is $40,000. Do these figures give you any reason to suspect the histogram may be skewed? Explain.

More Challenging Exercises

45. **Real options.** A large pharmaceutical firm is considering a deal to purchase the rights to a drug, still in development, from a small biotechnology lab. There is much uncertainty about the potential performance and market for the drug, and the firm must make substantial up-front investments to finish the development research. To help forecast its potential profit, the firm uses a computer simulation to create hundreds of possible outcome scenarios and computes the profit for each: these are in the file `option.xls` (an excerpt is in Table 2.17).

    (a) The histogram for the simulated profits can be viewed as describing where the actual profit is likely to fall, and the average of the simulated profits is viewed as the overall expected profit. Looking at the histogram and summary statistics, does this seem like a good deal?

    (b) The firm is negotiating an option (called a **real option**) to exit from the deal before all money is paid if in later studies the drug does not live up to expectations. The simulated profits with this option included are also given on the second sheet of the file. How does this option affect profits? How much value is it expected to add to the deal?

46. **Bidding on past-due accounts.** A collection agency is considering bidding on the past-due accounts from one of two different companies. Both companies have the same number of past-due accounts. For the first company the median amount due per account is $70, the average is $65, and the SD is $30. For the second company the median is $60, the average is $80, and the SD is $25. Which company has more past-due in terms of total dollars? Explain briefly.

47. **Sales territories.** Below is a histogram for sales (in thousands of dollars) of a certain product in several hundred US cities. Cities that have sales in the range from $20,000 to $25,000 are handled by Salesperson A, and cities that have sales in the range from $50,000 to $55,000 are handled by Salesperson B. Which of these two (A or B) has a larger territory, in terms of total overall sales dollars? Explain briefly.

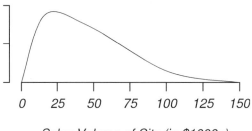

Sales Volume of City (in $1000s)

48. **How often do people move?** Market researchers for a moving company interview a representative sample of adults and ask them how recently they have moved. About 20 percent of them had moved in the last year, 5 percent hadn't moved for more than 35 years, and 20 percent had moved between 15 and 35 years ago. Would the SD for this data be closest to 2 years, 8 years, or 25 years?

49. **Comparing salaries.** Two companies have the same number of employees, but the total payroll is larger in the first company. Does this imply

that the median salary is also greater in the first company? Explain, drawing histograms if needed.

50. **Comparing sales.** You have 130 stores in Region 1 and 290 stores in Region 2. Total sales of all stores combined in Region 1 last month was $3.8 million and in Region 2 was $5.6 million. The median sales per store was the same in each region: $29,000 per store. Draw a reasonable sketch for the histogram of sales per store last month in each region.

51. **Corporate income.** An article in *Journal of Accounting and Economics* describes how researchers computed the annual change in net income reported by a representative sample of firms and created a histogram looking similar to the one below. The shape has an interesting spike in the middle which the article says is "revealing." What do you think this spike could be revealing?[20]

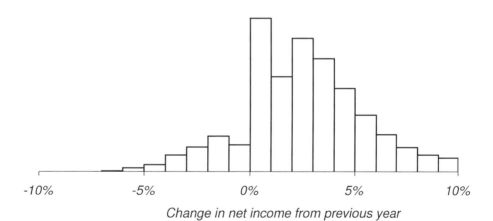

Change in net income from previous year

52. **Market volatility.** The NASDAQ and Dow Jones are stock indices that each move to follow the value of a corresponding portfolio of stocks. An analyst downloads the value of each index at the opening of each of the several hundred trading days during the period 2003–2004. Then she draws a histogram for the data corresponding to each index and the results are sketched below.

[20]D. Burgstahler and I. Dichev, "Earnings management to avoid earnings decreases and losses," *Journal of Accounting and Economics*, Volume 24, Issue 1, December 1997, Pages 99–126. Thanks to Brian Mittendorf (Yale University) for suggesting this article.

(a) One of the histograms shows a standard deviation of around 80 and the other around 300. Which one is which? Explain.

(b) Which histogram shows a larger coefficient of variation?

(c) If you invested the same amount of money in two mutual funds that followed each of the stock indices, which investment would show more variability from day to day?

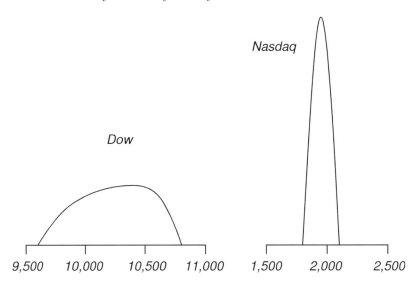

53. **MBA program rankings.** Each year *U.S. News & World Report* ranks graduate MBA programs based on factors both reported by the schools and gathered independently. Although its accuracy is controversial, the ranking is extremely influential. One factor that weighs heavily in the ranking is the undergraduate GPA of the incoming MBA class, and a difference of even 1/10 of a grade-point (for example, between 3.3 and 3.4) can mean a difference of several positions in the rankings. The average GPAs are computed by the individual schools themselves and are then sent to the magazine. Figure 2.37 is a histogram for the average GPA reported by the top 80 schools in the 2005 rankings. Each bar includes GPAs at the left endpoint of its range but not the right endpoint. What could explain the fact that the bars alternate high and low here? Does this have any implications for a school reporting its GPA to the magazine?

54. **Foster care.** The newspaper reports that "the average time spent by children in foster care [in New York City] more than doubled, to 4.4

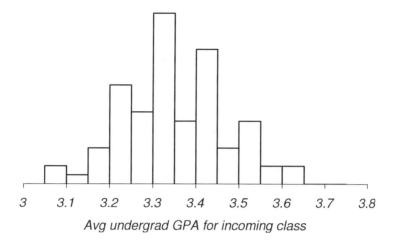

Avg undergrad GPA for incoming class

Figure 2.37. The average GPA reported by the top 80 schools in the 2005 US News & World report rankings.

years in 1996 from 2.1 years in 1990."[21] Does this mean the typical foster child spent more than twice the time in foster care in 1996 than in 1990? What additional summary statistics or charts might help you answer the question in the last sentence and how might they affect your conclusions about the change in time in foster care?

55. **Age distribution.** The city of Wahoo, Nebraska (population 4,000) is listed as the location of the "home office" that produces the Top-10 list for David Letterman's Late Show television program. On the city's Web page it gives a histogram of the ages of its residents.[22] Figure 2.33 is a sketch of the histogram given for women. What story does the shape tell about the city?

56. **Wall Street bonuses.** A news article writes "The average windfall for each individual at the five largest U.S. securities firms will be enough to buy a $165,000 Bentley Continental GT, the two-door coupe favored by Paris Hilton and Cher [see Figure 2.39]...Never in the history of Wall Street have so many earned so much in so little time. Goldman Sachs Group Inc., Morgan Stanley, Merrill Lynch & Co., Lehman Brothers Holdings Inc. and Bear Stearns Cos. are about to reward their 173,000

---

[21] *The New York Times*, August 12, 1997
[22] http://www.city-data.com/housing/houses-Wahoo-Nebraska.html

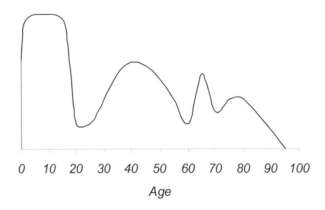

Figure 2.38. A histogram of the ages of women in Wahoo, Nebraska.

Figure 2.39. The 2007 Bentley Continental GT.

employees with $36 billion in bonuses.[23]

(a) What is the average employee bonus?

(b) Does this mean most employees could buy the Bentley with their bonus?

57. **Corporate debt.** The Securities and Exchange Commission (SEC) was concerned about a practice called "window dressing," when a firm slashes debt shortly before its quarterly reports in order to make financial state-

---

[23] "Wall Street's wild windfall," Tuesday, November 07, 2006, *Bloomberg News*. Photo from `http://www.bentleymotors.com/Corporate/`.

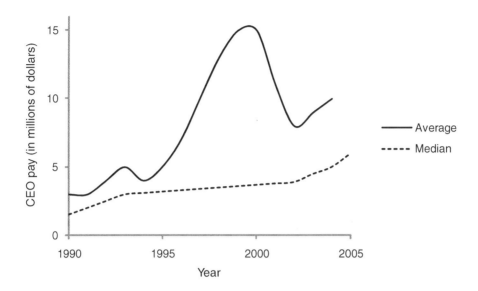

Figure 2.40. Average and median CEO pay from 1990-2005 (in 2005 dollars).

ments look healthier.[24] To investigate, the SEC gathered data over a period of time at a selected number of firms. Each time one of the firms slashed its debt, the SEC recorded the number of weeks that had elapsed since its most recent quarterly report was released; the average was 6.7 weeks. Is this evidence that "window dressing" was not a widespread practice among these firms? Explain briefly by sketching a carefully labeled histogram.

58. **CEO pay.** Figure 2.40 shows the average and median annual pay that American CEOs received between 1990-2005.[25]

   (a) What could explain why average pay rises so suddenly and then falls so sharply in the middle of the graph?

   (b) The average pay in 2000 was around $15 million. Were there more CEOs above or below this average?

59. **Batting averages.** Figure 2.41 shows a histogram of the end-of-season batting average for major league baseball players over the period from

---

[24] United SEC Votes to Propose Pulling Back Drapes on Debt, by Kara Scannell and Michael Rapoport, *The Wall Street Journal*, September 18, 2010

[25] From `www.demos.org/inequality/numbers.cfm`.

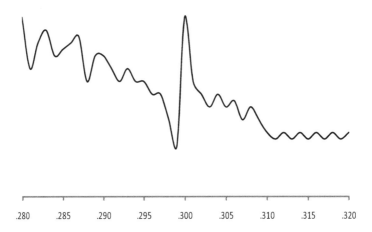

Figure 2.41. Histogram of end-of-season batting average for major league baseball players over the period from 1975 to 2008.

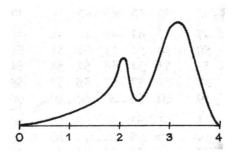

Figure 2.42. A histogram for the grade point average (GPA) reported by a sample of students interviewed at University of California, Berkeley.

1975 to 2008. What could explain the peculiar zigzag in the center of the graph?[26]

60. **GPA distribution.** In a survey carried out at the University of California Berkeley, a sample of students were interviewed and asked what the grade point average was. A histogram for the result is shown in Figure

---

[26]Sniffing .300, Hitters Hunker Down on Last Chances, by Alan Schwarz, *The New York Times*, October 2, 2010.

2.42. (GPA ranges from 0 to 4, and 2 is a bare pass.) What accounts for the spike at 2?[27]

61. **Smoking and health.** A study on the effects of smoking in a large sample of representative households showed that for men and women in each age group, those who had never smoked were on average somewhat healthier than current smokers, but the current smokers were on average much healthier than those who had recently stopped smoking. The lesson seems to be that you shouldn't start smoking, but once you've started, don't stop. Comment briefly.[28]

62. **Estimating standard deviation.** For each of the following pairs of lists, which has the larger standard deviation? Or are they the same?

 (a) i. 10, 20, 30
      ii. 10, 20, 30, 20
 (b) i. 1, 2, 3
      ii. 1, 2, 3, 2, 3, 1
 (c) i. 1, 2, 99, 100
      ii. 1, 25, 75, 100

63. **Comparing weather.** For the two cities Boston and San Francisco, one has an average daily temperature over the course of an entire year of 57 degrees and the other 52 degrees. Which city is which? Can you guess the standard deviation for each city?

64. **Market volatility.** Stock market volatility is often measured using the standard deviation. Figure 2.43 shows the standard deviation computed over a moving window of 100 days versus time. What explains the peak?

65. **Firm profitability.** The file `industries.xlsx` contains performance data from firms in different industries for 2012. The average profit margin for firms in the pharmaceutical industry is about -8% and in the semiconductor industry it is about +.2%. Does this mean that in firms in the pharmaceutical industry are typically less profitable than those in the semiconductor industry? Explain briefly using summary statistics and histograms.

---

[27] Taken from *Statistics*, 3rd edition, by D. Freedman, R. Pisani and R. Purvis. New York: W. W. Norton and Company, 1998, page 53.

[28] Take from *Statistics*, 3rd edition, by D. Freedman, R. Pisani and R. Purvis. New York: W. W. Norton and Company, 1998, page 25.

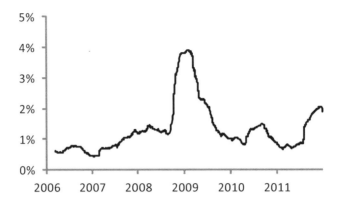

Figure 2.43. Standard deviation of average daily return over a 100 day period versus time.

66. **Industry statistics.** The file `industries.xlsx` contains performance data from firms in different industries for 2012. Use summary statistics and histograms to answer the following questions.

    (a) Do pharmaceutical firms generally have higher total revenues than semiconductor firms?

    (b) For each industry, how would you describe the mix of firms listed in terms of their total revenues.

    (c) The average return on assets (ROA) for semiconductor firms is negative. Does this mean the semiconductor industry is in bad shape?

    (d) How do companies in the pharmaceutical industry differ from companies in the food industry with respect to ROA?

    (e) In which industry do firms tend to have the highest ROA? The lowest?

    (f) In which industry do firms tend to have the most similar profit margins (PM)?

    (g) When looking at asset turnover (AT) ratios in an industry, are you more likely to see extreme values on the high side or the low side? How does this compare to PMs?

67. **MBA salaries at Boston University.** The file "BU MBA salaries.xlsx" contains data from four recent years for the MBA graduates at Boston University. The data includes their incoming GMAT score, their undergraduate GPA, the number of years of work experience they had, the

MBA program they are applied to, their graduating GPA, a column called "OA at graduation" indicating whether or not they had accepted a job offer at graduation (1=yes, 0=no), a column called "OA at 90 days" indicating whether or not they had accepted a job offered by 90 days after graduation (1=yes, 0=no), a column indicating their graduating year, and a column with the salary of the job they took after graduation if they reported it. Salaries are listed as 0 if the student did not report their salary. Answer the following questions using R.

(a) Draw histogram and compute the average salary for people who reported a salary and had a graduating GPA over 3.5.

(b) Draw a histogram and compute the average salary for people who reported a salary and had a graduating GPA of 3.0 or below.

(c) Using histograms, averages and standard deviations, do you see any interesting stories in these data?

Fun Problem: Size-biasing

A ring is hanging from the ceiling by a string. Someone will chop the ring in two completely random places on the circumference and you get to have the piece of the ring that falls on the floor. I get the piece that stays attached to the string. Whoever gets the bigger piece wins the game. Is this a fair game, or does one of us have an advantage over the other?

To illustrate the game, here is the ring hanging by a string from the ceiling:

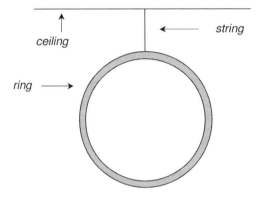

Next, we chop the ring:

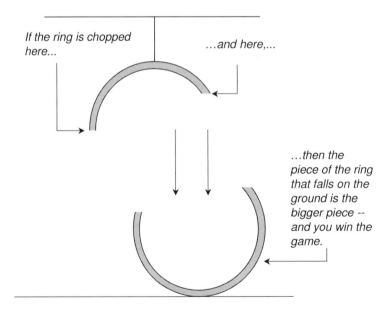

Is this a fair game?

Solution.   This is not a fair game—I have a big advantage over you. Suppose instead that before we start the game, the string is attached at random to the ring. Obviously this doesn't change the game at all. Now imagine that even before this happens, the random places where the ring will be chopped are marked in pencil—and then afterwards the string gets randomly attached. Since there will always be one penciled-off piece of the ring that is bigger than the other, if we attach the string randomly it is more likely to get attached to the bigger piece—because the bigger piece takes up more space around the ring than the smaller piece. Thus I am more likely to win.

Discussion.   This phenomenon is called **size-biasing** because the size of something biases the chance you choose it. This interesting effect appears in a lot of surprising places. For example, did you know that in most books, magazines and newspapers, the first word on a line of text tends to be longer on average than the last word on the line? Check it yourself with this book. To understand why this is true, suppose you drop your pencil onto a random word on the page of this book. Your pencil is more likely to land on a longer word than it is to land on a shorter word—just because the longer word takes up more space on the page. A word like "a" would be very difficult to land on, whereas a longer word like "phenomenon" would be easier to land on. This means that the word you land on tends to be longer than the average word. The same thing would happen for the same reason if you drew a line down the center of the page: the words that crossed the line would tend to be longer than the average word. The first word on each line of a book is essentially the word that would have crossed over the right margin and that's why it tends to be longer than the average word.

Size-biasing also turns up in surveys. If you ask random workers how many employees there are at the company they work for, you tend to get a number higher than the average company size. This is because the larger companies have more employees, so it is easier for you to randomly bump into one of them for your survey.

Size-biasing also turns up in transportation. Suppose buses come by a stop every half an hour on average. If you show up at random to the stop, you will tend to have to wait much longer than average. This is because of size-biasing—you are more likely to show up during a long time interval between buses than you are during a short time interval between buses. This is often called the **inspection paradox** since an inspector tends to see something very different than average. This also explains why the bus you get on often seems to be more crowded than the average bus.

This bias also turns up in medicine. If you show up at a hospital at random and take a sample of the patients you find there, you are more likely to see patients who are sicker than the average patient treated at the hospital. This is because the sicker patients tend to stay longer and thus are more likely to be there when you show up at random. Someone who is only there for a very short time is very unlikely to bump into you.

Finally, evaluation of some medical tests can run into problems from size biasing. Some cancer screening methods are, by size biasing, more likely to detect slower growing cancers (and these can be the easier-to-treat cases) than they are to detect the faster growing cancers. This can make the benefits of cancer screening seem better than they really are, since you are more likely to detect the least fatal cases anyway.

Chapter 3

The Normal Curve

1. Introduction

Although there are many different shapes histograms for data can take, one particular shape is surprisingly common in the business world. This famous "bell curve" or "**normal curve**" histogram shape looks like this:

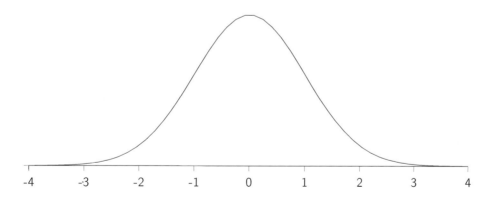

Statisticians have many different mathematical theories for why this shape turns up so often in practice, and for the business person it's important to be aware that this curve is a fact of life.

A normal curve-shaped histogram usually shows up when variation in the data is primarily caused by small variations in a large number of independently operating contributing sources of variation. A histogram for weekly sales figures over one year for an established brand of cereal might be expected to follow a normal curve, since weekly variation is primarily caused by small variations in the purchasing behavior of thousands of consumers. A histogram for sales figures during a single week across all the different brands of cereal might not be expected to follow the curve quite as well because the variation across brands is primarily caused by a smaller number of big factors like taste, advertising and brand image. Similarly, IQ, height and weight of people each tend to follow a normal curve probably because they are each influenced by variations in a large number of different genetic and behavioral contributing factors. Home values, on the other hand, usually do not follow a normal curve since they are greatly influenced by a smaller number of big factors such as number of bedrooms, square footage and location.

> A normal curve-shaped histogram usually appears when variation in the data is primarily caused by a large number of independently operating small contributing sources of variation.

It can be difficult to predict in advance whether or not a data set will follow a normal curve, so it's a good idea to take a look at the histogram shape if it's available.

Exercises

1. **Where do you expect a normal curve?** For each data set described below, say whether or not you would expect the data to follow a normal curve and explain why:

   (a) The number of chocolate chips contained in each bag of cookies produced by a bakery

   (b) The total miles driven with each full tank of gas for a delivery truck on its usual route

   (c) the number of calls received to 411 directory assistance each week over a year

   (d) The number of cell phone customers living in each city in Massachusetts

   (e) The number of cars sold in a year for each of the different car makes

(f) The salaries of each employee at a typical large corporation

Once you recognize that a histogram follows a normal curve, there are a number of useful shortcuts and "rules of thumb" you can use when analyzing your data. But the first step in using them is to start thinking about your data in what are called "standard units." We will describe these next.

2. Standard Units

Trying to directly compare data values from very different data sets can be difficult. Here we will give some examples of these difficulties and show you how the concept of standard units can help you.

On which exam did you do better?   Suppose you take two exams in a course. On the first exam you score 60 points, the class average was 50 points and the standard deviation (SD) was 20 points. On the second exam you score 70 points, the class average was 65 points and the SD was 5 points. On which exam did you score better relative to the rest of the class?

	Class Average	Class SD	Your Score
Exam 1	50	20	60
Exam 2	65	5	70

Because 70 points is larger than 60 points, it may seem that you scored better on Exam 2. But since you scored only five points above the average on Exam 2 and ten points above the average on Exam 1, it may seem you scored better on Exam 1. But notice that you scored one standard deviation above average on the second exam (5 points above average equals 1 SD above average) and only one-half a standard deviation above average on the first exam. This means on Exam 2 you scored further above the rest of the pack than on Exam 1. This reasoning is usually sensible if the histogram shape follows a normal curve, which is very often the case with exam scores.

Example.   How well did your mutual fund perform compared to the others? Mutual fund managers are often rewarded each year based on how well

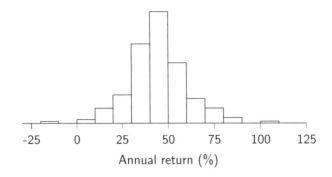

Figure 3.1. Return of mutual funds in 2003.

their fund performs compared with other funds. The Wellington Balanced mutual fund gave 20% return to its investors in 2003 and the Needham Growth mutual fund gave a 16% return in 2005. Which fund performed better relative to the other funds?[1]

Although a return of 20% is higher than a return of 16%, the economy was much different in 2003 compared with 2005: in 2003, the average mutual fund gave a 34% return and the standard deviation was 16%. In 2005 the average mutual fund gave a 9% return with a standard deviation of around 5%. In other words, the Wellington fund performed almost a full standard deviation below average in 2003 and the Needham fund performed almost a full standard deviation above average in 2005. The Needham fund performed much better relative to the others (and, in fact, the Needham fund gave almost a 50% return in 2003 while the Wellington fund gave almost no return in 2005). Again, this reasoning is usually sensible if the histogram roughly follows a normal curve, which Figure 3.1 shows it did.

A sensible way of comparing scores from data sets that have very different standard deviations and averages is to first convert the data into **standard units**. This means finding out how many standard deviations above or below average each score was, and then comparing these. For example, in the example of the two exams above, on the second exam you scored one standard deviation above average (70 is 5 points above 65), so on the second exam you scored a +1 in standard units. Similarly on the first exam you were one-half of a standard deviation above average (60 is 10 points above 50, and 10 points is half of the standard deviation of 20), so your score in standard

---

[1]See www.fund-track.com

units was $+1/2$.

 Standard units tell you how many SDs above or below average a data value is.

In general, the formula for converting your data into standard units is

$$\text{standard units} = \frac{\text{actual score} - \text{average}}{\text{SD}}.$$

For example, on Exam 2 your score in standard units was $(70 - 65)/5 = 1$.

Similarly, you can work backwards to find your actual score from your score in standard units as follows:

$$\text{actual score} = \text{average} + (\text{SD} \times \text{standard units})$$

For example, if your score in standard units was -2 on the second exam, your actual score was $65 - (2 \times 5) = 55$.

Question.   A financial analyst computes the annual return for each mutual fund currently available for sale and then converts the data into standard units. Here is a selection of the converted data for eight random funds: -3.0, $+2.1$, -1.5, -2.8, $+0.5$, $+4.5$, -3.9, -2.5. Do you notice anything suspicious here or are these reasonable numbers?

Discussion.   There is something suspicious here. As we saw in the previous chapter, the majority of values should be within 1 standard deviation from average in most data sets. This means most scores in standard units should be between $+1$ and -1. Here all except one score ($+0.5$) is outside this range. Someone most likely made a mistake converting the data into standard units or they didn't choose the funds randomly.

Comment. Examining observations in standard units is one of the ways quality control analysts can monitor a production process. A plot of measurements of a manufactured product in standard units over time is a type of **control chart**. If analysts see that far too many of the measurements are outside the range from -1 to $+1$, they may have a problem that needs investigation.

Technical notation: Scores in standard units are often called **z-scores**. If z is a z-score and X is the corresponding actual score, we can relate them with the following formulas: $z = \frac{X-\mu}{\sigma}$ and $X = \mu + z\sigma$, where σ is the standard deviation and μ is the average.

Exercises

2. **Account balances.** The average of a list of account balances is $10 and the standard deviation is $2.

   (a) What would $14 be in standard units?

   (b) How many dollars corresponds to a score in standard units of -4?

3. **Outstanding firms.** A financial analyst studying one industry is trying to identify companies that stand out from others in the industry for one reason or another. He investigates two companies, looking at several company statistics and comparing them with industry averages. Looking at the data below, does either of the two companies appear to stand out from others in the industry? Explain.

	Profits (millions of $)	Number of employees	CEO's salary (millions of $)
Company #1	14	150	4
Company #2	10	190	6
Industry average	20	130	8
Industry SD	10	15	4

4. **MBA rankings.** Each year *U.S. News & World Report* ranks MBA programs based on a number of different numerical factors and these rankings are highly influential. Two important factors are the percentage of students with a job lined up at graduation and the percentage with a job within three months of graduation. Below are the figures for Boston

University in 2003 and 2004 as well as the average and SD for the top 50 schools overall. For which of these two factors did Boston University make the biggest gain relative to the other schools?

	% employed at graduation		% employed within 3 months of graduation	
	2003	**2004**	**2003**	**2004**
Boston University	62	71	87	89
Average of top 50 schools	70	66	81	87
SD of top 50 schools	9	10	7	5

5. **High-yield fund.** One mutual fund is designed to give an average return of 5% with a standard deviation of 1% and another is designed to give an average return of 4% with a standard deviation of 2%. Which fund would you guess is more likely to give a 7% return? Explain briefly.

3. The Normal Curve

The standard **Normal Curve**, or bell curve, is a very famous histogram that uses standard units as its unit of measure. Very many data sets encountered in real life have a shape similar to this curve and data are often put into standard units to see the similarity. The curve was discovered almost 300 years ago by the French mathematician Abraham de Moivre and is shown in Figure 3.2. The equation of this curve is

$$y = \frac{1}{\sqrt{2\pi}}e^{-x^2/2},$$

which interestingly contains both e and π—two very famous numbers that have mystified mathematicians over the ages.[2] We only show you the equation because it's interesting that this curve with this particular formula turns up in a wide variety of very practical situations in real life.

---

[2]Incidentally the number e (which equals approximately 2.71) may have been first used around 400 years ago in Egypt, and π may have first been estimated almost 4,000 years ago in Italy. An estimate of π actually appears in the Bible. For more on this history see the excellent book *History of Pi*, by Petr Beckmann.

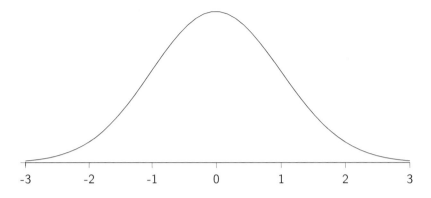

Figure 3.2. The standard normal curve.

Important properties of the curve are that it is symmetric around zero, the area between -1 and $+1$ is roughly 68% of the total area (illustrated in Figure 3.3), the area between -2 and $+2$ is roughly 95% of the total area, and the area between -3 and $+3$ is roughly 99.7% of the total area. The curve theoretically continues on forever in both directions, but it gets so close to being flat that the area under the curve beyond $+3$ or -3 can usually be considered negligible for all practical purposes.

Approximately **68%** of the area under the normal curve is between -1 and $+1$.

Approximately **95%** of the area under the normal curve is between -2 and $+2$.

Approximately **99.7%** of the area under the normal curve is between -3 and $+3$.

Question.   Find the following area to the left of $+1$ under the normal curve:

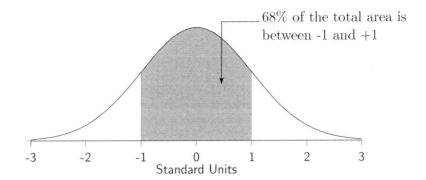

Figure 3.3. About 68% of the area is between -1 and +1 for the standard normal curve.

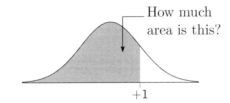

Solution. We know that 50% of the area must be on one side of the Normal Curve, because it's symmetric. Also, we know that 68% of the total area is between -1 and $+1$. This means that the area between 0 and $+1$ must be half of 68%, or 34%. We can use this to get a final answer of 84% by breaking down the area as follows:

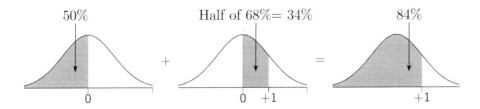

So the answer is $50\% + 34\% = 84\%$.

Question. Next, let's find the area for a range in the middle of the curve. Find the area between −2 and +1 under the normal curve.

Solution. We know that the area between −2 and +2 under the curve corresponds to 95% of the total area, and this means that the area between −2 and 0 corresponds to half of this, or 47.5%. Applying the same reasoning as before, we see that the area between 0 and +1 must be half of 68%, or 34%. We add 47.5% to 34% to get a final answer of 81.5%. We illustrate this below:

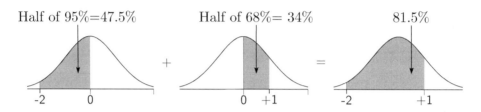

Question. Next, let's find the area under the upper part of the curve. Find the area to the right of +2 under the normal curve.

Solution. Here we have to subtract two areas. We know that to the right of zero we find 50% of the area. We also know that 95% of the area can be found from −2 to +2, and that therefore 47.5% of the area can be found from 0 to +2. Subtracting 47.5% from 50% gives 2.5%. This can be illustrated as follows:

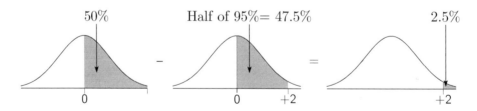

Question. So far we've been dealing with round numbers of standard units. Let's move away from this now. Find the area to the left of +1.5 under the normal curve:

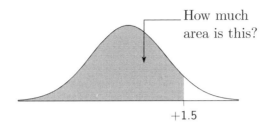

Solution.   It is not possible to answer this question using the same method we have used so far. It is important to realize that the normal curve is not a rectangle, so you can not find the area corresponding to +1.5 if you know the area corresponding to +1 and +2—it is not one-half of the area between +1 and +2. Also, for example, the area from 0 to +2 is not twice the area from 0 to +1:

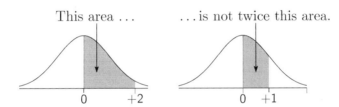

There are two ways you could find the area we are looking for: the easy way and the hard way. The hard way would be to start with the equation of the normal curve given above and use calculus to compute the area.

There is an easier way. Since the normal curve is so important, we have computed the areas for a large number of possible values you might be interested in and have listed them in Table 3.1   towards the end of this chapter. Most statistics books contain some sort of table like this, and we will show you later how you can use Excel to easily compute numbers in the table.

To find the area to the left of +1.5 standard units under the normal curve, look in the column of the table where it says "number of standard units z" using a value of $z = +1.5$, and read across to find the corresponding "percent of area below z." Figure 3.4 is an excerpt from the table. Notice there is no value listed for exactly $z = 1.5$, so we look up the closest value of 1.48 (using the closest value will give us the correct answer to the nearest percentage,

Number of standard units z	Percent of area below z
1.34	91%
1.41	92%
1.48	93%
1.55	94%
1.64	95%

Look up 1.48 because it's the closest value listed to 1.5,...

...and it tells you about 93% of the area is below 1.5.

Figure 3.4. An excerpt from the normal table.

but if you need more accuracy you could interpolate or use Excel—which we describe below). This tells us that approximately 93% of the area is to the left of +1.5 under the normal curve.

By breaking down an area into pieces as we did and using this table, you can estimate the area in just about any range under the normal curve. If you are looking for a number that is far off the range of the table, you can usually just use the closest number listed and you won't be too far off. You can also use this table in reverse: you can look up a percentage in the second column and find the corresponding number of standard units in the first column. We will do an example of this in the next section.

Using Excel. If you want more precision or don't want to bother with the normal table, you can use the Excel command "=Normsdist(z)", which tells you the fraction of the total area to the left of z under the normal curve. For the last problem of this section we can apply the Excel command "=Normsdist(1.5)" to get an answer of .933193, or about 93%. In practice you will not usually need this much precision but it is convenient to compute these percentages in Excel. To work backwards to find the value of z that corresponds to a given fraction of area, you can use the Excel command "=Normsinv(x)" to get the number of standard units corresponding to the fraction x of the total area. For example, the command "=Normsinv(.933193)" gives an answer of 1.5. To save the trouble of first converting to standard units, you can use the command

"=Normdist(x,m,s,1)" if you want excel to convert x into standard units using a mean of m in the standard deviation of s and then give the area under the normal curve to the left.

Exercises

6. Find the area to the right of $+1.2$ under the normal curve.

7. Find the area between -2.5 and $+2.5$ under the normal curve.

4. Using the Normal Curve

If you have a histogram with normal curve shape, you can easily estimate the percentage of your data values that fall in various ranges without having to keep looking back at the area under the histogram. All you need to know is the average and the standard deviation, and then you can approximate the histogram using the normal curve. It's a great way to summarize a set of data that follows a normal curve.

Graduate Management Admission Test (GMAT) Scores.   The   GMAT scores in 2001–2002 had an average of 529 with an SD of 113 and the scores roughly followed a normal curve.[3] If you scored a 642 on the exam, approximately what percentage of test-takers scored below you?

Solution.   If you got a 642 you were exactly 113 points above average (since $642 - 529 = 113$) or, in other words, exactly one standard deviation above average. This corresponds to a score in standard units of $+1$. The calculation we used is $(642-529)/113 = +1$. This means that to estimate the percentage of people who scored below 642, we can find the area to the left of $+1$ under the normal curve by looking up $+1$ standard units in the table. The closest thing to $+1$ in the table is $+.99$, and using this tells us that roughly 84% of the people scored below 642 points.

---

[3]See *Guide to the Use of GMAT Scores* at the web site http://www.gmac.com/gmac/ TheGMAT/Tools/GuidetotheUseofGMATScores.htm

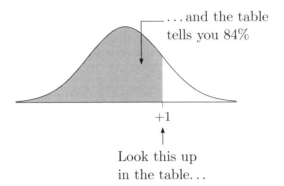

Percentiles. It's common terminology to say that 642 points is the 84th percentile for the GMAT. If you are at the 84th percentile, it means that 84 percent of the people scored lower than you did. The concept of a percentile applies to any distribution, not just the normal distribution.

If you are at the Xth **percentile**, it means that $X\%$ of the people are below you.

Question. In some graduate programs you may be given a scholarship if you score above the 90th percentile. For the previous example, compute the 90th percentile for the GMAT.

Solution. If someone scored at the 90th percentile, he or she did better than 90% of the people taking the test. By looking up 90% on the normal table, we see that the corresponding score in standard units must be 1.28. This means that the actual score would be around $529 + (113 \times 1.28) = 673.6$. So we can say that the 90th percentile for the GMAT here is somewhere around 670 or 680 points (GMAT scores are in multiples of 10).

Inventory planning. Your company estimates demand for a product will be around 800 units with a standard deviation of 200 and past experience has shown that demand follows a normal curve. How many units should you order from the factory if you want only around a 5% chance of running out?

Solution. To get a 5 percent chance of running out, you would want there to be a 95 percent chance that demand is less than what you order. Looking up 95% in the normal table gives you a score in standard units of $+1.64$. This corresponds to demand of $800 + (1.64 \times 200) = 1{,}128$ units of product. You should order 1,128 units.

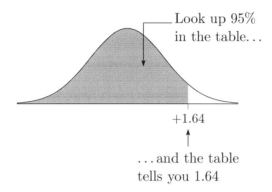

Look up 95%
in the table...

$+1.64$

...and the table
tells you 1.64

Comment. You may remember by now that 95 percent of the area under a normal curve falls between $+2$ and -2, but in this problem we instead wanted 95 percent of the area to be below some cut off. This means we needed to use 1.64 standard units instead of 2 standard units.

Value at Risk. Financial analysts commonly use an approach called **value at risk** to compare the risk of stocks or portfolios. The analyst tries to estimate a low percentile (such as the 5th or 10th percentile) of the forecasted range of possible returns for a given stock or portfolio and that percentile is used as a rough measure of the worst-case scenario for that investment. Different investments can then be compared.

Suppose an analyst is trying to compare two investments: Investment A has a forecasted average return of $10 million with a standard deviation of $3 million, and Investment B has a forecasted average return of $12 million with a standard deviation of $5 million, and investment returns typically follow a normal distribution. The analyst wants to avoid risk and invest in the one that has a larger 5th percentile for the return distribution. Which investment should she make?

Solution.   Even though Investment B has the higher forecasted average return, it may not necessarily be the better investment. To find the 5th percentile of the return distribution for each investment, we look up 5% in the normal table and we see this corresponds to -1.64 in standard units, or 1.64 standard deviations below the average.

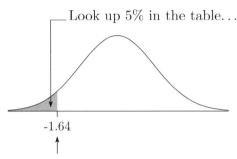

Look up 5% in the table...

-1.64

...and the table tells you -1.64

This gives a 5th percentile of $10 - (1.64 \times 3) = 5.08$, or \$5.08 million for Investment A and $12 - (1.64 \times 5) = 3.8$, or \$3.8 million for Investment B. Based on her stated goal, she should invest in A. Even though the forecasted average return for investment B is higher, its higher standard deviation makes it much riskier and its 5th percentile comes out to be lower than A's. This means you should not always pick the investment that's expected to do best—you should consider the risk as well.

Home Prices.   Suppose home prices in a city average \$150,000 with a SD of \$100,000. Does this mean approximately 50% percent of homes are priced over \$150,000?

Solution.   The answer would be "yes" if the data followed a normal curve, but there is something suspicious in the numbers that indicates home prices do not follow a normal curve. With a normal curve, you expect there will be some data one or two standard deviations below the average. In this case homes that are 2 standard deviations below average would be valued at $\$150,000 - (2 \times \$100,000) = -\$50,000$. Does the negative value mean someone should pay you \$50,000 to take it off their hands? Obviously there should be no homes valued below \$0, so these data most likely do not follow a normal curve. To find the percentage of homes over \$150,000, we would need to see the histogram for home values.

The moral is that if going 1 or 2 standard deviations away from average takes you outside of the range of the data, the histogram probably does not follow a normal curve very well. Don't just blindly use the normal curve for every single data set you come across; think if you have reason to believe it's appropriate or not.

> If going 1 or 2 standard deviations away from average takes you outside of the range of the data, the histogram probably does not follow a normal curve very well.

Technical note

A normal distribution with mean μ and SD σ is usually written in shorthand notation as $N(\mu, \sigma)$. The standard normal distribution is $N(0, 1)$.

5. Exercises

8. (a) Estimate the area under the standard normal curve to the right of -1.5.

(b) Estimate the area under the normal curve between -2 and $+0.5$.

(c) Fill in both blanks with the same number to make the following statement true: the area under the normal curve between positive _____ and negative _____ equals approximately 40%.

9. GMAT percentiles. GMAT scores in 2001–2002 nationwide had an average of 529 with an SD of 113 and scores roughly followed a normal curve.[4]

(a) If you got a 600 on the exam, approximately what percentage of test-takers scored below you?

(b) Compute the 75th percentile for this exam.

---

[4]See *Guide to the Use of GMAT Scores* at the web site `http://www.gmac.com/gmac/TheGMAT/Tools/GuidetotheUseofGMATScores.htm`

Areas under the normal curve

Number of standard units z	Percent of area below z	Number of standard units z	Percent of area below z
−3.00	0.13%	0.03	51%
−2.58	0.5%	0.05	52%
−2.33	1%	0.08	53%
−2.05	2%	0.10	54%
−1.96	2.5%	0.13	55%
−1.88	3%	0.15	56%
−1.75	4%	0.18	57%
−1.64	5%	0.20	58%
−1.55	6%	0.23	59%
−1.48	7%	0.25	60%
−1.41	8%	0.28	61%
−1.34	9%	0.31	62%
−1.28	10%	0.33	63%
−1.23	11%	0.36	64%
−1.17	12%	0.39	65%
−1.13	13%	0.41	66%
−1.08	14%	0.44	67%
−1.04	15%	0.47	68%
−0.99	16%	0.50	69%
−0.95	17%	0.52	70%
−0.92	18%	0.55	71%
−0.88	19%	0.58	72%
−0.84	20%	0.61	73%
−0.81	21%	0.64	74%
−0.77	22%	0.67	75%
−0.74	23%	0.71	76%
−0.71	24%	0.74	77%
−0.67	25%	0.77	78%
−0.64	26%	0.81	79%
−0.61	27%	0.84	80%
−0.58	28%	0.88	81%
−0.55	29%	0.92	82%
−0.52	30%	0.95	83%
−0.50	31%	0.99	84%
−0.47	32%	1.04	85%
−0.44	33%	1.08	86%
−0.41	34%	1.13	87%
−0.39	35%	1.17	88%
−0.36	36%	1.23	89%
−0.33	37%	1.28	90%
−0.31	38%	1.34	91%
−0.28	39%	1.41	92%
−0.25	40%	1.48	93%
−0.23	41%	1.55	94%
−0.20	42%	1.64	95%
−0.18	43%	1.75	96%
−0.15	44%	1.88	97%
−0.13	45%	1.96	97.5%
−0.10	46%	2.05	98%
−0.08	47%	2.33	99%
−0.05	48%	2.58	99.5%
−0.03	49%	3.00	99.87%
0.00	50%	3.72	99.99%

Table 3.1. Areas under the normal curve.

Figure 3.5. A histogram of hourly wages paid to employees at a firm.

10. **GMAT scores.** GMAT scores of applicants to a certain university average 600 points with an SD of 50 points and the scores follow a normal distribution.

    (a) If you look at only the group of applicants scoring in the range 550 to 650, the average of this group will be (i) around 600 (ii) above 600 (iii) below 600. Choose one and explain.

    (b) If you look at only the group of applicants scoring in the range 600 to 700, the average of this group will be (i) around 650 (ii) above 650 (iii) below 650. Choose one and explain.

11. **Hourly wages.** Figure 3.5 is a histogram of hourly wages paid to employees at a firm. The data are spread evenly within each bar. Does the 80th percentile for this data equal twice the 40th percentile? Explain.

12. **Weight data.** A large group of people are part of a study. Thirty people have weights in the range 110–130 pounds, seventy are in the range 130–150 pounds, twenty are in the range 150–170, and thirty are in the range 170–190. The weights are spread fairly evenly in these ranges.

    (a) True or false and explain: the SD for this data is larger than 39 pounds.

    (b) Give the 80th percentile for this data and explain.

13. **GMAT percentiles.** GMAT scores for 1000 applicants to a business school closely follow a normal distribution with an average of 600 and an SD of 100. Nationwide GMAT scores average 550 with a SD of 100 and

also closely follow a normal distribution. The school decides to admit its top 300 scoring applicants. What percent of its admitted applicants will be above the 90th percentile nationally?

14. **Call centers.** You are responsible for the staffing of a service center that responds to customer telephone inquiries. During normal times, customers wait an average of 30 seconds before being served (with a standard deviation of 6 seconds), and during busy times, customers wait an average of 50 seconds before being served (with a standard deviation of 10 seconds). Waiting times roughly follow a normal distribution.

   (a) During normal times what is the chance a customer has to wait more than 40 seconds?

   (b) Marketing wants to say "even during busy times 95% of our calls are answered within _____ seconds." What number should go in the blank?

15. **Picking investments.** Investment A has an expected return of $25 million and investment B has an expected return of $5 million. Market risk analysts believe the standard deviation of the return from A is $10 million, and for B is $30 million (negative returns are possible here).

   (a) If you assume returns follow a normal distribution, which investment would give a better chance of getting at least a $40 million return? Explain.

   (b) How could your answer to part (a) change if you knew returns followed a skewed distribution instead of a normal distribution? Explain briefly.

16. **Forecasting inventory.** The monthly demand for a company's product follows a normal curve and has an average of 10,000 units with a standard deviation of 3,000 units. Currently the company has 12,000 units in inventory available to satisfy demand.

   (a) What is the chance the company sells its entire inventory?

   (b) What is the chance it has at least 9000 units left unsold at the end of the month?

17. **Picking investments.** An investor has three possible investments to consider: the first has a forecasted return of 40% with a standard deviation of 25%, the second has a forecasted return of 20% with a standard deviation of 10%, and the third has a forecasted return of 30% with a standard

deviation of 5%. The investor wants to be conservative and choose the one with the largest 10th percentile, and all returns are expected to follow a normal curve. Which one should he choose?

18. **Manufacturing problems.** A machine filling potato-chip bags is designed to put in 8 ounces of chips with a standard deviation of 0.1 ounces. If a bag comes out with 9 ounces in it, is this reasonable, or a sign of a possible problem? Explain.

19. **Product lifetimes.** The lifetime of a particular product is normally distributed with a mean of four years. About 16% of the products have a lifetime of at least five years. What percent of the products have a lifetime of at least six years?

20. **Forecasting demand.** The weekly demand for a product follows a normal curve with an average of 1,000 units and a standard deviation of 200 units. In how many weeks during a typical 52 week year is the demand expected to be above 1,200 units?

More Challenging Exercises

21. **Delivery time guarantees.** A manufacturing supplier offers delivery time guarantees. It promises delivery within 60 days, or it will pay a penalty to the customer. Actual delivery times follow a normal distribution with an average of 52.5 days and an SD of 5 days.
    (a) What percent of deliveries are delivered within 50 days?
    (b) If the company budgets enough to pay penalties for 4% of deliveries, should it expect to stay within budget?
    (c) What is the median delivery time for the deliveries that incur a penalty?

22. **High-speed production.** You are trying to get an important new customer to buy a product that you produce. Their decision to buy the product from you will depend primarily on the speed with which you can produce the product once they have placed an order. Currently, it takes you 70 hours on average to produce the product with a standard deviation of 8 hours. From historical data, you have determined that actual production times follow a normal distribution.
    (a) What percent of the time can you produce the product within 80 hours?

Table 3.2. An excerpt from the file `funds2005.xls`.

Mutual Fund Name	% Return 2004	% Return 2005
ABN AMRO Chic Capital Bal N CHTAX	5.02	-5.14
ABN AMRO/Veredus Aggr Grth VERDX	20.69	2.51
Acadian Emerging Markets AEMGX	33.53	25.14
Aegis Value AVALX	13.42	-6.41
Alpine Realty Income & Growth AIGYX	30.95	5.74
Alpine U.S. Real Estate Equity Y EUEYX	39.32	5.95
Amer Cent Global Growth TWGGX	15.11	17.05
Amer Cent Intl Bond BEGBX	10.1	-12.14
Amer Cent Real Estate REACX	28.85	2.62
Amer Cent Small Cap Quant ASQIX	28.26	-4.12

(b) You want to promise that 95% of the time you will deliver the product in under _____ hours. What number should you put in the blank?

(c) Assuming the standard deviation stays the same, how much do you have to reduce your average production time so that 95% of the time you can deliver the product in under 75 hours?

23. **Mutual fund returns.** The file `funds2005.xls` (an excerpt is in Table 3.2) contains the annual return for mutual funds in 2005. One of the funds, the Cullen Value fund, had return of 11.46%.

(a) What percentage of the funds listed had a return below this?

(b) Use the normal curve to estimate the percentage in (a). Are the two answers close? (Hint: first compute the average and standard deviation.)

24. **Mutual fund returns.** The file `funds2005.xls` (an excerpt is in Table 3.2) contains the annual return for mutual funds in 2005.

(a) Do the data appear to follow a normal curve?

(b) The standard deviation is a lot larger than the average. How is this possible? Is this evidence the data don't follow a normal curve?

25. **Just-in-time ordering.** A factory has a "just-in-time" ordering policy for an important component in the product it manufactures and its supplier guarantees delivery of ordered components within 30 days. Historical data shows that components arrive on average within 24 days from

ordering, the standard deviation is 4 days, and delivery times follow a normal curve closely.

(a) What percent of components are received within 30 days of ordering?

(b) Orders are "flagged" for special monitoring on the 24th day after ordering if they haven't yet arrived. What percent of these flagged orders arrive within 30 days of ordering? What percent of these flagged orders arrive more than 30 days after ordering?

26. For people who have studied calculus. Without doing any calculations, approximately evaluate the following integral:

$$\int_{-1}^{+1} \frac{1}{\sqrt{2\pi}} e^{-x^2/2} \, dx$$

27. Is there a shortage of women in high-aptitude professions? In 2005 Harvard University president Larry Summers was intensely criticized for an explanation he proposed for an apparent shortage of women in high-aptitude professions; he shortly thereafter resigned. One of several explanations he gave was "different availability of aptitude at the high end" and he used the following example to illustrate his point: suppose an equal number of men and women take an exam where each gender gets the same average score, but the SD for women is 20% lower than the SD for men. What percentage of people above the 95th percentile will be women? Assume the data follow a normal curve within each gender.[5]

28. Physician salaries. The percentiles for salaries of physicians in a particular specialty are computed. The 25th percentile equals $200,000, the 50th percentile equals $270,000 and the 75th percentile equals $340,000. The salaries follow a normal curve.

(a) Approximately what percentile would a salary of $120,000 be?

(b) Repeat part (a) for $420,000.

29. Physician salaries. The percentiles for salaries of physicians in a particular specialty are computed. The 25th percentile equals $200,000, the 50th percentile equals $250,000 and the 75th percentile equals $400,000. Does it look like the salaries could follow a normal curve?

---

[5]See "Remarks at NBER Conference on Diversifying the Science & Engineering Workforce," Lawrence H. Summers, Cambridge, Mass., January 14, 2005. http://www.president.harvard.edu/speeches/2005/nber.html

30. **Account balances.** An analyst records a large number of account balances and draws a histogram. The 20th percentile is $400, the average is $500, and the median is $600.

    (a) Sketch a histogram having these three features and label the horizontal axis.

    (b) A different set of account balances has the same 20th percentile and average as in part (a), but this time the median equals the average and the histogram follows a normal curve. Using just this information is it possible to determine how many standard deviations above average a balance of $800 would be? Explain briefly.

31. **Accounts receivable collection.** You are analyzing data on the time it takes to collect accounts receivable. Data from the last 3 years shows that times are approximately normally distributed. You have separately identified two groups of receivables: ones having times below the 20th percentile, and ones having times above the 80th percentile. To summarize the difference between the two groups, you examine two ratios: (a) the median collection time in the high group to the median collection time in the low group; (b) the mean collection time in the high group to the mean collection time in the low group. Which ratio (a) or (b) will be larger? Briefly justify your answer.

32. **SAT gender differences.** A newspaper article reports the following: "Although girls consistently earn higher high school grades, their SAT scores continue to lag behind boys', with the gap reaching 36 points in math."[6] (The SATs are a standardized test students applying to college are usually required to take). Overall the average is 500, the standard deviation is 100, and scores are normally distributed. Assume girls score on average 482 and boys score on average 518, each has a standard deviation of 85, and each are normally distributed.

    (a) What percentage of girls score higher than 75% of the boys?

    (b) Among those girls who score higher than 75% of the boys, what percentage score higher than 90% of the boys?

    (c) What is the median score of girls who score higher than 75% of the boys?

33. **Overseas delivery times.** When an order is placed with an overseas manufacturer, the product takes an average of 90 days to arrive with a

---

[6] *The Los Angeles Times*, August 27, 1997.

standard deviation of 15 days. Orders from a local manufacturer take an average of 70 days with a standard deviation of 25 days. Delivery times in both cases follow normal distributions.

(a) What percent of orders placed with the overseas manufacturer arrive within 70 days?

(b) When one particular retailer places an order for product that is needed by a certain date, its goal is to order as late as possible but still be 90% sure that the product will arrive by that date. With this goal, which manufacturer would be better to use? Explain briefly.

(c) If we look at orders from the local manufacturer where 95 days have passed since product was ordered and the order has not yet arrived, what percentage of these orders should arrive within the next 25 days? Explain briefly.

Fun Problem: Monty Hall

Question. There are three doors in front of you. Behind one of the doors is a fantastic prize and behind the other two doors is nothing. You get to choose a door and you can have what's behind it. But before we open your door, I will show you what's behind one of the remaining two doors, which I know does not have a prize. Then I will give you the opportunity to keep your initial door choice or you may instead have what's behind the remaining unopened door. Should you keep your original choice or switch? Or does it really make any difference?

Should you keep your initial choice, or switch?

Solution. This is often called the "Monty Hall" problem because this type of situation arose in a television game show hosted by Monty Hall. Although it might seem as though it shouldn't make any difference, you should always switch to the unopened door. It makes a big difference and actually doubles your chance of winning.

Suppose Door #1 is your initial choice. Now there are three possibilities for the location of the prize: behind Door 1, Door 2 or Door 3. Now if you keep Door #1, you'll only win if the prize is actually behind Door #1—you'll lose in the other two scenarios. This means you win in one out of three equally likely scenarios, so your chance of winning is 1 out of 3.

What if Door #1 is your initial choice and you decide to switch to the unopened door? If the prize is behind Door #1 you will lose if you switch doors. If the prize is behind Door #2, you will win because I will open up Door #3 and the remaining unopened door will be Door #2 with the prize. The same reasoning applies if the prize is behind Door #3: I will open Door #2 and the remaining unopened door will have the prize. So if you switch when the prize is behind Door #3 you will win also. We can summarize this with the table below:

	Prize behind #1	Prize behind #2	Prize behind #3
Keep Door #1	Win	Lose	Lose
Switch doors	Lose	Win	Win

This means that if you decide to switch doors, you will win in two out of the three equally likely scenarios. This gives you a two out of three chance of winning. So you should always switch!

Comment.   This problem appeared in a column by Marilyn vos Savant in *Parade Magazine*, where she gave the correct solution.[7] Since it is so counterintuitive, many people wrote in saying she was wrong—even some college math professors. In the end, they were all wrong and she was right. Incidentally, if I did not know where the prize was and just opened a random door (and by luck it did not have any prize behind it) then it would not make any difference whether or not you switch. The fact that I know where the prize is, interestingly, makes a big difference.

More comments.   Puzzles like this are routinely used in "pressure" job interviews for competitive industries such as banking, law, consulting, insurance, airlines, media, advertising and even the Armed Forces. In the "old days" they may have used questions such as "describe your worst fault," but the trend now seems to be towards these types of brain-teasers. A recent *Harvard Business Review* article discusses this trend and gives a few more examples:[8]

1. Suppose you are tied to a chair. I pick up a gun with six chambers—all empty. Now watch me as I put two bullets in the gun in adjacent chambers. I close the cylinder and spin it. I put the gun to your head and pull the trigger. Click. You are still alive. Lucky you! Now, before we discuss your resume, I'm going to pull the trigger one more time. Which would you prefer, that I spin the cylinder first or that I just pull the trigger?

2. If you are in a boat and toss a suitcase overboard, will the water level rise or fall?

3. If you could remove any of the 50 states, which would it be?

4. Why do mirrors reverse left and right instead of up and down?

---

[7]Marilyn vos Savant, *Parade magazine,* September 9, 1990. See also the front page article in *The New York Times,* July 21, 1991.

[8]"Beware the Interview Inquisition," by William Poundstone. *Harvard Business Review,* May 2003, pages 18–19.

5. How would you design Bill Gates' bathroom?

6. How many dump trucks would it take to cart away Mount Fuji?

Chapter 4

How to Tell a Statistical Story with a Graph

1. Introduction

Making a graph tell the right story on its own is very important. Since you usually will not be around to explain the story while people are looking at the graph, it should be able to speak for itself. Don't rely on Excel to make your story clear—Excel is not thinking about your audience. It's also best if your audience can believe the story just by looking at the graph without having to trust your analysis or credibility. For this it can often be better to show raw data from an insightful angle than to show the results of a fancy analysis or summarized final conclusions. People will believe their own eyes much more easily than they will ever believe you. And don't make the audience search for your story—it should jump right off the page.

> It can often be better to show raw data from an insightful angle than to show the results of a fancy analysis or summarized final conclusions.

Finally, don't rely on the choices Excel gives you to help decide what type of graph would be best. Although Excel floods you with dozens of choices of graphs it can make for you, the best graph to tell your particular story is very often not one of the choices; you usually must imagine and create it yourself. Before you start pointing and clicking, decide on your own what type of graph would be best.

> Decide on your own what type of graph would be best before you start pointing and clicking.

2. Examples

Example.   Pie charts are one of the most common statistical graphs, and also among the most misused. Here we will give an example of the right way and the wrong way to use them.

It gets very cold in Boston during January and many people take warm weather vacations then. Below are the average high temperatures (in degrees Fahrenheit) for the month of January in a selection of tourist destinations.[1]

City	Temperature
Acapulco (Mexico)	87
Amsterdam (Netherlands)	41
Athens (Greece)	54
Auckland (New Zealand)	73
Bangkok (Thailand)	89
Beijing (China)	35
Belgrade (Yugoslavia)	38
Berlin (Germany)	35
Bombay (India)	83
Cairo (Egypt)	65
Calcutta (India)	80
Cape Town (South Africa)	69
Caracas (Venezuela)	75
Copenhagen (Denmark)	36

For this data, suppose a tourism analyst creates the corresponding pie chart

---

[1] http://www.infoplease.com/ipa/A0004587.html

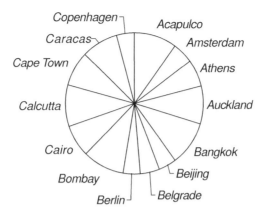

Figure 4.1. Pie chart for temperatures in a selection of tourist destinations. What story does this tell?

in Figure 4.1 to summarize the data. The size of each slice is equal to the temperature. What story does this tell?

Well, this pie chart very clearly tells the story that the analyst does not know how to use pie charts. This is not the right type of data to summarize using a pie chart and we learn almost nothing looking at it. Pie charts are best used when it's important to look at each data value as a percentage of the total of the data values. The data here are better summarized using the bar chart shown in Figure 4.2. From this bar chart it's quite easy to immediately see that Bombay, Bangkok, and Acapulco are the warmest, and Berlin, Copenhagen, and Beijing are the coldest.

When you have data values that each represent a fraction of the total, a pie chart can be informative. For example, Figure 4.3 is a pie chart illustrating how the federal government plans to spend its money in 2007.[2]

Example. You have historical price data for two different products (Product A and Product B) and you would like to tell the story that they tend to have similar price movements. Take a look at the graph in Figure 4.4.

What story does this graph tell? We can vaguely see that the price movements tend to be somewhat similar, but we really have to look carefully and we probably need someone to tell us what we are supposed to be looking for.

---

[2]data taken from http://nationalpriorities.org/

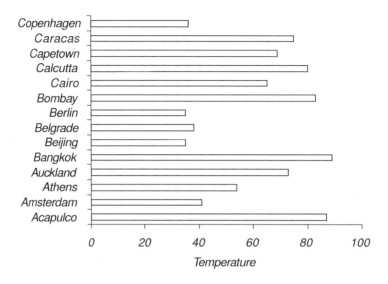

Figure 4.2. A bar chart for temperatures for average January temperatures for selected tourist destinations.

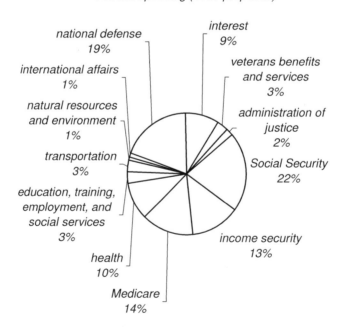

Figure 4.3. The allocation of the federal budget in 2007.

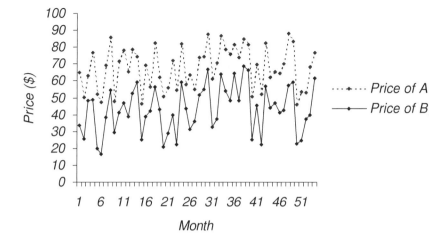

Figure 4.4. The prices of two products over time.

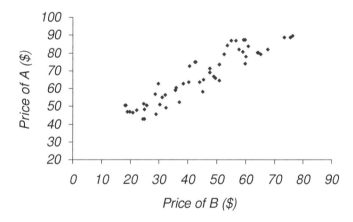

Figure 4.5. The price of one product versus the price of another product.

It is a nicely drawn graph, but it's not so great for telling the story that the prices move together. This graph may be better at telling the story that A's prices tend to be higher than B's prices—but that's not the story we wanted to tell. Figure 4.5 is a much better graph where the story that prices move together just jumps right out.

See the difference? Nobody has to explain the correct story to you this time—you see it and believe it yourself.

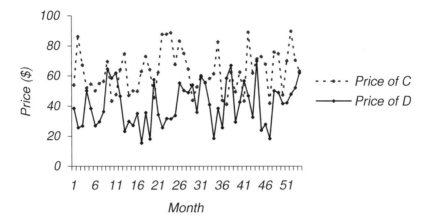

Figure 4.6. The price of two products versus time.

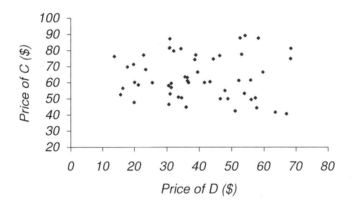

Figure 4.7. The relation between C's price and D's price.

To show you how things would look if the products did not have similar price movements, suppose instead we have two different products (C and D) and we want to tell the story that there is no relation between their prices. When we draw the first type of graph, shown in Figure 4.6, it is not so clear whether or not prices move together – the graph looks somewhat similar to the previous one.

But when we draw the second graph, shown in Figure 4.7, the correct story jumps right out. We immediately see that there is no relation between the prices.

Table 4.1. A portion of the file `abprices.xls`.

Purchase #	A Price	B Price	Product Purchased
1	$0.07	$0.12	A
2	$0.10	$0.10	A
3	$0.07	$0.04	B
4	$0.05	$0.13	B
5	$0.06	$0.05	B
6	$0.07	$0.12	A
7	$0.08	$0.07	B
8	$0.09	$0.14	A
9	$0.05	$0.10	A
10	$0.05	$0.11	A

Figure 4.8. A bad way to tell the story that the cheaper product is more likely to be purchased.

Example.   You have data on the price for two products (A and B) at all times when someone purchases either of them. And suppose you want to tell the story that the cheaper product is more likely to be purchased. The data is in the file `abprices.xls` and a portion is shown in Table 4.1. First, Figure 4.8 shows a very bad way to tell the story. This graph tells us that B was priced lower on average than A, and that more people bought B than A. It does not tell the story we want.

Incidentally, people often think that charts with fewer bars are easier to understand or believe. But it can be just like saying "because I told you so!" when someone asks you why to believe something. Consider the clunker of a graph in Figure 4.9 that unfortunately turns up in too many presentations.

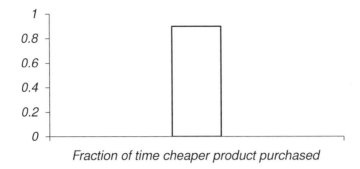

Figure 4.9. A clunker of a graph.

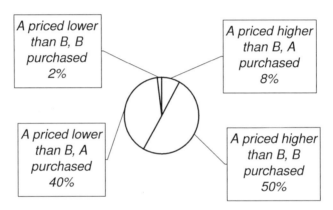

Figure 4.10. Another bad graph.

Yes, this graph tells the story that the cheaper product is purchased most of the time. But it makes me very suspicious. I have no way of knowing if it is telling the whole story or only what someone wants me to know. It also raises more questions than it answers: is one product usually priced lower than the other anyway, and how much data is this graph really based on? This graph doesn't really convince me that a price difference is responsible for what gets purchased.

Another bad graph is shown in Figure 4.10. You might see the story that most of the time a sale is made for a product when it is cheaper, but you really have to look for it. There is so little information being displayed, and it takes so long to figure out what is going on.

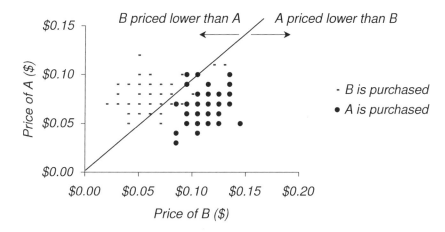

Figure 4.11. A better way to tell the story that the cheaper product is more likely to be purchased.

Finally, Figure 4.11 shows a much better way to tell the story. There is a lot of information presented and we can form our own opinion since it's all raw data. We can see that prices don't seem to be correlated and are scattered all over the place, but it just so happens that the cheaper one is much more likely to get purchased. This would convince me that we should keep our price just slightly below the competition's.

3. Exercises

1. **Price variability.** The file `prices1.xls` (an excerpt is in Table 4.2) contains historical price data for two products A and B. Create a graph that tells the story that the price of A tends to be more variable than the price of B.

2. **Price changes.** The file `prices1.xls` (an excerpt is in Table 4.2) contains historical price data for two products A and B. Show that price changes for product B tend to be followed one month later by a similar price change for product A.

3. **Monitoring pharmaceutical news.** The pharmaceutical company Merck hadn't received much media attention until after it withdrew its block-

Table 4.2. An excerpt from the file `prices1.xls`.

Month #	A's Price	B's Price
1	80.00	50.00
2	82.00	56.90
3	96.56	55.33
4	83.43	45.42
5	67.76	45.72
6	65.80	46.59
7	68.76	46.11
8	66.26	38.33
9	58.33	40.47
10	61.99	47.63

Table 4.3. An excerpt from the file `Vioxx.xls`.

Date	Number of negative stories	Number of neutral stories	Number of positive stories
Jan-02	13	1	4
Feb-02	9	6	7
Mar-02	9	2	6
Apr-02	4	2	5
May-02	8	1	9
Jun-02	4	5	1
Jul-02	4	10	5
Aug-02	13	0	10
Sep-02	7	6	12
Oct-02	0	4	9

buster drug Vioxx from the market on September 30, 2004.[3] The file `Vioxx.xls` (an excerpt is in Table 4.3) shows approximately the number of positive, neutral, and negative news stories appearing in print during each month of several years around that time. What story does this data tell?

4. **Can foreign operations act as a hedge against slumping domestic performance?** It is commonly believed that the foreign operations of a

---

[3] "Reputation and Its Risks," *Harvard Business Review*, February 2007, page 104, by Robert G. Eccles, Scott C. Newquist, and Roland Schatz.

Table 4.4. An excerpt from the file `margins.xls`.

Year	Domestic	Foreign
1991	10	10
1992	8.5	7.5
1993	9.5	6.5
1994	9	8.5
1995	10	8.4
1996	9.5	8.3
1997	10.3	9
1998	9	7.5
1999	10	8
2000	10.5	8.7

large company can act as a "hedge" against slumping domestic results.[4] In other words, profit margins from foreign and domestic operations tend to move in opposite directions over time. This would reduce risk since it would mean that in years when profit margins were low from domestic operations, they tend to be offset by high margins for foreign operations—and vice versa. The file `margins.xls` (an excerpt is in Table 4.4) gives foreign and domestic operating margins averaged over a large number of companies from 1990–2001. Do margins tend to move in opposite directions, or in the same direction? Create a graph to illustrate the story.

5. **IQ gender differences.** The file `salary.xls` (an excerpt is in Table 4.5) contains data on salary, IQ, experience, and gender for a group of people employed at a company. (a) Does salary differ by level of experience? IQ? Gender? (b) After controlling for differences in IQ and experience, does salary differ by gender?

6. **Product quality.** A *Harvard Business Review* article discusses a study of many products over many years that compares actual quality, as measured by *Consumer Reports*, and perceived quality, as measured by consumer surveys.[5] The file `quality.xls` contains data for products rep-

---

[4] "The Forgotten Strategy," by P. Ghemawat, *Harvard Business Review*, November 2003, pages 76–87.

[5] "Quality Is in the Eye of the Beholder," by D. Mitra and P. Golder, *Harvard Business Review*, Apr2007, Vol. 85 Issue 4, p26–28.

Table 4.5. An excerpt from the file `salary.xls`.

ID	Experience	Gender	Male	IQ	Earnings
1	20	M	1	107	55419
2	20	F	0	110	57056
3	10	M	1	139	66260
4	20	F	0	145	65385
5	20	M	1	124	64111
6	10	F	0	145	63975
7	20	F	0	121	56934
8	20	M	1	114	59868
9	20	M	1	127	62944
10	20	F	0	125	60960

resentative of the study's findings. Draw a graph to illustrate the story you see in the data. What strategic implications does this have?

7. **Earnings deception.** An industry analyst would like to know if division heads at large corporations are more likely, compared with small corporations, to deceptively collaborate and move money between them to help each other meet their earnings forecasts. She has data on forecasted and actual earnings for a number of small firms, as well as for separate divisions in large firms. Below are histograms for the ratio of actual to forecasted earnings. What story do the graphs tell?

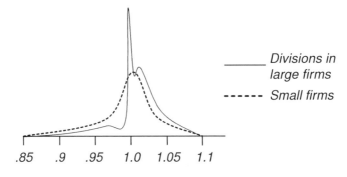

Actual earnings / Forecasted earnings

8. **Stock price momentum.** Does the price of Yahoo stock exhibit momentum? In other words, when the price starts going down one day is it more likely to continue going down the next day, and when it starts

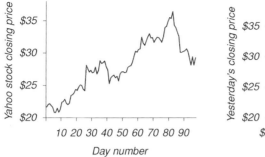 

Figure 4.12. Yahoo's stock price. Does it exhibit momentum?

going up one day is it more likely to continue going up the next day? In Figure 4.12 on the left is a graph of the closing price of Yahoo stock over a ninety day period, and this data is also in the file `stock.xls`. Also, on the right is a graph of the current day's price versus the previous day's price. Does it look like there is momentum in the graph on the left? Does the strong correlation in the graph on the right indicate momentum? Explain. If not, create a graph which can tell the story of whether or not there is momentum.

9. **Forecasting earthquakes.** The United States Geological Survey reports that worldwide since 1900 there has been an average of 19.4 earthquakes of magnitude greater than 7.0 per year.[6] The data are given in the file `earthquake.xls` (an excerpt is in Table 2.6). Does there appear to be any relation between the number of such earthquakes in consecutive years?

---

[6]See http://neic.usgs.gov/neis/eqlists/7up.html.

Chapter 5

Correlation

1. Scatter plots

A scatter plot can be used to see the relationship between two different variables. Below is a **scatter plot** of approximately 200 mutual funds you can buy; each dot here represents a different mutual fund. The percentage return in 2003 is graphed along the X-axis, and an estimate of the standard deviation of this return—labeled as "risk"—is graphed on the Y-axis.[1]

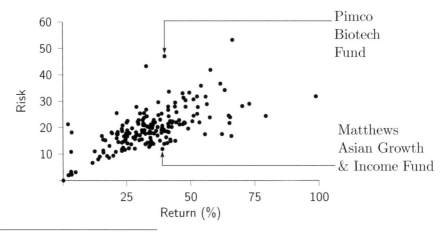

---

[1]Data are from www.fund-track.com. Interestingly, for the first half of 2004 there didn't seem to be any clear relationship between risk and return.

176

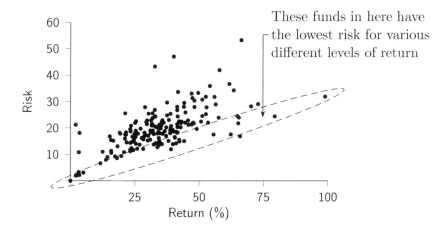

Figure 5.1. A scatter plot of risk versus return for mutual funds.

Which funds are the best? Points that are high vertically correspond to high-risk funds, and points that are to the right horizontally correspond to high-return funds. As you can see, the graph generally slopes upward, meaning that funds that have higher return tend to also have higher risk associated with them. The Pimco Biotech Fund had about the same return as Matthews Asian Growth and Income Fund but had much more estimated risk associated with it. If your investment goal is to find the highest return with the lowest risk, you should buy funds along the bottom edge of the graph circled in Figure 5.1. A scatter plot like this can reveal interesting associations between variables and help you identify data points of interest.

Exercises

1. **Mutual funds.** In Figure 5.2 we've identified Fund A and Fund B. Which of the two funds had a higher return? Which of the two funds had a lower risk? Which of the two funds do you think would be a better investment? Explain.

2. **Hire or promote from within?** It is claimed that it's better to hire a manager from outside a firm than promote one from within. A human resources analyst at a firm justifies this claim by showing that average performance evaluations tend to be significantly higher for managers hired from outside the firm compared with those promoted from within.

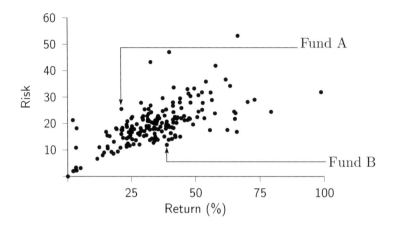

Figure 5.2. A scatter plot of risk versus return for mutual funds and we've identified Fund A and Fund B. Which of the two funds do you think would be a better investment?

Look at the graph of salary versus performance evaluation in Figure 5.3 for managers at the same level at the firm. Is it true that average performance evaluation is higher for managers hired from outside the firm compared with those promoted from within? What other stories does this graph tell?

2. The Correlation Coefficient

The **correlation coefficient**, usually represented by the letter "r," is a measure of the strength of the linear relationship between two variables. The correlation coefficient is a number between -1 and $+1$ that tells how tightly clustered points in the corresponding scatter plot are around a line. A correlation of $+1$ means all the points lie perfectly on a line sloping upwards and a correlation of -1 means all the points lie perfectly on a line sloping downward. Figure 5.4 gives an example of a correlation of $+1$ as well as an example of a correlation of -1, and Figure 5.5 gives examples of values of correlations between -1 and $+1$.

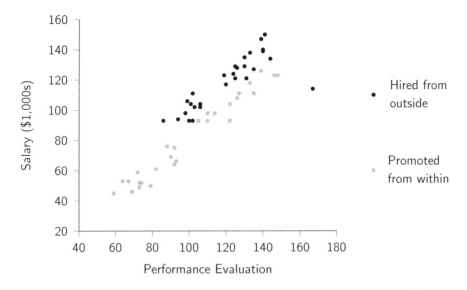

Figure 5.3. Performance evaluation versus salary for employees at a firm. The highest possible performance evaluation score is 200 points.

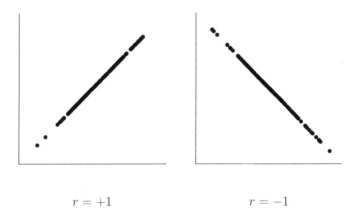

$$r = +1 \qquad\qquad r = -1$$

Figure 5.4. Correlations of +1 and -1.

The **correlation coefficient** measures the strength of the linear association between two variables.

Curved relationships. Interestingly, two variables can have a strong relationship and still have a correlation of zero: Figure 5.6 shows an example

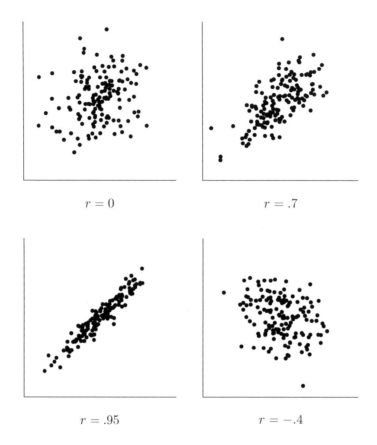

$r = 0$ $\qquad\qquad$ $r = .7$

$r = .95$ $\qquad\qquad$ $r = -.4$

Figure 5.5. Scatter plots and associated correlation coefficients.

of this. Although there is a strong relationship between X and Y, there is no linear relationship. In this case the correlation coefficient actually equals zero. The point here is that you shouldn't always rely on just the correlation coefficient to tell you what is going on—a scatter plot can be very informative.

A **positive correlation** between two variables generally means that when one variable tends to be large the other also tends to be large. On the other hand, when one variable tends to be small the other also tends to be small.

A **negative correlation** between two variables generally means that when one variable tends to be large, the other tends to be small—and vice versa.

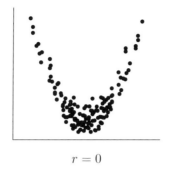

$$r = 0$$

Figure 5.6. A curved relationship.

Car age and value.   If we gather data on the ages of cars on the road and their book values we will find a negative correlation. Older cars tend to be worth less (antiques excluded) and newer cars tend to be worth more. Some older cars are worth a lot, so the correlation is not a perfect -1, but rather is most likely close to -1.

Height and weight.   If we look at the weight and height of people, most likely we will find a positive correlation. Taller people tend to be heavier than shorter people.

Ages and dating.   If we look at the ages of men and women who are dating each other, most likely we will find a positive correlation. Young people tend to date other young people and older people tend to date other older people. This is not guaranteed, but is just a general tendency.

Question.   Suppose that each man only dated a woman who was exactly four years younger than he was. What kind of correlation would you expect to find then?

Discussion.   In this case you would find a perfect $+1$ correlation. Even though the women are younger than the men they date, the ages of men and women tend to be large together and small together.

Question.   Suppose the correlation between X and Y is positive. What kind of correlation would you expect between Y and X?

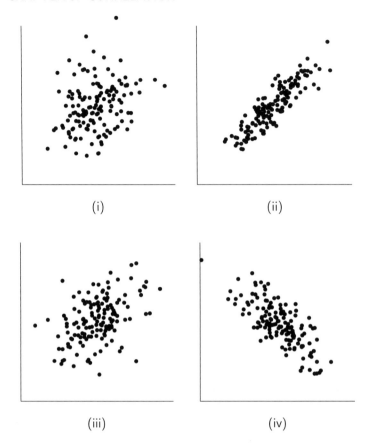

Figure 5.7. Four scatter plots. Can you guess the correlations?

Discussion.   You would expect the exact same positive correlation. A positive correlation means both variables tend to be large together and small together. The correlation between X and Y is the same as the correlation between Y and X.

Exercises

3. **Estimating correlations.** Figure 5.7 shows four scatter plots. Estimate the correlation for each using the following choices: $-.8$, $+.2$, $+.5$, $+.9$

4. Suppose the correlation between two variables is almost zero. What does this say about the link between them?

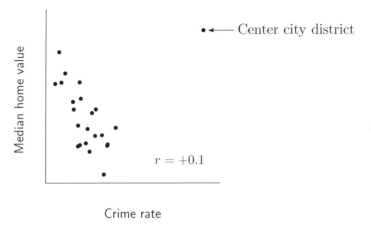

Figure 5.8. Crime rates versus home values.

5. Over the past few years, do you think the correlation between the stock prices of General Motors and Exxon was positive or negative? Why?

3. Some Tricks the Correlation Coefficient Can Play On You

The correlation coefficient is a widely used statistical measure. There are a few things to watch out for when using it. We will give a few examples in this section.

Crime rates and home values.   If you look at the crime rates and the median home value in various districts of one large metropolitan area, the correlation turns out to be about $+0.1$. Does this mean the poor neighborhoods have less crime than the rich neighborhoods? How can you explain this?

Discussion.   Just looking at the positive correlation it may seem as though higher crime rates are associated with higher home values, but let's take a look at the scatter plot. We see in Figure 5.8 that most of the districts follow a downward sloping trend, but a single district, the Center City district, has much higher crime rates and home values than any of the other districts.

We call this type of point an **outlier** since it lies significantly outside the range of the other points. Such outliers have a surprisingly strong impact on

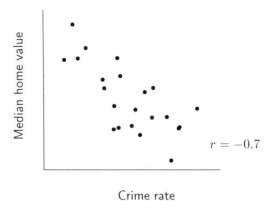

Figure 5.9. Crime rates versus home values excluding the Center City district.

the correlation, especially if they are outliers in both the X direction as well as the Y direction. Because of this, it is not very appropriate to summarize this type of data set with the correlation coefficient. If we remove the Center City district outlier, we get a correlation of -0.7 and the scatter plot in Figure 5.9.

It may seem as though removing a data point is "cheating." But I would summarize the results of this analysis as follows: there is a negative correlation between crime rates and median home value in all the districts except the Center City district. This district does not fit the trend of all the other points—it has a very high crime rate as well as a very high median home value.

In order to tell if an outlier is going to cause a problem, try removing it and see if the correlation changes dramatically. If it does not, then it was not a problem. It usually only causes a problem if it is an outlier in both the X and Y directions – this is summarized in Figure 5.10. Removing too many points from your data can also lead to misleading conclusions.

The correlation coefficient is really only appropriate when your data roughly follow an upward sloping or a downward sloping line. If you remove an outlier so that the correlation becomes an appropriate tool to use, you should think about what removing it says about your overall conclusion. Don't just remove it and quietly "sweep it under the rug."

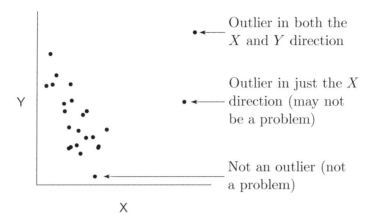

Figure 5.10. Outliers in a scatter plot.

Outliers can have a surprisingly strong impact on the correlation coefficient.

Quality and time spent.   For a large number of American and Japanese companies that all manufacture the same product, there is a positive correlation between the speed at which units are produced and the resulting quality of these units. Does this mean faster production also means better quality? Figure 5.11 shows the scatter plot.

Discussion.   American companies in this industry have lower quality than Japanese companies and tend to also have slower production. If you break the points down by country, Figure 5.12 shows that within each country there is actually a negative correlation: among American companies there is a negative correlation and among Japanese companies there is also a negative correlation. When you combine the two data sets together, interestingly you get a positive correlation. The story here is that higher production speed tends to be associated with lower quality products, but the speed and quality levels are different in Japanese and American companies. Just because two variables have a positive correlation, it does not necessarily mean that in a subset of the data they will also have a positive correlation.

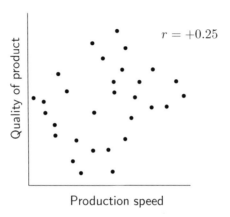

Figure 5.11. Quality versus time spent manufacturing a product.

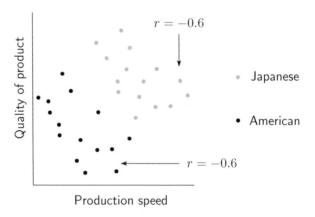

Figure 5.12. When the positive correlation is broken down into two subgroups, we see two negative correlations.

> When you combine data sets or look at a subset of a data set, the correlation coefficient may change in surprising ways.

Theater ticket prices and cancer death rates. For countries worldwide, there is a positive correlation between national cancer rates and national average theater ticket prices. Does this mean that a country could lower its cancer rate by legislating price controls for theater tickets?

Discussion. This correlation is most likely caused by some other factor that is perhaps linked to economics, and theater ticket prices are also linked to economics. Just because there is a correlation it does not mean one causes the other—they could both be caused by some other factor. Don't fall into this trap. It may be true that one influences the other, but you cannot tell this from the correlation coefficient. The correlation coefficient measures association, not causation.

> The correlation coefficient measures association, not causation.

Crime rates and home values, continued. In the example above, there was a negative correlation between crime rates and home value (excluding the Center City district). Does this mean that if the crime rate can be reduced in a town, we could expect the home prices to go up?

Discussion. Suppose a town would like to determine how much property tax revenue could be gained from a given reduction in the crime rate, and if this increase in tax revenue could actually pay for the extra policing necessary. But unfortunately it is impossible to tell what portion of the negative correlation is because the crime rate has an impact on the home values and what portion of the correlation is because home values have an impact on the crime rate. Or perhaps both factors are influenced by some other factor—such as the economy. You cannot tell if one factor causes the other to change just because there is a correlation. Certainly crime rates have an impact on home values, but the correlation here most likely would greatly overestimate the true impact.

Changing the scale and correlation. Below are two scatter plots. Which shows a higher correlation?

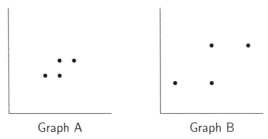

Graph A   Graph B

Discussion. Even though the points in Graph A appear to be more tightly clustered around a line, both correlations are actually the same. The reason for the optical illusion here is that the variables in both graphs have very different standard deviations. To visually compare the correlations in two different scatter plots, both must have approximately the same standard deviation for the X values as well as the same standard deviation for the Y values. If we shrink Graph B down in size so the standard deviations are on the same scale as in Graph A, then we see that both will have the same correlation. Changing the scale of your data does not change the correlation coefficient.

Scaling all your data up or down by some positive constant does not change the correlation coefficient.

---

4. Using Excel

Computing correlations using standard Excel

To have Excel compute the correlation coefficient you can use the following command:

$$=\texttt{CORREL}(X\text{-values},\ Y\text{-values})$$

By "X-values" we mean the range of cells containing the X-values. For example if the X-values are in the range A1 to A10 and the Y-values are in the range B1 to B10, you can use the following command:

	store	adv	miles	sqfeet	own	income	prom	sales	prom1	prom2
store	1.0									
advert	0.2	1.0								
miles	0.1	0.5	1.0							
sqfeet	0.0	0.5	0.4	1.0						
own	0.2	0.7	0.3	0.5	1.0					
income	−0.1	0.6	0.4	0.6	0.7	1.0				
prom	0.0	−0.2	−0.3	−0.1	−0.2	0.0	1.0			
sales	0.1	0.6	0.5	0.7	0.6	0.6	−0.2	1.0		
prom1	0.0	0.1	0.1	−0.2	−0.1	−0.1	−0.7	−0.2	1.0	
prom2	0.1	0.2	0.2	0.5	0.3	0.1	−0.2	0.6	−0.5	1.0

Figure 5.13. Computing pairwise correlations using the `Correlation.xlsx` worksheet.

$$=\text{CORREL(A1:A10, B1:B10)}$$

In addition, most versions of Excel come with the "Analysis Toolpak" which has a feature to compute correlations. In Excel 2010 you can go to the "Data" ribbon and click "Data Analysis" and then "Correlation." If you don't see it there, you may have to install first it by going to the "File" ribbon and clicking "Options/Add ins/Go" and checking off "Analysis Toolpak."

Computing correlations using the Correlation Worksheet included with this book

You can use the "Correlation" worksheet included with this book to compute a large number of correlations easily. First open the file `salesdata.xlsx` which contains some sample sales data. Copy the data and then open the file `correlation.xlsx` and paste the data into the "Data" sheet. Check to make sure that the top row contains the variable names, the first column of data is in Column A and that there are no blank rows or columns. Then go to the "Correlations" sheet and you will see all the pairwise correlations as shown in Figure 5.13.

Exercises

6. **Hospital visit costs.** Medical records show strong positive correlations between the age of a patient, the number of days the patient stays in the hospital, and the total cost of the visit.

    (a) Does this mean more efficient management practices that reduce the length of hospital stays could reduce the age of patients? Explain.

    (b) Does this mean such practices could reduce the total cost of the hospital visits? Explain.

7. **Incomes and home prices.** An analyst finds a positive correlation between median incomes in a city and median home prices in the city, using data from 200 cities nationwide. A second analyst decides to compute the correlation using only data from the top 20 most expensive cities. Would you expect the second analyst's correlation to be higher or lower than what the first analyst got? Explain.

8. **Commodity trading.** A commodity trader finds a positive correlation between wheat prices and soybean prices, when both are measured in terms of dollars per bushel.

    (a) Would you expect the correlation to be higher, lower or the same if prices were instead measured in dollars per pound?

    (b) What if prices for soybeans were in dollars per bushel and prices for wheat were in dollars per pound?

5. Computing the Correlation Coefficient

You will not usually need to compute the correlation coefficient by hand, since the computer does such a good job at that. But there are some insights you can get by understanding how the formula works. Below are the steps for computing the correlation coefficient between X and Y:

1. Convert the data to standard units
2. Multiply each pair of X and Y values together
3. Average the numbers you get in Step 2

We illustrate with an example.

Example. Compute the correlation coefficient for the following data set of paired X- and Y-values.

X	Y
12	20
8	16
9	14
10	8
6	12

Step 1. To convert to standard units, we must first compute the two averages and SDs:

$$\text{Average of } X = \frac{12 + 8 + 9 + 10 + 6}{5} = 9$$

$$\text{Average of } Y = \frac{20 + 16 + 14 + 8 + 12}{5} = 14$$

$$\text{SD of } X = \sqrt{\frac{(12-9)^2 + (8-9)^2 + (9-9)^2 + (10-9)^2 + (6-9)^2}{5}}$$

$$= 2$$

A similar calculation tells us that the SD of Y equals 4.

Now we convert the data to standard units by subtracting the corresponding mean from each and dividing by the corresponding SD. The first X value is 12, and converting this into standard units gives $(12 - 9)/2 = +1.5$. The second X value is 8, and converting this into standard units gives $(8-9)/2 = -0.5$. Repeating this calculation all the way down for the X values and then again for the Y values gives us the following:

		Standard Units	
X	Y	X	Y
12	20	1.5	1.5
8	16	−0.5	0.5
9	14	0.0	0
10	8	0.5	−1.5
6	12	−1.5	−0.5

Steps 2 and 3. Multiply each pair of X- and Y-values in standard units and

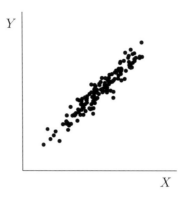

Figure 5.14. This is a scatter plot.

averaging them together gives the following:

$$r = \frac{(1.5) \times (1.5) + (-.5) \times (.5) + (0) \times (0) + (.5) \times (-1.5) + (-1.5) \times (-.5)}{5}$$
$$= 0.4$$

This means the correlation equals $+0.4$.

How the formula works

You may think it is odd that a mathematical formula can tell if points are sloping upward or downward on a scatter plot. To see how the formula for the correlation coefficient works, consider the scatter plot in Figure 5.14.

The first step in the correlation coefficient formula is to convert all the data into standard units. Since in standard units the point $(0,0)$ is at the average of the X values and the average of the Y values, converting to standard units essentially moves the axes to the center of the scatter plot – this is shown in Figure 5.15.

The next step is to multiply each pair of X and Y values together. There are four different quadrants here to consider. In the upper-right quadrant, we have positive X-values multiplied by positive Y-values, and this will give positive products. In the lower-left quadrant we have negative X-values multiplied by negative Y-values, and this will also give positive products. In the upper-left quadrant we have negative X-values multiplied by positive Y-values, and this gives negative products. Finally, in the lower-right quadrant

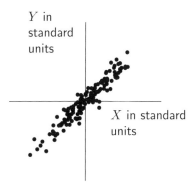

Figure 5.15. The standard units axes for a scatter plot.

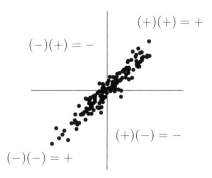

Figure 5.16. The four quadrants of a scatter plot in standard units.

we have positive X-values multiplied by negative Y-values, giving negative products. This is summarized in Figure 5.16.

When the graph slopes upward, most of the points are concentrated in the upper-right and lower-left quadrants—and these are the ones with positive products. This means the positive products usually outweigh the negative products, so when you take the average you get a positive number for the correlation coefficient. When the graph slopes downward, most of the points are in the quadrants that have negative products—so the correlation coefficient will come out negative.

6. A Couple More Things to Watch Out For

Correlations between averages.   Suppose we create a scatter plot for state average education level and state average income for each of the 50 states. In this scatter plot there is one point for each of the 50 states. We find there is a strong positive correlation. If we instead had data from all residents of these states, would we get a higher or lower correlation?

Discussion.   If you had hundreds of millions of points (one for each person), there would be much more scatter. Averaging all the people in the state together and representing them as a single point reduces a lot of the scatter and acts to increase the correlation.

Here is a simple illustration. Suppose there are three states (a, b, and c) and each state has five or six residents. For each person we find the income and education level and, instead of marking it with a dot in the scatter plot, we mark it with the letter of that person's state. The black dot in the middle of each cluster of letters represents the average income and education level for that state.

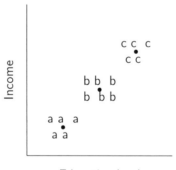

Education level

If you just look at the letters, you can see there is not a perfect correlation: the graph slopes upward but all the letters do not fall exactly on a line. If you now instead look at just the state averages—the three black dots—you see that they all fall almost perfectly on a line. This means that the correlation between state average education and state average income is much higher than the correlation between individual income and individual education. This illustrates how correlations between averages tend to be stronger than correlations for the underlying data.

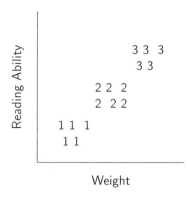

Figure 5.17. Reading ability and weight in elementary school children.

Correlations for averaged data tend to be stronger than correlations for the underlying data.

Weight and reading ability in children. Interestingly, there is a strong positive correlation between weight and reading ability in elementary school children. What explains this?

Discussion. This was a trick question. If you look at children in a single grade, there generally is no correlation. But if you combine all the grades together you see a positive correlation. Children tend to get bigger and heavier as they get older and their reading skills tend to get better as well. If you combine data sets that have no correlation at all, it is possible to get a positive correlation.

Figure 5.17 is an illustration with five or six students in each of grades 1, 2, and 3. Each child is marked with the number of his or her grade level. Notice that if you look at only a single grade by itself there is no correlation. But if you look at all the grades combined there is a positive correlation.

7. Technical Notes

The mathematical formula for the correlation coefficient is

$$r = \frac{1}{n} \sum_{i=1}^{n} \left(\frac{X_i - \mu_X}{\sigma_X} \right) \left(\frac{Y_i - \mu_Y}{\sigma_Y} \right)$$

where n is a number of data points, X_i is the ith X-value, and μ_X and σ_X are the mean and standard deviation of the X-values (for the Y values it's Y_i, μ_Y and σ_Y).

The **covariance** is also used to measure association between X and Y and it is defined to equal

$$Cov(X, Y) = \frac{1}{n} \sum_{i=1}^{n} (X_i - \mu_X)(Y_i - \mu_Y).$$

You can see that the covariance will equal the correlation if both standard deviations are equal to 1 or if the data have been converted into standard units. Another equivalent formula for covariance is

$$Cov(X, Y) = r\sigma_X\sigma_Y,$$

which is useful if you know the correlation already.

8. R lab: MBA graduate salaries

Start R-studio and click on "File/Import dataset/From Excel" and navigate to the data file "BU MBA salalaries.xlsx" and click "Import". Then create a copy titled "d" as follows:

```
d=data.frame(BU_MBA_salaries)
```

Let's take a look at correlations between some interesting variables and salary. The command

```
cor(d[d$Salary>0,c(1,2,3,5,9)])
```

prints out the correlations between all pairs of variables in columns 1,2,3,5, and 9 where the salary is greater than zero. Why did we choose only these columns? We chose these columns because they respectively correspond to GMAT, incoming GPA, work experience, graduating GPA, and salary. Column 4 is the name of the program, and columns six and seven are just variables that can be either zero or one. The correlation coefficient and scatter plots are not very informative if a variable only takes two values.

From this we see that the highest correlation with salary is .177 and corresponds to graduating GPA. We will look at a scatter plot of this in a moment but first we will take a brief diversion on selecting rows and columns.

In R to select a subset of rows and columns, you can use the command

```
d[c(2,4,6),c(1,3,5)])
```

to select rows 2,4,6 and columns 1,3,5. To select all rows and just columns 1,3,5 you can use

```
d[,c(1,3,5)])
```

You can select a range of rows and columns using

```
d[1:5,4:6])
```

to select rows one through five and columns four through six.

To draw scatter plot for graduating GPA and salary, type

```
plot(d$Graduating.GPA[d$Salary>0],d$Salary[d$Salary>0])
```

and you should see the plot in Figure 5.18 below. To quickly draw scatter plots for all pairs of variables 1,3,5,9, type

```
pairs(d[d$Salary>0,c(1,2,3,5,9)])
```

and you should see the plot in Figure 5.19 below. It doesn't appear that salary is strongly correlated with any of the other variables.

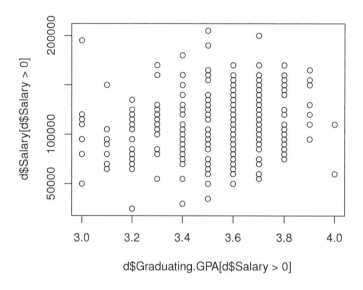

Figure 5.18. A scatter plot of graduating GPA and salary for people who report salaries.

9. Exercises

9. **Employee evaluations.** In evaluating employee performance and determining salary raises, a company uses a large number of subjective and objective criteria gathered from numerous evaluators that are then combined together to give each employee an overall score between 0 and 100. The raise an employee gets depends on the score as follows: people scoring under 41 points get no raise, people scoring between 41–50 get a 1% raise, 51–60: 2% raise, 61–70: 3% raise, 71–80: 4%, 81–90, 6%, 91–100: 8%. Below in Figure 5.20 is a histogram of the overall scores and in Figure 5.21 a scatter plot (with the correlation coefficient displayed above it) showing the score for each employee versus the length of time between when the evaluation was finished and when the employees learned the results from their managers.

(a) Rumors at the company suggest that the longer it takes to hear the results, the lower the score because managers may be postponing a

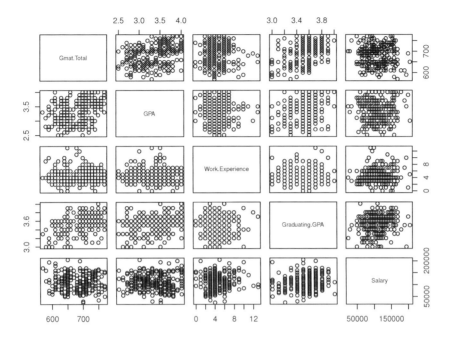

Figure 5.19. All scatter plots for columns 1,3,5,9 for people who report salaries.

difficult discussion with the employee about poor performance. Do the data support this concern?

(b) Top management is concerned about rumors that scores were unfairly bumped up a few points for many employees to help them get raises that weren't justified by their performance. Is there any evidence of this here?

10. **Exam scores.** A small class of five students takes two exams. The scores are shown below and the correlation coefficient between the two exam scores equals +0.7.

Student #	1	2	3	4	5
Exam 1 score	73	82	100	77	81
Exam 2 score	83	64	99	68	79

(a) On Exam 1, is the average higher than or lower than the median?

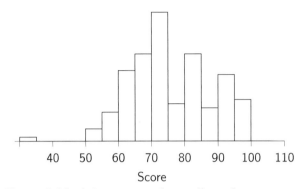

Figure 5.20. A histogram of overall employee scores.

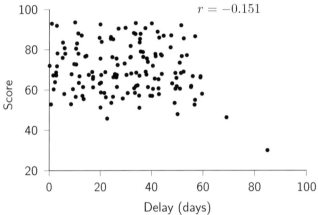

Figure 5.21. Score for each employee versus the length of time between when the evaluation was finished and when the employees learned the results from their managers.

    (b) Does this correlation of +0.7 here mean that people who scored well on one of the exams in general tended to score well on the other exam? Explain briefly.

11. **Salary and demographics.** A manager would like to know which of the following factors has the strongest link to the salary an employee earns: age, gender, experience, time at the firm, or education. Looking at the file employees.xls (an excerpt is in Table 5.1), what do you conclude?

12. **Order completion times.** In a manufacturing job shop, different orders

Table 5.1. An excerpt from the file `employees.xls`.

gender (1=F, 0=M)	Age	Yrs Exp	Yrs at Firm	Yrs Ed	Annual Salary
1	39	5	12	4	$38,450
0	44	12	8	6	$50,912
0	24	0	2	4	$29,356
1	25	2	1	4	$27,750
0	56	5	25	8	$109,285
1	41	9	10	4	$48,442
1	33	6	2	6	$40,207
0	37	11	6	4	$42,331
1	51	12	16	6	$87,489
0	23	0	1	4	$26,118

Table 5.2. An excerpt from the file `jobshop.xls`.

Order completion time (days)	Total number of other orders in shop
78	44
74	39
68	40
76	40
80	35
84	41
50	23
93	47
55	17
76	49

can require work to be performed in different sequences at work centers in the plant. As a result, there are constantly shifting bottlenecks at work centers. Management has found it difficult to predict the length of time within which an order can be completed. File `jobshop.xls` (an excerpt is in Table 5.2) shows the time to complete each of last year's orders. Based on this, the plant has been telling customers to expect that on average it will take 71 days to complete an order.

(a) How accurate does this prediction tend to be? Quantify your answer.

(b) Can you come up with a better prediction rule? How much more accurate is your prediction rule?

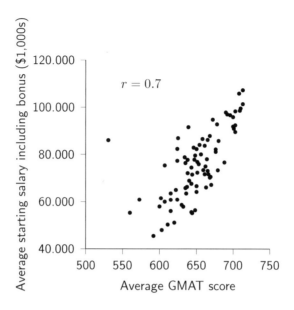

Figure 5.22. Average starting salary including bonus (in $1,000s) versus the average GMAT score for students at each of the top 80 ranked schools in the country.

13. **GMAT scores and salaries.** U.S. News & World Report is a popular magazine that annually ranks business schools. They publish the average starting salary and the average GMAT score for students at each of the top 80 ranked schools in the country. In 2004 the correlation was 0.7. The graph of the 80 data points is in Figure 12.

    (a) Does this mean that, in general, schools in this group with higher average GMAT scores tend to have higher average starting salaries? Explain.

    (b) If you instead looked at the data for all the individual enrolled students at these schools, the correlation should be (i) around .7 (ii) higher than .7 or (iii) lower than .7. Choose one and explain.

14. **Do immigrants improve a state's economy?** For each of the fifty states if you look at the state average income and the percentage of people in that state who are foreign-born immigrants, you will see a positive correlation. Does this mean immigrants tend to earn more than other people? Or does it mean immigrants improve a state's economy? If not,

what could explain the correlation?[2]

15. **Height and income.** There is a small but positive correlation between income and height for adults and a scatter plot tends to follow a gently sloping line.

    (a) Does this mean people who earn more tend to be a bit taller on average than people who earn less?

    (b) Does this mean people who are shorter tend on average to earn a bit less than people who are taller?

    (c) If you increase the salary for a large group of people, should you expect them to grow a little taller on average? Explain.

    (d) What could explain the correlation?

16. **Value at risk.** The **value at risk** for $X\%$ confidence is defined to be the $(100 - X)$th percentile of the forecasted distribution of returns for a portfolio. This value says you believe there is an $X\%$ chance you will receive at least this return. The **marginal value at risk** for a given investment in a portfolio tells you how much the value at risk would change for that portfolio if you removed the given investment from that portfolio. Suppose there are two investments in a portfolio: investment A has a forecasted average return of $10 million with a standard deviation of $10 million, investment B has a forecasted average return of $12 million with a standard deviation of $5 million, and investment returns typically follow a normal distribution. Also, the correlation between the two investment returns is estimated to be +0.4.

    (a) Calculate the value at risk for 95% confidence for the whole portfolio.

    (b) Calculate the marginal value at risk for 95% confidence corresponding to each investment, and interpret the results.

17. **Investment styles.** As part of evaluating its fund managers, an investment firm assesses the manager's investment style. If the manager's goal is to use a "growth" style (or a "value" style) of fund management, then his fund's return should follow "growth" stock indices (or "value" stock indices). The file `fund.xls` (an excerpt is in Table 5.3) contains returns for one manager's stock fund, as well as the returns over the same period

[2]D. A. Freedman, "Ecological Inference," in Neil J. Smelser and Paul B. Baltes, Editors-in-Chief, *International Encyclopedia of the Social & Behavioral Sciences*, Elsevier Science Ltd, Oxford, 2001, Pages 4027–4030, 0080430767. (`http://www.sciencedirect.com/science/article/B6WVS-4DN8TNP-VV/2/97f107752e403d5197484f8351c0d959`)

Table 5.3. An excerpt from the file `fund.xls`.

Week	Fund price	Growth index	Value index
1	51.35	1012.00	348.00
2	51.82	1019.19	348.77
3	52.42	1030.98	352.93
4	53.49	1038.39	355.58
5	54.65	1056.20	358.25
6	56.44	1068.96	364.75
7	56.11	1072.74	369.65
8	56.24	1087.36	373.24
9	56.89	1102.55	375.16
10	58.00	1126.94	376.34

for a standard "growth" index and a standard "value" index. Is it more likely the manager is using a "growth" style or a "value" style?

18. **Does the world need more pornography?** A recent newspaper article by Suzanne Reisman titled "Porn: it just might cure what ails us" argues that "the world needs more pornography" and writes that "studies find a positive correlation between a society's consumption of pornography and its level of equal opportunity for women in education, employment and politics. Residents of Sweden and Denmark view a large amount of porn, but the countries have high rates of gender equality and low levels of violence against women, especially when compared to the United States."[3] Can you think of some alternate ways to explain the correlation?

19. **Manufacturing costs.** You have overseas plants in both Latin America and Asia and the regional manager of your Latin American plants has been asking for more production volume. He argues that when weekly volume from the Latin American plants is higher, cost per item is lower. In fact, he says the correlation between weekly volume and cost per item is −.70 and has prepared a scatter plot showing that the relationship fits a line sloping downward quite well. You ask for a similar analysis on the Asian plants and the results are similar (a correlation of −.65

---

[3] "Porn: it just might cure what ails us," by Suzanne Reisman, November 16, 2006, *Boston Metro*, `http://boston.metro.us/metro/blog/my_view/entry/Porn_it_just_might_cure_what_ails_us/5727.html`

Table 5.4. An excerpt from the file `innovation.xls`.

City #	Population of city	Total innovation index
1	5,400,000	29,000
2	2,950,000	5,000
3	1,413,270	1,575
4	18,107	9
5	13,281	5
6	38,441	12
7	240,789	151
8	2,040,000	4,000
9	581,536	252
10	4,013,366	9,900

and a scatter plot showing the relationship fits a downward sloping line nicely). You are surprised because a previous scatter plot created for you that used the same data showed no correlation between volume and cost per item from all of the overseas plants combined. Explain how this could possibly happen.

20. **Innovation and firm size.** An article in the *Harvard Business Review* argues that the size of a company greatly affects the amount of innovation you see, not only in total, but also when measured as innovation per employee.[4] To illustrate the point, it discusses research on the relationship between the size of various cities and the total amount of innovation that occurs. Data illustrating the relationship is contained in the file `innovation.xls` (an excerpt is in Table 5.4). How would you describe the relationship and what implications does it have?

21. **Forecasting growth.** Having a reliable forecast of national economic growth can be important in making investment decisions. The file `growth.xls` (an excerpt is in Table 5.5) contains actual economic growth figures over a time period as well as the forecasted growth figures using three separate forecasting techniques. Which of the three forecasting techniques seems to be the best?

22. **Estimating correlations.** For the data shown in Figure 5.23, does the

---

[4] "Innovation and Growth: Size Matters," *Harvard Business Review*, February 2007, page 34, by Geoffrey B. West.

Table 5.5. An excerpt from the file `growth.xls`.

Month	Actual growth %	Forecast 1	Forecast 2	Forecast 3
1	2.34%	0.84%	1.82%	1.68%
2	1.11%	0.73%	1.51%	1.04%
3	2.48%	1.20%	2.87%	2.34%
4	0.48%	0.97%	-0.36%	0.37%
5	1.84%	0.63%	0.51%	1.43%
6	2.32%	2.73%	0.88%	2.41%
7	1.50%	1.33%	0.16%	1.53%
8	3.50%	2.66%	-0.39%	3.13%
9	2.21%	2.14%	0.63%	1.81%
10	0.84%	1.54%	0.01%	0.71%

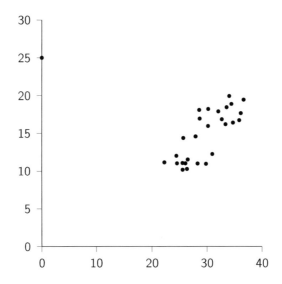

Figure 5.23. Does the correlation coefficient here appear to be closest to $+.7$, $-.7$, or 0?

correlation coefficient appear to be closest to $+.7$, $-.7$, or 0? Explain briefly.

23. **Four data sets.** The file `fourdatasets.xls` contains four sets of eleven pairs of data values.[5] (a) Compute summary statistics for each variable,

---

[5]From "Graphs in Statistical Analysis," F. J. Anscombe, *The American Statistician,*

and correlations for each of the four pairs of variables. How are they similar across the four pairs of variables? (b) How do the four relationships differ? Discuss briefly.

24. **Customer survey.** A company surveys its customers and asks them to evaluate the benefits of a number of different models in two different product lines. The average survey results along with the prices of the products are given in the file `survey.xls`. How well aligned are perceived benefits and prices in each of the product lines? What managerial implications does this have?[6]

25. **Heights and weights.** Data is gathered on the heights and weights of people in a large group of adults and children. Among adults the correlation between height and weight equals 0.6, and a similar correlation is observed among children. Would the correlation for both groups combined be higher or lower than 0.6? Justify your answer briefly, using a diagram if necessary.

26. **Height and income.** A book discusses an interesting relationship between height and income for people in the United States.[7] The file `height-income.xls` contains data on height, income, and gender for a large representative sample of people. What kind of relationship do you see and what could explain it?

27. **Happiness survey.** The city of Somerville, MA became the first city in the United States to conduct a well-being survey of its residents.[8] The survey included general happiness questions such as "On a scale of 1 to 10, how happy do you feel right now?" as well as questions about the city such as "Taking everything into account, how satisfied are you with Somerville as a place to live?" The file `Somerville.xlsx` contains the responses of residents who completed the full survey as well as a brief explanation of the variables gathered.

Vol. 27, No. 1 (Feb., 1973), pp. 17-21.

[6] "Charge What Your Products Are Worth," by Bala, Venkatesh and Green, Jason. *Harvard Business Review*, Sep2007, Vol. 85 Issue 9, p22-22.

[7] *Teaching Statistics: A Bag of Tricks*, by A. Gelman and D. Nolan, Oxford University Press, 2002.

[8] How Happy Are You? A Census Wants to Know, by John Tierney, *The New York Times*, April 30, 2011. http://www.nytimes.com/2011/05/01/us/01happiness.html?pagewanted=all

(a) What factors seem to be most correlated with a person's satisfaction with the city of Somerville?

(b) Do you see see any racial differences in general happiness?

(c) Do you see any other interesting stories in the data?

Fun Problem: Coin Flipping Patterns

Someone is going to flip a fair coin until either one of the following two patterns of heads and tails (H and T) appears in a row:

(1) T H H H

(2) H H H H

I win if Pattern (1) comes up first, and you win if Pattern (2) comes up first. Who has the advantage? Or is this a fair game?

To illustrate the game, suppose we start flipping and we get "H T H H T H T H H H." We would stop flipping at this point because the last four outcomes were "T H H H," and this means that Pattern (1) appeared before Pattern (2)—and I would win this time.

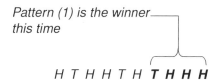

Pattern (1) is the winner this time

H T H H T H **T H H H**

For this game, who has the advantage here? Or is it fair?

Solution.   Pattern (1) is 15 times more likely to come up first than is Pattern (2). The only way Pattern (2) can win is if it comes up on the first four flips. If a "tails" comes up anywhere during the first four flips, then by the time we see "H H H H" we would have seen right before it an appearance of "T H H H." We will see later why the chance of getting "H H H H" on the first four flips is 1/16, so this means the chance of getting "T H H H" must be 15/16—which is 15 times larger.

Chapter 6

Regression

1. Introduction

When you have a scatter plot where the points appear to slope upwards or downwards in an approximate linear relationship, it is natural to describe the relationship by fitting a line to the points. The **regression line** is the best fitting line that attempts to go through the average Y value for each X value.

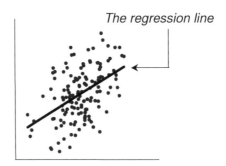

The regression line

As you may remember from algebra, the equation for a line is

$$Y = mX + b$$

where m is the **slope** (which, as illustrated in Figure 6.1, equals the "rise" divided by the "run" for the line, or the change in Y for a unit change in X), and b is the **Y-intercept** (which is where the line crosses the Y-axis).

210

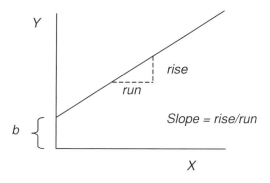

Figure 6.1. The slope of a line.

The variable on the X-axis is usually called the **independent variable** and the variable on the Y axis is usually called the **dependent variable**.

The equation of this line is often used for summarizing the relationship between two correlated variables. For example, suppose the regression line for predicting the value of a home in a city from the square footage is

$$Y = .25X - 50,$$

where Y is the value of the home (in \$1,000s) and X is the square footage. You can then answer questions such as "what is an estimate of the average value of a home with 2,000 square feet?" Using the line we get an answer of $Y = .25 \times 2000 - 50 = 450$, which corresponds to \$450,000. We can interpret the slope .25 as "each extra square foot is associated with an extra \$250 on average." (The \$250 comes from the fact that the units of Y are in thousands of dollars here, so .25 corresponds to \$250.)

> The slope of the regression line tells you how much of a change in the Y variable is associated with a unit change in the X variable.

Notice that if we plug in 10 for X we get a predicted value of -47.5, which corresponds to a home value of negative \$47,500. This clearly does not make sense as a home value and the reason for this nonsense is that we are extrapolating far outside the range of the data. It's not appropriate to use a regression line too far outside of the range of the data. The intercept -50 does not have a useful interpretation here.

As a second example, suppose a different line for predicting the home value in the same city is

$$Y = 60X + 250,$$

where Y is the value of the home (in \$1,000s) and X is the number of sinks in the home (the kitchen sink, the bathroom sink, etc.). Using this line, we can estimate that the average value of homes with three sinks is around $Y = 60 \times 3 + 250 = 430$, or \$430,000. We can interpret the slope 60 by saying that "each extra sink is associated with an extra \$60,000 in value on average."

What would happen to the value of a home if we installed an extra sink? Would we expect the home value to go up by around \$60,000? Since a sink obviously costs a lot less than \$60,000 to install, does this mean that we should start installing extra sinks in every room of the house? The reality is that this regression line does not tell us what would happen to the value of the home if we came in and increased the number of sinks; the regression line only tells us the average value of homes that already have that number of sinks. The reason why the number of sinks in a home can be a reasonably good predictor of the value of the home is because an extra sink usually corresponds to an extra bathroom, and an extra bathroom often corresponds to an extra bedroom and thus more square footage overall. So it would be unwise to interpret the coefficient as meaning "each extra sink adds an extra \$60,000 to the home value." The better way to interpret the coefficient is "each extra sink is associated with an extra \$60,000 in home value."

The regression line does not describe the causation (what would happen to Y if you intervened to change X) but merely describes the association between X and Y. There may truly be a causal link between X and Y but you can't tell this from the equation of the line—you must have other reasons for believing this.

> The regression line describes how two variables are naturally associated, but it does not tell you how one variable would change if you intervened to change the other variable.

Technical Note: Computing the Slope and Intercept

Although the computer handles these details, computing the slope and intercept of the regression line is fairly easy. The formulas are the following:

$$\text{Slope} = r \times \frac{\text{SD of } Y \text{ values}}{\text{SD of } X \text{ values}},$$

where r is the correlation coefficient, and

$$\text{Intercept} = (\text{Avg of } Y \text{ values}) - \text{Slope} \times (\text{Avg of } X \text{ values}).$$

This means the slope and correlation will actually be equal if both the X and Y values have the same standard deviation, which is the case if they are both put into standard units. Otherwise the slope of the regression line will be different from the correlation coefficient.

2. Some Things to Watch Out For

The two regression lines.   To get the regression line for predicting square footage from home value, can we just solve for X in the first equation above?

Discussion.   No. You may think if you have an equation for predicting Y from X that you can just solve for X to get an equation for predicting X from Y—but that doesn't work very well. You need to use a completely different line to predict X from Y. When predicting Y from X, you want the line to go through the average Y for each value of X. When predicting X from Y, you want the line to go through the average X for each Y. This gives two different lines with different slopes. In short, you can't just solve for X to get a prediction for X. Figure 6.2 gives a graphical illustration.

Correlation coefficient versus regression coefficient.   Suppose   someone gives you one regression line for predicting variable Y from variable A, and another for predicting Y from variable B: $Y = .02A+7$, and $Y = 1{,}500B+2$. Since 1,500 is so much bigger than .02, does this mean B is much more highly correlated with Y than A is? Does this mean B has much more of an impact on Y than A does?

This line forecasting Y from X goes through the middle of each vertical slice -- so it gives the average Y for each X,...

...whereas this steeper line for forecasting X from Y goes through the middle of each horizontal slice -- so it gives the average X for each Y.

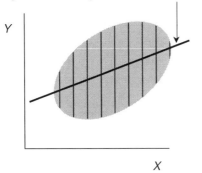
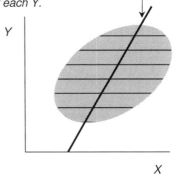

Figure 6.2. The regression line for predicting X from Y is steeper than the one for predicting Y from X.

Discussion. No. There is a difference between the correlation coefficient and the regression coefficient. Just because something has a larger regression coefficient, it doesn't mean it has a stronger link with Y. It could be that A and B are in very different units and have very different standard deviations. To see the strength of the link, look at the correlation coefficient—not the regression coefficient. And, again, just because two things are linked doesn't mean one has any causal impact on the other.

3. More Examples

The beta of a stock. The "beta" for a stock is a way of measuring how risky it is compared with the rest of the market. It is usually defined as the slope of the regression line when you put weekly returns for your stock on the Y axis and weekly returns for a market index, like the Standard and Poor's 500 (S&P 500), on the X axis and use data over a period of several years. It tells you how much return you expect to see from your stock for each 1% of return in the market index. If the beta value is greater than 1, it means you expect to see larger than 1% return. This means that small fluctuations in the market are associated with large fluctuations in your stock

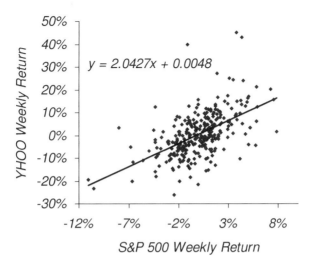

Figure 6.3. Calculating the beta value for Yahoo stock.

price and your stock is more risky than the overall market. If the beta value of your stock is less than 1, it means that a large fluctuation in the market corresponds to a small fluctuation in your stock price and your stock is less risky than the overall market.

For example, we look at Yahoo stock (symbol YHOO) over a few hundred weeks (from 1999–2004) and compute the return for each week. We put this on the Y axis. Then we look at the returns from the S&P 500 index and put this is on the X axis. Since the slope of the regression line comes out to about 2, the "beta" for Yahoo stock is around 2; it is much more risky than the market. Figure 6.3 is the graph of what we get. Notice that Yahoo's stock price can change by up to 50% in a single week whereas the market index is much more stable.

Optimal hedging.   Suppose you own shares in an S&P 500 index fund (designed so the value closely follows the S&P 500 stock market index) and you want to reduce the risk of this investment by buying shares of a gold index fund. How many shares of the gold fund should you buy to best reduce your risk? Figure 6.4 is a graph of the monthly closing price of each fund as well as the equation of the regression line. You can assume the price of one share of each index fund equals the value of the index.

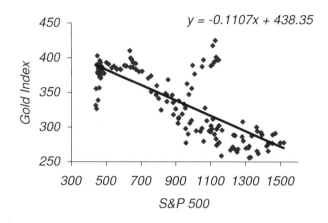

Figure 6.4. The link between gold prices and stock market prices.

The regression line only roughly fits the data. This coefficient "−0.1107" says that when the S&P 500 fund goes down by $1 we expect to see the gold fund go up by about 11 cents on average. This means that if we owned $1/.11 \approx 9$ shares of the gold fund, when the S&P fund goes down by $1 we would expect to see the gold fund go up by about $(9) \times (.11) \approx \$1$ on average. This is a great way to reduce the risk of our investment in the S&P 500 fund and means losses we suffer with one of the funds tend to get canceled out by gains from the other fund. This means that for every share we have in the S&P 500 fund we should own 9 shares of the gold fund. The reciprocal of the slope tells you how much you should own.

4. Standard Error For a Regression Line

The **standard error for a regression line** is a measure of how far in general points are (vertically) from the line. Usually around two-thirds of the points will be within one standard error from the regression line. This also tells you the typical amount of error you would see when making forecasts using this regression line. An approximate formula is as follows:

$$\text{Standard error for regression line} \approx \text{SD of } Y \text{ values} \times \sqrt{1 - r^2}$$

This means that if you are trying to predict a Y value having a standard deviation of 10 and you use an X variable having a correlation with Y of .7,

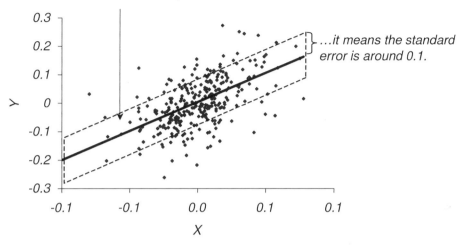

Figure 6.5. The standard error for a regression line.

the typical amount of error you'll see using this line would be approximately $10 \times \sqrt{1 - .7^2} \approx 7.1$.

> The **standard error for a regression line** is a measure of how far in general points are vertically from the line.

Figure 6.5 is an illustration of the standard error for a regression line. Since about two-thirds of the points there are within 0.1 from the line, the standard error is about 0.1. This means that the typical forecast for Y using this line will be off by around 0.1 up or down from the actual Y value.

5. Regression Toward the Mean

The "regression towards the mean effect," or the **regression effect**, says that in situations where there is some sort of test and then some sort of retest, the people who score extremely high on the first test tend to score a bit lower and closer to the mean (but still above the mean) on the second test. Similarly, people who score extremely low on the first test tend to score a bit higher and closer to the mean on the second test. This phenomenon is caused purely by random fluctuations and not by any more interesting psychological theory of test taking. The main reason for this phenomenon is that there are two types of people who score extremely high on a test: the extremely smart and the extremely lucky. When you give the test a second time, the people who were extremely lucky the first time tend to have more ordinary luck the second time and this tends to pull the overall average for this group down a bit and closer to the overall mean. The same type of reasoning applies to the people who score extremely low on the first test.

This interesting effect explains why movie sequels are usually not as good as the original movie. A sequel is usually only made when the original movie is extremely good and the regression effect says that the quality of the sequel tends to be a little closer to ordinary. Below are two more examples where this phenomenon arises.

Business writing skills.   A business school would like to improve the writing style of its graduates and decides to institute intensive writing assignments in a number of courses. To help monitor the effectiveness of this approach, each student has one assignment graded for style by a team of outside experts at the beginning of their freshman year and again near the end of their senior year. Each assignment is given a score from 1 to 10.

If you look at the students who scored below 5 points during their freshman year, the vast majority of them scored above 5 points during their senior year. This looks like evidence of a big benefit for the lowest scoring incoming students. But the surprising news is that if you look at the students who scored above 7 points during their freshman year, the vast majority of them scored *below* 7 points during their senior year. Does this mean the writing assignments can help improve the students with low ability but may actually hurt the students with high ability?

Solution.   Both could be caused by the regression effect. In Figure 6.6 we plot each pair of scores for the 95 students who were studied, so each point represents one student.

The first surprise here is that there does not appear to be much of a correlation between freshman and senior year, nor does there appear to be much

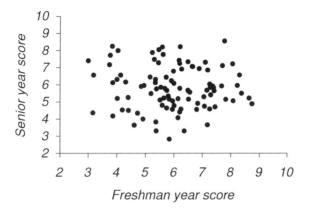

Figure 6.6. Freshman and senior year scores for a writing assignment.

of an increase in the overall average from the freshman year to the senior year. And, even so, we can still observe a rise in the low group and a drop in the high group. This is illustrated in Figure 6.7. This apparent rise and drop can be explained by the regression effect.

> If you are trying to draw a conclusion in a situation with a test and a re-test, check to see if the regression effect plays a role.

Next we give another more subtle but important example from health-care management.

Health Care Cost Management.   A company is in the business of managing treatment for the most expensive types of patients in an effort to reduce costs. They offer their services to a large health plan. They say they will take the plan's most expensive patients this year and through careful management reduce their average cost for next year (without reducing quality) from the average for the current year, and will take a portion of the cost reduction as their fee. Is this a reasonable arrangement for the health plan?

Discussion.   The health plan should be very careful here because the regression effect will play a large role. If you take people who were extremely

These people scored below 5 points as freshmen,...

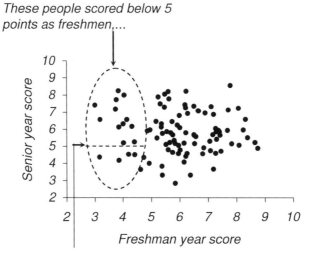

...but notice that the vast majority of them scored above this dotted line (5 points) during their senior year.

These people, on the other hand, scored above 7 points as freshmen,...

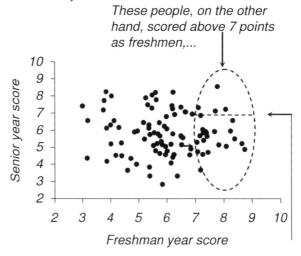

...but this time the vast majority scored below the dotted line (7 points) as seniors.

Figure 6.7. Illustration of the regression effect in a test/retest writing assignment.

expensive this year, the regression effect says that on average costs in this group will tend to naturally come down the following year—even without any form of special management. So the health plan might be paying for something that would be happening anyway. They need to consider the regression effect and make sure they only pay for cost savings beyond what the regression effect would explain.

6. Using Excel

In Excel to get the slope and intercept of the regression line you use the formulas

$$=\text{SLOPE}(\textit{Y-values},\ \textit{X-values})$$

and

$$=\text{INTERCEPT}(\textit{Y-values},\ \textit{X-values}),$$

where by "*Y-values*" and "*X-values*" we mean you should type in the ranges where the Y values and X values are stored separated by a comma ",". In other words, if your Y-values are in the range A1:A100 and your X-values are in the range B1:B100, you should use the formulas

$$=\text{SLOPE}(\textit{A1:A100,B1:B100})$$

and

$$=\text{INTERCEPT}(\textit{A1:A100,B1:B100}).$$

7. Exercises

1. **Traffic lights.** A city would like to install improved traffic lights at intersections to reduce traffic accidents. For a preliminary one-year test they randomly select a large number of intersections and install the new lights. They observe that intersections having the highest accident rates (accidents per thousand cars passing through the intersection) on average had significantly lower accident rates during the year of the test. A spokesman for the light vendor says this means their product saves lives—the lights are larger and brighter, and drivers notice them more easily. But a spokesman for the city points out that the intersections with the lowest accidents rates in the year prior to the test on average

had significantly higher accident rates during the year of the test. The spokesman also says that this means the lights can cause accidents—the new larger lights can distract drivers from noticing other cars entering the intersection. Which spokesman's claim is better justified here? Or are both wrong?

2. **Traffic lights, continued.** In the previous problem, an analyst points out that the overall average accident rate at the selected intersections during the year of the test was significantly lower compared with the previous year. Which spokesman's argument does this support? Or can this be explained by the regression effect?

3. **Purchasing and distance-to-store.** Store records show that people who live closer to a store tend to purchase more. The relationship between the distance to a store in miles (X) and the amount a customer spends over the year in dollars (Y) fits the following regression line well: $Y = -100X + 1000$. Give an interpretation for the coefficient -100.

4. **Height and weight.** The regression equation for predicting the height of a man Y (in inches) from his weight X (in pounds) is given by $Y = .2X + 40$. (a) Does this mean heavier men tend to be taller than lighter men? (b) Does this mean that if a man gains five pounds he is expected to grow an inch taller on average? If not, what does the .2 coefficient mean?

5. **GMAT score and salary.** U.S. News & World Report publishes the average starting salary and the average GMAT score for MBA graduates at each of the top 80 ranked schools in the country. The data fit the following regression line nicely: $Y = 288X - 111,408$, where X is the average GMAT score for students at a school and Y is the average starting salary plus bonus (in $) for graduates from that same school.

   (a) Estimate the average starting salary plus bonus for a school having an average GMAT of 700 points.

   (b) Give a managerial interpretation of the coefficient 288.

   (c) If you plug in the lowest possible GMAT score of 200 for X, you get $-\$53,808$. Explain what this number represents and what it being negative means here.

6. **Production cost forecasting.** Production costs for a large number of previous orders of varying sizes for a product are in the file `production.xls`

Table 6.1. An excerpt from the file `production.xls`.

Order size (units)	Cost ($)
983	51275
798	40808
527	38195
1049	49989
780	37299
831	55707
857	43168
1316	58435
1737	76411
1511	65020

Table 6.2. An excerpt from the file `jobs.xls`.

Job #	Billable hours	Total cost including materials
1	680	123757
2	450	85920
3	520	98450
4	740	122800
5	880	128544
6	650	122957
7	250	82746
8	270	84262
9	770	121967
10	570	116384

(an excerpt is in Table 6.1). An analyst computes the production cost per unit in each order and averages these to get $50. Using this he gives a cost estimate of $24,000 for a new order for 500 units. Is this a reasonable cost estimate? Explain why or why not. Looking at the data in the file, can you give a better cost estimate?

7. **Billable hours and total cost.** A manager would like to know the relationship between the billable hours spent on a job and the total cost of the job including materials. The file `jobs.xls` (an excerpt is in Table 6.2) contains the data. Summarize the relationship.

8. **Hiring poor performers.** A human resources director, on learning about

the regression effect, decides to hire people who have been fired by their previous employer for poor performance. He argues that the regression effect says people who perform very poorly in their previous job tend to perform well in their next job. Is this what the regression effect says? Explain.

9. **High performing salespeople.** Sales figures for each sales associate in a large company are calculated for the current year and last year. Last year the average sales per associate was $500,000 and the average for this year is about the same. True or false and explain briefly: the regression effect says that the number of associates who sold over $1 million in the current year should be somewhat less than the number who sold over $1 million last year.

10. **Workplace morale.** In order to assess work environment and morale, a human resource department surveys all employees every 2 years. Based on information from the survey 2 years ago, a series of special programs were implemented in those business units with the worst work environment (as assessed by the employees) as an attempt to improve morale. When the survey was repeated two years later, the work environment and morale in these units had improved. Does this mean the programs worked or could there be an alternate statistical explanation for this?

11. **Accounting fraud detection.** An accounting firm specializing in fraud detection has developed an index that should be correlated with the net income for a company. They have found in the past that the relationship between their index and actual net incomes follows the regression equation $Y = 2X + 10$, where X is their index and Y is net income in millions. The standard error for this regression line is 10. If a company's reported net income is much higher or lower than what the prediction from this line says, it is justification for a detailed audit. (a) One client firm reports a net income of $280 million and has an index of 120. Is this justification for a detailed audit? (b) Another client firm reports a net income of $200 million and has an index of 100. Is this justification for a detailed audit?

12. **The risky business of hiring superstars.** A recent article in *Harvard Business Review* titled "The Risky Business of Hiring Stars" argues that "Odds are, the superstars you eagerly and expensively recruit will shine much less brightly for you than for their previous employers. Research

shows why—and why you're usually better off growing stars than buying them." The article argues that primary reasons include the difficulty in making the transition from one company to another, the effect on morale in the new company, and shareholder perception of the transfer. Can you think of a statistical explanation for this phenomenon?[1]

13. **Forecasting growth.** The price-earnings ratio (PE ratio) for a stock is a commonly used measure of how over-priced or under-priced a company's stock is. There are a number of different statistics about a company that are available that might explain why this ratio differs for different companies. One of these statistics is a measure of future growth. To examine the relationship between PEs and the measure of future growth (FG), you run a simple regression and get the equation

$$PE = 3 + .9FG.$$

The R^2 for this model is 18% and the standard error is 5. Another model was run using a measure of dividends (D) to explain the PE. This gives the equation

$$PE = 1.6 + 13.2D.$$

(a) Give a managerial interpretation for the coefficients 3 and .9

(b) A particular company has a value of 15 on the measure of future growth. Its PE ratio is 4.5. What would you conclude about this company's PE? Briefly explain.

(c) Since 13.2 is greater than .9, can you conclude the PE ratio has a stronger relationship to dividends than future growth? If not, what would you need to know to conclude which variable has a stronger relationship to the PE ratio? Briefly explain.

(d) A company looking at this equation, concludes that if it increases its dividends 0.1, its PE is expected to increase by 1.32. Is this a sensible interpretation? Briefly explain.

(e) A set of industries was identified as high growth (HG) industries. To examine the relationship between the PE ratio and whether or not the company comes from a high growth industry, the following

[1] "The Risky Business of Hiring Stars," by Boris Groysberg, Ashish Nanda, and Nitin Nohria, *Harvard Business Review*, May 2004. For the alternative statistical explanation, see "The Risky Business of Hiring Stars," Letter to the Editor, Stephen M. Robinson, *Harvard Business Review*, September 2004.

model was developed (where $HG = 1$ if the company is from a high growth industry and 0 otherwise):

$$PE = 7.8 + 23.2HG.$$

Give a managerial interpretation for the coefficients 7.8 and 23.2

14. **Product pricing.** You head the marketing department of a company that makes a product sold in two different types of stores. You would like to understand how sales receipts are related to the product price you set. An analyst has prepared a report for you using weekly average price (in $) and sales (in number of units) data over the past year. In his report he gives the below graph and regression line:

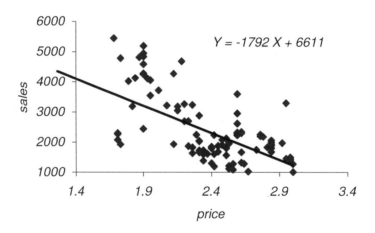

Since the slope of the line is $-1,792$, the report says that "a ten cent decrease in price would increase weekly sales by roughly 179.2 units." Is this report telling the whole story? For each week and store type during the past year you have data on average price, sales, and which of the three promotion methods were used. Below are the first six rows of the data you have; the data are in the file `marketing.xls`.

Store (1 or 2)	Week	Sales (# units)	price ($)	Promotion method (1, 2, or 3)
1	1	2305	1.71	3
1	2	2084	1.71	3
1	3	1928	1.73	3
1	4	2446	1.9	3
1	5	1825	2.26	3
1	6	2049	2.31	3

15. **Explaining sales.** You own similarly-sized stores widely scattered across the United States. To examine the relationship between sales and income of people living near the store, you create the regression equation

$$S = 30 + .7I,$$

where S = yearly sales of a store (in $100,000$'s) and I = median income (in $1,000$s) of people living near the store. The R^2 for this model is 14% and the standard error (SE) is 20.

(a) Give managerial interpretations for the numbers 30, .7, and 20.

(b) If a new industry opening near a given store raises median income for the people living near the store by $1,000$, what can you say about how this would affect sales? Discuss briefly.

(c) Half the stores are in rural areas and half are in urban areas. You decide to create separate models and get

$$S = 50 - .25I, R^2 = 30\%, SE = 18$$

for the rural areas and

$$S = 90 - .22I, R^2 = 31\%, SE = 17$$

for the urban areas. How is it possible that the regression coefficients are negative in the urban and rural equations, but positive in the overall equation above? Explain with a diagram, if necessary.

(d) You then run the following model:

$$S = 42 + 40Z,$$

where Z is a "dummy" variable which equals 0 if the area is rural and 1 if it is urban. Give managerial interpretations for the numbers 42 and 40.

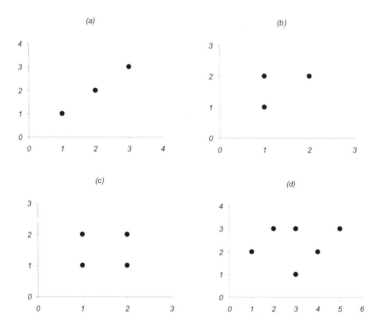

Figure 6.8. Above are four scatter diagrams. Sketch the regression lines for each.

16. Figure 6.8 shows four different scatter diagrams labeled (a) through (d). Sketch the corresponding regression line in each scatter diagram.[2]

17. **Backing the wrong horse.** A news article discusses a study of companies featured on the covers of major business magazines that tries to test "the long-held impression that the media usually back the wrong horse."[3] The study found that companies with positive cover stories had outperformed the market index 42.7% (after adjusting for sector and size) but on average fell to an outperformance of just 4.2% after the stories appeared. Can you think of an explanation for this drop? What implications does this have for investing?

18. **Is there something wrong with the scientific method?** A magazine

---

[2]From *Teaching Statistics: A Bag of Tricks*, by A. Gelman and D. Nolan, Oxford University Press, 2002.

[3]"Covered in shame: Investors, like magazines, have a terrible tendency to extrapolate," May 3, 2007, *The Economist*. http://www.economist.com/finance/displaystory.cfm?story_id=9122878

article titled "The Truth Wears Off: Is there something wrong with the scientific method?" discusses the alarming trend in medicine that "many results that are rigorously proved and accepted start shrinking in later studies." For example, one popular antipsychotic drug showed dramatic benefits in its initial trials, but follow-up studies 20 years later showed an effect less than half of that documented in the first trials. What could explain this type of decline?[4]

19. **Imports and healthcare spending.** An analyst creates a scatter plot for the value of goods imported into the United States in the years between 1990 and 2001 versus the total private spending on health care in the US during those years. The correlation equals 0.975, and the slope of the regression line for predicting health care spending from imports equals 0.8. If tariffs are used to reduce imports by $50 billion dollars, approximately what type of reduction in health care spending is to be expected from this? Explain briefly.

20. **Price and quantity sold.** The file `pricequantity.xlsx` contains data on weekly price and quantity sold for five different brands of cereal over period of 40 weeks.

   (a) On the worksheet titled "graphs" there are number of scatter plots for price and quantity. Quantity (in millions) is on the horizontal axis and price is on the vertical axis. What would you estimate the demanded quantity to be if Honey Nut Cheerios set its price at $3.25? Explain briefly.

   (b) If you compute a regression line using quantity as the dependent variable and price as the independent variable, economists define the "price elasticity" to be the slope of this regression line divided by the ratio of the average quantity to the average price. Give an interpretation of the slope of this regression line and compute the price elasticity for each brand. The averages have been computed for you on the sheet titled "means."

---

[4]The Truth Wears Off, *The New Yorker*, December 13, 2010, by Jonah Lehrer.

Fun Problem: Betting on Red

Here are two games you can play:

Game 1: A dealer holds a regular shuffled deck of cards. He will turn the cards over one at a time for you to see and you are allowed to watch him turn over as many cards as you like. As soon as you think the next card is going to be a red card, you place your bet on red. If you guess correctly, you win a prize. But if you're wrong, you lose and are kicked out of the game. You only get to play this game once and you are only allowed to bet once on the color red—you are not allowed to bet on black. If you don't bet on any card by the time the dealer gets through the whole deck, you lose your chance at the prize. That's Game #1.

Game 2: Someone flips a coin and if it's heads you win a prize. If it's tails you lose.

Which game has a better chance of winning? If you pick Game 1, which strategy is best to use?

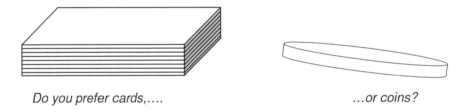

Do you prefer cards,....   *...or coins?*

Solution. Believe it or not, both games give you the exact same chance of winning—no matter what strategy you use for Game 1. Obviously this second game gives you a 50–50 chance of winning. I will show you why the first game always gives you this same chance of winning no matter what strategy you use.

In the first game, suppose you decide to bet on the very first card. In this case your chance of winning would be 50 percent, since half the cards are red. Now instead suppose you decide in advance to wait and bet on the very last card, no matter what happens. If you are betting on the last card no matter what happens, the dealer does not need to turn over every single card one at a time and can just immediately turn over the bottom card. Again, since half the cards are red, your chance of winning with this strategy would also be 50 percent.

Now suppose you have a very complex strategy of waiting until a certain number of black cards have been removed or until whenever you feel the time is right. When you finally decide to place your bet on red, does it give you an advantage if the dealer lets you pick from the remaining cards in the deck the one to be turned over next? Since each of the remaining cards has the same chance of being red, it shouldn't matter whether or not the dealer just turns over the next card in the stack or lets you pick the one to turn over. Right? And since it really doesn't matter which card you pick to turn over, you might as well just pick the bottom card. But this means you are just going to bet on the bottom card no matter what happens. Then there is no need to even go through each of the cards one at a time—you can just go straight to the bottom card, and your chance of winning is 50 percent.

In fact, there is no way to do any worse than 50 percent either—no matter how hard you try. Interesting, isn't it?

Chapter 7

Introduction to Multiple Regression

1. Introduction

La Quinta Motor Inns is a mid-sized hotel chain headquartered in San Antonio, Texas. As a growing chain, hotel executives spent much time considering where they should open new hotels. One year the chain had consultants develop an equation for forecasting future profitability for potential new hotel sites under consideration. This approach was so successful that the company president said that he did not feel obliged to personally select the new hotel sites anymore.[1] The profit forecasting equation the consultants came up with used only the state population for the potential site, the planned price of rooms at the hotel, the median income in the area, and the number of college students within four miles of the hotel. That's it.

How did the consultants use such limited data to get an accurate profit forecast? To combine several variables together to forecast another variable, they used the statistical tool called "multiple regression." Simple regression uses a single X variable to forecast a Y variable; multiple regression, in contrast, uses multiple X variables to forecast a Y variable. For forecasting hotel profitability, the Y variable was profitability and the X variables were

---

[1]Kimes, S. and J. Fitzsimmons, "Selecting profitable hotel sites at La Quinta Motor Inns," *Interfaces* 20, no. 2 (1990): 12–20.

things such as income and population in the area, room price, and so forth.

Our discussion of multiple regression will take place over two chapters. In this chapter, we will discuss where you can use a multiple regression equation and then how to interpret it. We will consider in broad terms what makes a good equation. Once we have this grounding, in the final chapter we will turn to how one can choose the best predictor variables.

2. The Multiple Regression Equation

As we saw in an earlier chapter, simple regression equations are of the form

$$Y = b + mX$$

where Y represents what you are trying to forecast, b is the Y-intercept, and X is the variable used to make your forecast.

Multiple regression forecasting equations, in contrast, look something like

$$Y = b + m_1X_1 + m_2X_2 + m_3X_3 + \cdots$$

where Y represents what you are trying to forecast and X_1, X_2, and X_3 are the different X variables you are using to make your forecast. Here b is still the Y-intercept. The big difference, of course, is that multiple regression equations have more than one X variable in them.

The coefficients m_1, m_2, and m_3 can tell you the marginal relationship between each of the X-variables and the Y variable, and these can reveal managerially relevant facts about your data. They tell you the difference in the Y variable you expect to see associated with a unit difference in a given X variable, when all other X variables stay the same.

> The coefficients in a multiple regression equation tell you the difference in the Y variable you expect to see associated with a one unit difference in a given X variable, when all other X variables stay the same.

It is important to realize that these coefficients measure the association between variables and can be very misleading about the direction of causation.

Table 7.1. An excerpt from the file `brooklinehomes.xls`.

Parcel-Id	Value	Sqft	Age	Bedrooms
046199 001000000	1255300	3900	106	10
046067 000300000	1384400	3229	100	7
046413 001400000	623100	2280	54	4
046085 002800000	1302200	2778	83	4
046224 001100000	1537200	3704	106	6
046395 000200000	1332400	2374	66	3
046111 000400000	1993900	4374	86	7
046433 001900000	1618700	3945	81	6
046029 000900000	333200	2760	96	4
046278 000800000	1626000	2482	46	4

For example, if consultants had used the number of housekeeping employees as one of the X-variables in the hotel profit forecasting equation we mentioned above, they probably would have seen a positive coefficient for this variable. This would tell us that hotels with a larger housekeeping staff tend to have higher profits—but this does not tell us that profit would rise if we hired more housekeepers. The causation probably goes in the other direction: profitable hotels hire more housekeepers to keep up with high demand. The bottom line is that to determine the direction of causation you need to rely on your business common sense—not just the regression coefficients the computer gives you.

Let's get into a specific example.

Home values   When we study the value of homes in the town of Brookline, Massachusetts using the computer, we see the data fit the following multiple regression equation:

$$Y = 230{,}040 + 478X_1 - 6{,}400X_2,$$

where Y is the value of the home in dollars, X_1 is the square footage of the home and X_2 is the age of the home in years. You can find the data in the file `brooklinehomes.xls` (an excerpt is in Table 7.1).[2] Programs such as Excel can take data and generate a multiple regression equation and we will

[2]Data file from `http://www.town.brookline.ma.us/Assessors/`

show you later how to do this. Before we try to interpret the coefficients, let's first use this equation to make a basic forecast.

Question.   Estimate the average value of homes that are 50 years old and have 3,000 square feet.

Answer.   We plug in 3,000 for X_1 and 50 for X_2 in the equation. This gives us

$$\begin{aligned} Y &= 230{,}040 + 478(3{,}000) - 6{,}400(50) \\ &= \$1{,}344{,}040. \end{aligned}$$

This tells us that the average value of these types of homes is about \$1.3 million. That makes it a pretty expensive town to live in.

> To get forecasts from a multiple regression equation, just plug in X values and what you get out is the forecast.

We mentioned above how the coefficients can give important managerial insights. In this case, we can interpret the coefficient 478 as follows: for houses of the same age, with each extra square foot we see an additional \$478 in value on average. We can interpret the coefficient $-6{,}400$ as follows: in houses having the same square footage, we see a decrease in value of \$6,400 for each additional year in age. This makes sense because we usually expect larger houses to be more valuable and older houses to be less valuable. This equation fits our general intuition of what likely should be the case.

It's a good idea to try to intuitively make sense of the regression coefficients before you start using the equation to make forecasts. Doing this can teach you something new about your data and can sometimes reveal problems with your data. There may be a few extreme outliers or missing values (sometimes the computer treats these as "0") in a data set that can throw all the coefficients off. As we discussed earlier with correlation and simple regression, it may be best to remove such extreme outliers and treat them as special cases.

Suppose we decided to add in another variable, X_3, that equals the number of bedrooms. It seems reasonable that a house with three or four bedrooms is likely to be more valuable to a growing family than a house with only two bedrooms. When we use the computer to re-calculate the regression equation with this new variable we get the following:

$$Y = 231,485 + 480X_1 - 6,365X_2 - 2,682X_3.$$

You may be surprised to see that all the coefficients and the intercept have changed even though we're looking at the same town and the same houses. Why has everything changed?

The answer is that the interpretation of each of the coefficients has now completely changed. To see this, let's first take a look at the coefficient for X_3, the number of bedrooms. The interpretation of the coefficient $-2,682$ is that for houses of the same age and the same square footage, with each additional bedroom we see a *decrease* in value of around $2,700.

But this last interpretation seems puzzling—each extra bedroom *decreases* the value? We stated above that we expected an extra bedroom to definitely *increase* the value. The answer to this puzzle is that the coefficient now tells you what happens to the value when you have an extra bedroom but still have the same age and same square footage overall; houses with an extra bedroom but without any additional square footage have *lower* home value because the other rooms in the house would have to be smaller and the house more crowded. So the negative coefficient makes sense after all. It is important to understand that the coefficient tells you what happens to the Y variable when all the other variables in the equation stay constant and you look at a change in only a single X variable.

In summary, each of the coefficients in this new multiple regression equation now represents something completely different: what happens to the home value when you look at a difference in a variable with the restriction that all the other variables in the equations stay the same. The moral is that you should watch out when adding or removing X variables from a multiple regression—all the coefficients can change in surprising ways as the interpretations change. This also means that the interpretation of the coefficient for some variable depends on which other variables are in the equation.

> When adding or removing X variables from a multiple regression, the coefficients can change in surprising ways as the interpretations change.

Incidentally, to get a better picture of the value of an extra bedroom we could try this alternate approach. When you add a bedroom to your home you naturally expect that the square footage would rise. This means we should leave out square footage from the multiple regression equation and predict home value using only the other variables: the age variable, X_2, and the number of bedrooms, X_3. Running the multiple regression equation with only these variables, we get the following:

$$Y = 1{,}011{,}313 - 5{,}464X_2 + 179{,}955X_3.$$

In this equation we can interpret the coefficient 179,955 as follows: for houses of the same age, for each additional bedroom we see an increase in value of around \$180,000. Without square footage in the equation, this may be a more reasonable picture of the value of an extra bedroom.

Question.   If we create a multiple regression equation to predict home value using the age of the home, number of bedrooms, and the number of full bathrooms, we get a coefficient of \$165,000 for the number of full bathrooms. But if we then take out the number of bedrooms, so we have only the age of the home and the number of full bathrooms, the coefficient for the number of full bathrooms increases to \$277,000. What explains this increase?

Discussion.   The \$165,000 tells us the change in value when we look at homes with an extra full bathroom but the same number of bedrooms and age. When we leave out the number of bedrooms, the \$277,000 tells us the change in value when we look at homes of the same age with an extra full bathroom. Since homes with more bathrooms also tend to have more bedrooms, we are now looking at homes having extra bedrooms as well. So the larger figure \$277,000 also includes the value of extra bedrooms that tend to be associated with the extra bathrooms.

What have we done in this first section? We introduced the multiple regression equation and contrasted it with the simple regression equation. Then we went on to give you a little practice using a multiple regression equation. We showed you that it is simple to plug in values to get a forecast, but the interpretation of coefficients takes some thought. Further, we showed you that the addition of one new variable can change the coefficients of all the other variables: an entirely new equation with a new interpretation is the result, not the old equation with a new variable tacked on.

Exercises

1. **Real estate.** A real estate market analyst is interested in estimating the value a garage parking space adds to a home. He computes the multiple regression equation $Y = 200{,}000 + 500X_1 + 50{,}000X_2$, where Y is the value, X_1 is the square footage of the living area, and X_2 is the number of garage parking spaces. The analyst also computes an equation without using square footage to be $Y = 700{,}000 + 175{,}000X_2$. (a) Estimate the value of a home with 1,000 square feet and one garage parking space. (b) Give managerial interpretations for the coefficients. (c) Can you think of a reason why the two coefficients 50,000 and 175,000 might be so different? (d) Is it more reasonable to estimate the value added by a single garage parking space to be \$50,000 or \$175,000? Explain.

2. **Employee salaries.** A human resources analyst at a firm computes the following regression line for employee salaries: $Y = 60 + 3X_1 + 2X_2$, where Y is the employee salary in thousands of dollars, X_1 is the number of years of experience the employee has, and X_2 is the number of people the employee supervises. (a) Give managerial interpretations for the co-efficients here. (b) The analyst also computes the following regression equation omitting work experience: $Y = 70 + 4X_2$. What could explain why the coefficient for the number of people the employee supervises goes up from 2 to 4?

3. **Job costs.** A regression line to predict the cost of a job is given by $Y = 1{,}000 + 1{,}500X_1 + 40X_2$, where Y is the cost of the job in dollars, X_1 is the number of tons of raw material required, and X_2 is the number of labor hours required. (a) Give managerial interpretations for the coefficients here. (b) How much do you expect a job requiring 2 tons of raw materials and 10 labor hours to cost?

3. The Standard Error for the Regression Line

How good are forecasts using a multiple regression equation? When a regression equation is used to forecast something, the **standard error for the regression line** is a measure of how far in general forecasted values tend to be from actual values in your data. Usually around two-thirds of the actual values will be within one standard error of the forecasted values and almost all will be within two standard errors.

> The **standard error for the regression line** is an
> overall measure of how accurate forecasts using the
> regression line tend to be on your current data.
>
> Usually around two-thirds of the forecasted values
> are within one standard error of the actual values.
> Very few are off by more than two standard errors.

For example, the standard error for the first equation above $Y = 230{,}040 + 478X_1 - 6{,}400X_2$ for predicting home values equals \$386,361. This means home values tend to be off by around \$390,000 from the forecasts you get from this regression equation. This may seem like a lot of error, but it's not so bad when you realize that homes values in the town range up into the several millions of dollars and that we're only using square footage and age to make the forecast.

After you compute the standard error, how do you know if it's high or low? The answer to this question depends on the application you have in mind. For example, if you are trying to forecast automobile values in the thousands of dollars, a standard error of \$1 million is useless. But if you are trying to forecast corporate profits which are up in the hundreds of millions of dollars, a standard error of \$1 million could be great. This means there are no general rules for assessing how good a standard error is, and you must evaluate it yourself using a common sense estimate of the type of accuracy you need for the business application you have in mind.

To see how the standard error is calculated for this regression equation, imagine the forecasting error (called the "**residual**") is computed for each home in the town by taking the actual value and subtracting the forecasted value you get using the regression equation. The standard error for the regression line equals the standard deviation of this list of forecasting errors, and thus tells you how variable the forecasts are from the actual values. This also means, by the empirical rule, approximately two thirds of the actual values are within the standard error from the forecasted values. We illustrate this in Figure 7.1, where we plot the forecasted value for each home using the above regression equation versus the actual value. Note that if a home is on the diagonal line its actual value equals its forecasted value.

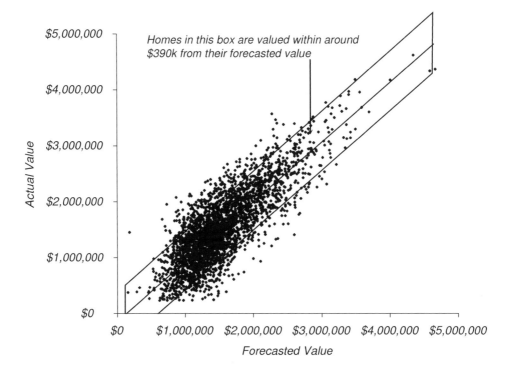

Figure 7.1. An illustration of the standard error for a regression line.

Technical note: A formula for the standard error for the regression line. A formula for the standard error for the regression line is

$$\text{SE of regression line} = \sqrt{\text{sum of (observed } Y \text{ - forecasted } Y)^2} \times \sqrt{\frac{n}{n-k-1}}$$

where n is the number of data points in the data set, and k is the number of different X variables being used.

Exercises

4. **Gas mileage.** An analyst creates a multiple regression equation to predict the gas mileage for a new vehicle that is still in the design stage using the weight, horsepower and other relevant factors about the design. This will help determine the appropriate potential market size for the vehicle. The analyst finds that the standard error for the regression equation equals 25 miles per gallon. Explain why this equation is not

going to be so useful.

4. Using Excel

Using the Regression Worksheet included with this book

Here we will show you how to use the Regression worksheet that comes with this book to compute a multiple regression equation. This works with any version of Excel for Windows or for the Mac. It can handle up to 5000 rows and 50 columns of data; up to 15 variables can be in the regression equation at one time.

Suppose we would like to build a multiple regression equation for predicting price from square footage for the home price data file (the data are in the Excel file `brooklinehomes.xls`). The first step is to copy the data from there and paste it into the "Data" worksheet of the Excel file `regression.xls`.

Next, go to the "Regression" worksheet and put the letter "y" to the left of "Value" and put the letter "x" to the left of both "AGE" and "SQFT." This indicates that we want "Value" to be the dependent variable and "AGE" and "SQFT" to be the independent variables. Also, remove any other "x" or "y" that may already be there next to other variables by selecting each such cell and pressing the "Delete" key. You should see the following on the "Regression" worksheet:

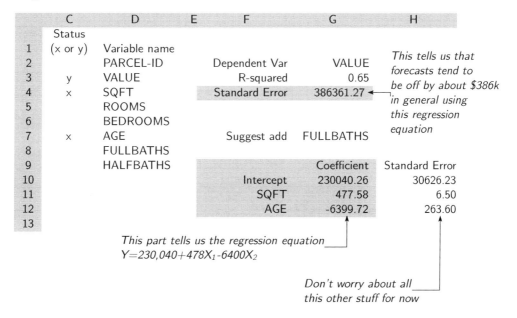

	C	D	E	F	G	H
		Status				
1		(x or y)	Variable name			
2			PARCEL-ID		Dependent Var	VALUE	*This tells us that forecasts tend to be off by about $386k in general using this regression equation*
3		y	VALUE		R-squared	0.65	
4		x	SQFT		Standard Error	386361.27 ◄	
5			ROOMS				
6			BEDROOMS				
7		x	AGE		Suggest add	FULLBATHS	
8			FULLBATHS				
9			HALFBATHS			Coefficient	Standard Error
10					Intercept	230040.26	30626.23
11					SQFT	477.58	6.50
12					AGE	-6399.72	263.60
13							

This part tells us the regression equation _____
$Y=230{,}040+478X_1-6400X_2$

Don't worry about all _____
this other stuff for now

There is a lot of information there, but just focus for now on the two areas which are shaded above. The multiple regression coefficients and intercept are given as well as the standard error of the regression line.

Using the Analysis ToolPak included with Excel

Using special Excel add-ins to compute a regression equation can be more convenient than using what comes standard in Excel. But since Excel is so widely available without such an add-in, we will show you how to do it directly in Excel. Versions of Excel for Windows usually come with the "Analysis ToolPak," which you may need to activate by going to Tools/Add-ins in Excel 2003 (or Office Button/Excel Options/Go in Excel 2007). Excel for the Macintosh may not have this add-in; in this case you should use the "Regression" worksheet discussed below.

Below we have the first 10 rows of the home price data file (the data are in the Excel file `brooklinehomes.xls`—the first column "Parcel ID" is the ID code for the property):

	A	B	C	D	E	F
1	PARCEL-ID	VALUE	SQFT	ROOMS	BEDROOMS	AGE
2	046360 001200000	$ 383,000	687	4	2	56
3	046305 001900000	$ 242,000	728	4	2	106
4	046305 001800000	$ 249,600	728	4	2	106
5	046093 000500000	$ 523,200	775	6	3	76
6	046305 002100000	$ 252,300	785	5	2	106
7	046304 001300000	$ 254,500	794	5	2	106
8	046321 001300000	$ 515,900	800	7	3	39
9	046305 001700000	$ 337,400	832	5	2	106
10	046214 000900010	$ 338,300	848	4	2	116
11	046307 000500000	$ 528,600	884	6	2	54

Suppose we would like to build a multiple regression equation for predicting price from square footage and age. In Excel, the columns of data you want in your regression equation must all be adjacent. Therefore first rearrange the columns so that they are adjacent:

	A	B	C
1	VALUE	SQFT	AGE
2	$383,000	687	56
3	$242,000	728	106
4	$249,600	728	106
5	$523,200	775	76
6	$252,300	785	106
7	$254,500	794	106
8	$515,900	800	39
9	$337,400	832	106
10	$338,300	848	116
11	$528,600	884	54

Now we are ready to run the regression analysis. To do this, choose "Data Analysis" from the "Data" ribbon and, as we show below, scroll down and select "Regression" and then click "OK." (you may need to install the "Analysis Toolpak" first by going to the "File" ribbon and clicking "Options/Add ins/Go" and checking off "Analysis Toolpak.")

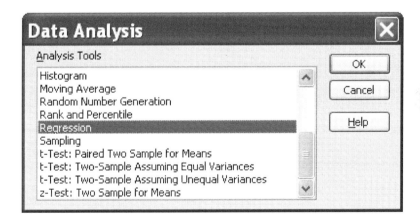

Then fill in the blanks with the references to the Y values, the references to the adjacent columns of X-values and don't forget to check off "Labels" if the first row of your data holds the names of the columns.

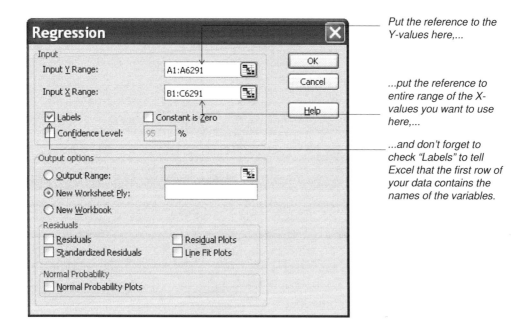

Put the reference to the Y-values here,...

...put the reference to entire range of the X-values you want to use here,...

...and don't forget to check "Labels" to tell Excel that the first row of your data contains the names of the variables.

Excel will then give you a printout that looks something like the one below. There is a lot of information printed out there, but for now just focus on the areas that are shaded. The multiple regression coefficients and intercept are given as well as the standard error of the regression line.

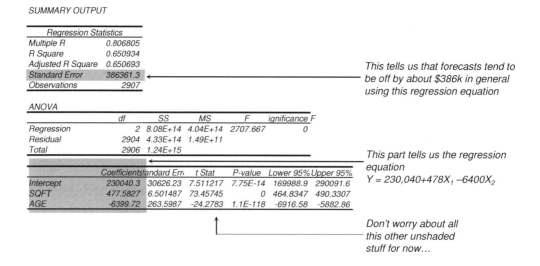

SUMMARY OUTPUT

Regression Statistics	
Multiple R	0.806805
R Square	0.650934
Adjusted R Square	0.650693
Standard Error	386361.3
Observations	2907

This tells us that forecasts tend to be off by about $386k in general using this regression equation

ANOVA

	df	SS	MS	F	ignificance F
Regression	2	8.08E+14	4.04E+14	2707.667	0
Residual	2904	4.33E+14	1.49E+11		
Total	2906	1.24E+15			

	Coefficients	tandard Err	t Stat	P-value	Lower 95%	Upper 95%
Intercept	230040.3	30626.23	7.511217	7.75E-14	169988.9	290091.6
SQFT	477.5827	6.501487	73.45745	0	464.8347	490.3307
AGE	-6399.72	263.5987	-24.2783	1.1E-118	-6916.58	-5882.86

This part tells us the regression equation
$Y = 230{,}040 + 478X_1 - 6400X_2$

Don't worry about all this other unshaded stuff for now...

Using R

Start R-Studio and click on "File/Import dataset/From Excel" and navigate to the data file "brooklinehomes.xls" and click "Import". Then create a copy titled "d" as follows:

```
d=data.frame(brooklinehomes)
```

Then to predict value from square footage and age, type

```
summary(lm( d$VALUE ~ d$SQFT + d$AGE ))
```

and you should see the following:

```
Call:
lm(formula = d$VALUE ~ d$SQFT + d$AGE)

Residuals:
Min      1Q   Median      3Q      Max
-1215056  -273290   -4246   258148  1292589

Coefficients:
             Estimate   Std. Error t value Pr(>|t|)
(Intercept) 230040.261  30626.232    7.511 7.75e-14 ***
d$SQFT          477.583      6.501   73.457  < 2e-16 ***
d$AGE         -6399.716    263.599  -24.278  < 2e-16 ***
---
Signif. codes:  0 *** 0.001 ** 0.01 * 0.05 . 0.1   1

Residual standard error: 386400 on 2904 degrees of freedom
Multiple R-squared:  0.6509,Adjusted R-squared:  0.6507
F-statistic:  2708 on 2 and 2904 DF,  p-value: < 2.2e-16
```

The multiple regression coefficients for square footage and age are respectively given as "477.583" and "-6399.716," and the intercept is "230040.261." The standard error of the regression line is given as "386400" labeled as "Resudual standard error." The R package uses slightly different formulas for the standard error compared with Excel, but they can be interpreted roughly the same.

Technical Note

Calculating multiple regression coefficients by hand is quite difficult and there is no easy formula for them. The coefficients are computed by finding values for m and b that minimize the average squared forecasting error

$$\frac{1}{n} \sum_{i=1}^{n} (Y_i - m_1 X_1 - m_2 X_2 - \cdots - b)^2.$$

For this reason the regression equation is sometimes called the "least squares regression line"—the squares of the errors are being minimized to find the coefficients of the line. There are several ways to do this minimization and the details are usually handled by software such as Excel. If you want to do the minimization by hand, as it was done in the old days before computers, you need to use techniques from calculus. The convention of "squaring" the errors instead of taking the absolute value makes it possible to take derivatives of the forecasting error to do the minimization using calculus.

Exercises

5. **Real estate.** Using the data file `brooklinehomes.xls` (an excerpt is in Table 7.1), create the appropriate multiple regression equation and use it to estimate the value of a home having 3 full bathrooms, 2 half bathrooms, and 3,200 square feet overall.

5. The Multiple Correlation Coefficient

The standard error is one way to summarize how good forecasts are from a multiple regression equation. Here we describe another very commonly used method.

If you have a Y variable and a single X variable, the correlation coefficient is a good way to summarize how well the data fit a line on a scale from -1 to $+1$. When you have multiple X variables there are many different correlation coefficients you might look at: one for each pair of variables. The common practice for summarizing how well data fit a multiple regression equation is to compute the correlation between the forecasted Y values from the regression equation and the actual Y values. This is called the **multiple**

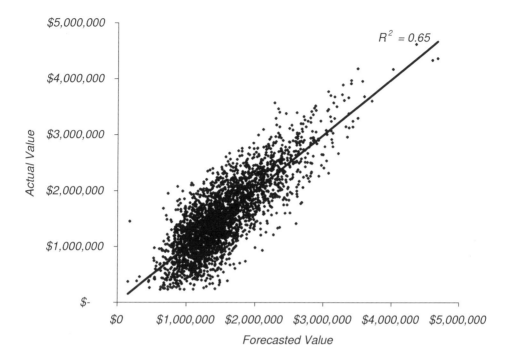

Figure 7.2. The correlation between actual and forecasted values is the multiple correlation coefficient.

correlation coefficient and is often denoted with the capital letter R. Usually people take the square of this and call R^2 the **coefficient of determination**.

> The **multiple correlation coefficient** is the correlation between the actual Y values and the forecasted Y values (forecasted using a multiple regression equation).

In the example of home prices above, when we plot forecasted prices and actual values we get correlation of .807. This means that the coefficient of determination R^2 equals $(.807)^2 = .65$. This is illustrated in Figure 7.2.

Forecasting anything in real life is difficult and analysts on Wall Street get paid lots of money for forecasts that give people even a tiny edge over other investors. Thus, even an equation with a small R^2 can be useful. When forecasting something difficult like sales figures or home values, you will see that R^2 values usually will be quite low. When forecasting something easy like the cost of postage from the weight of a package, you will see a much higher R^2 value.

Some properties of R

If you only have a single X variable that is positively correlated with Y, the multiple correlation coefficient will be the same as the correlation between X and Y. This is because in a simple regression the forecasted Y value is a linear function of the X variable—and so is perfectly correlated with it. Also notice that the multiple correlation coefficient is never negative even if your Y variable is negatively correlated with every X variable. This means it ranges from 0 to 1 rather than from -1 to $+1$ like the usual correlation coefficient does. This is because even if X and Y are negatively correlated, the forecasted Y value and the actual Y value will still be positively correlated.

If the multiple correlation coefficient equals $+1$, this means that every Y value is exactly equal to its forecast—and so the multiple regression equation fits the data perfectly. If, however, the multiple correlation coefficient equals 0, it means there is no correlation between the forecasted and the actual Y values—and so the multiple regression equation is essentially useless. As with simple correlation, just because a multiple correlation coefficient equals 0 it doesn't mean that there is no relationship—just no linear relationship. There might be a more complicated nonlinear relationship hidden in the data.

An interpretation of R^2

When you use the standard error to summarize how good forecasts are from a multiple regression equation, it's important to keep in mind the overall variability of what you are trying to forecast. For example, suppose we see that the weather forecasts in Boston are accurate only half of the time, but in Honolulu they are accurate almost all the time. Does this mean weather forecasting in Honolulu is better? Not necessarily, since the weather in Boston is much more unpredictable than the weather in Honolulu—and we need to take this variability into consideration when assessing the quality of the forecasting method.

To take into account the variability of what we are trying to forecast, people usually look at the standard deviation of the Y variable. Then if you compute the ratio of the standard error to the standard deviation, this then tells you something about how big your forecasting error is compared with the overall variability. The tradition is to square this and then subtract from 1, and this turns out to be almost equal to the coefficient of determination R^2. This is summarized with the following formula:

$$R^2 = 1 - \frac{\text{sum of (observed } Y \text{ - forecasted } Y)^2}{\text{sum of (observed } Y \text{ - overall mean of } Y)^2}.$$

Thus when R^2 is large it means that forecasting errors are a small fraction of overall variability. For this reason people often say that the R^2 value tells you the "fraction of variation in the Y variable that is explained by variation in the X variables." This is a mouthful so here is another way to look at it. In the example of home prices we saw that prices were not completely unpredictable—they could be somewhat explained by the size of the house and the age of the house. So variation in the size and the age of houses "explains" variation in prices. This does not mean it "causes" it, but just "explains" it. Since here R^2 was .65 we could say "variation in the size and age of houses explains about 65% of the variation in home values." If R^2 was 1.0, we would say "variation in age and size of the home completely explains all the variation in home values."

> The **coefficient of determination**, R^2, tells you the fraction of variation in the dependent variable that is explained by variation in the independent variables.

Exercises

6. **Sales forecasts.** An analyst at a firm creates one multiple regression equation to forecast domestic sales and another one to forecast international sales. The equation for domestic sales has a standard error of $5 million and $R^2 = .8$ and the one for international sales has a standard error of $1 million and $R^2 = .05$. The analyst says that since $1 million is smaller than $5 million, the multiple regression approach is more useful for international sales forecasting. Is this sensible reasoning?

6. Using Multiple Regression Models

Here we will show you a few of the wide variety of situations where multiple regression models are useful.

Do men get paid unfairly more than women?   A manager in human resources would like to know the answer. Company records show that the average salary paid to women is higher than the average salary paid to men. Does this mean that the answer is "no?" The University of Michigan in 2001 conducted a study to see if there were gender inequities in the salaries paid at that university.[3] We will use a hypothetical company to show you the approach they used and then afterwards we will tell you the conclusions of their study.

Discussion.   Suppose in our company we gather the necessary data and compute the regression equation

$$Y = 30 + 2X_1 - 3X_2,$$

where Y is salary (in thousands of dollars), X_1 is years of experience, and X_2 equals 1 if the person is female and equals 0 if the person is male. This type of variable is called a **dummy variable** and it can be used to be able to put a non-numerical variable, such as gender, into a regression formula.

You might at first think of just creating separate regression lines for each gender. But then if you want to incorporate other non-numerical factors, such as employee experience, department and position title, you would end up with separate regression lines for each of the dozens of possible combinations. And you would get dozens of different coefficients for every variable, each based on a very small set of data. Using the approach of incorporating gender as a dummy variable, you get a single equation and a single set of coefficients that are estimated using the whole set of data.

Finally, the standard error for the regression line here equals $15,000 and $R^2 = .3$.

Before we answer the question about gender unfairness, let's interpret the coefficients, the standard error and R^2.

---

[3] "University of Michigan Gender Salary Study: Summary of Initial Findings," M. Corcoran, P. Courant, and P. Raymond. `http://www.provost.umich.edu/reports/U-M_Gender_Salary_Study.pdf`

To interpret the "2": for people of the same gender, each extra year of experience adds $2,000 to the average salary they are paid.

To interpret the "−3": for people with the same years of experience, on average women earn $3,000 less than men.

The interpretation of the standard error is that people's actual salaries tend to be about $15,000 off from what the regression equation would predict. And the interpretation of $R^2 = .3$ is that about 30% of the variation in salary can be explained by looking at gender and years of experience. Since we are not interested in forecasting an individual person's salary here, and instead only the average salary, these measures of fit are not so important to us.

The insight revealed by this regression equation is that, even though records show that the men get paid on average lower than women, the coefficient "−3" tells us that women on average earn less (by about $3,000) than men who have the same years of experience as they do.

In fact, in the study commissioned by the University of Michigan in 2001 using this approach, university records showed that the average salary paid to women was around $72,000 and the average to men was around $88,000. This salary difference of $16,000 was troubling to university administrators. But when researchers ran a multiple regression to predict salary and included X variables such as the number of years of experience, gender, education level and dummy variables for the areas of specialization, the coefficient for gender became nearly equal to zero. The researchers concluded that the salary difference between men and women could be explained by variables other than gender.

Example. Which features of a sport-utility vehicle (SUV) make it sell well? A large number of consumers are asked to separately rate the perceived fuel economy, style, and safety on a scale from 1 to 10 for a large number of different SUVs. Analysts then relate these ratings to annual sales figures in order to get a better sense of what makes an SUV sell well.

Suppose an analyst conducting a simple regression analysis gets the following equation:

$$Y = 4 - 3X_1$$

where Y is annual sales (in thousands of units), and X_1 is the average miles per gallon for the SUV. Does the negative coefficient (-3) mean that consumers actually prefer worse fuel economy?

The answer is no. Although there is a negative coefficient for fuel economy, it could be that SUV's with low fuel economy rate high on the other factors— and this makes it more desirable overall. To find out how fuel economy is linked to sales when all the other factors stay the same we must look at the multiple regression coefficient—not the simple regression coefficient above. The analyst then computes the multiple regression equation to be

$$Y = 5 + 4X_1 + 8X_2 + 2X_3,$$

where X_2 is the average style rating, and X_3 the average safety rating.

How do you interpret the coefficients now and what does this tell us about consumers' view of fuel economy? Now the positive coefficient "4" tells us that for SUVs with the same style and safety rating, higher fuel economy is linked with higher sales. This is in contrast to the negative coefficient we saw before in the simple regression. The explanation is that SUVs with higher fuel economy tend to have a lower style rating. It does not mean that customers prefer lower fuel economy—they actually prefer higher fuel economy when all else is the same. The multiple regression analysis reveals this whereas the simple regression analysis was misleading.

> If you would like to know how two variables are related to each other when other variables stay the same, you need to do a multiple regression analysis—not a simple regression analysis.

You might also ask "which factor has the biggest link to sales?" Since the coefficient 8 for style is the largest, an increase in style is linked with a bigger increase in sales than an equivalent increase in either of the other factors. This means style is very important to consumers. We can say each extra unit in average style rating is associated with an extra 8,000 units sold, when the other two other factors stay the same (if the standard deviations of the variables are very different, people sometimes compare importance by comparing the ratio of each coefficient to the standard deviation of the corresponding variable).

Is a company's stock undervalued or overvalued?   The price-earnings ratio (PE ratio) for a stock is a commonly used measure of how over-priced or under-priced a company's stock is. If this ratio is too high, the stock price is viewed as being over-priced and likely to decrease in the future. For example on May 10, 2005, Yahoo finance reports that Ford Motor Co.'s PE ratio was 6.9 and Toyota's ratio was 17.78. If all else were the same for these two companies, most analysts would say that buying shares of Ford is a better bet than buying shares of Toyota.

But all else isn't the same and there are many differences between these companies that could explain the different PE ratios. A multiple regression analysis can help adjust for some differences between companies so you can more fairly compare the PE ratios. Financial analysts call this technique "relative valuation."[4]

Suppose analysts use published data for a group of comparable companies in one industry to compute the regression equation

$$Y = 14 + .9X_1 - .05X_2 + .02X_3,$$

where Y is the PE ratio, and X_1, X_2, and X_3 are three different statistics about the company that are readily available. For example, these X-variables might include measures of future growth, dividends, and risk. And suppose this regression equation has $R^2 = .2$ and a standard error of 20. This R^2 value may not seem very large but in finance even a small R^2 may be enough to give you a small edge over other investors.

Now suppose analysts would like to evaluate one of these companies that has $X_1 = 30$, $X_2 = 10$, $X_3, = 1.25$ and a PE ratio of 80. Is this company's stock overpriced?

Solution.   Plugging in the X-values into the regression equation gives a predicted Y value of

$$Y = 14 + .9(30) - .05(10) + .02(1.25) = 43$$

which is our estimate of the average PE ratio for comparable companies with the same values of X_1, X_2, and X_3. Since the actual value of Y is 80 and is almost twice the forecasted value of 43, the company appears to have a PE

---

[4]see   http://pages.stern.nyu.edu/~adamodar/New_Home_Page/datafile/MReg97. html and http://pages.stern.nyu.edu/~adamodar/pdfiles/darkside/ch8N.doc

ratio much higher than other comparable companies. It seems over-priced compared to comparable companies and you should be cautious purchasing this stock now.

Reducing risk exposure.   Suppose you own a large stock portfolio and you would like to purchase some gold and silver to help reduce your risk. How much of an investment in each would best help reduce your overall risk exposure?

Questions like these can be answered using multiple regression. The intuitive idea is that you would like to make side investments with values that move in the opposite direction from your stock portfolio. This means if the value of your stocks goes down, the value of your other investments will go up to offset some of this loss.

Suppose an analyst gathers weekly historical data over a recent period and fits the multiple regression equation

$$Y = 100{,}000 - 50X_1 - 60X_2,$$

with $R^2 = .4$, and where Y is the value of your stock portfolio at the start of a given week, X_1 is the corresponding value of one ounce of gold, and X_2 is the same but for silver.

Since the multiple regression equation attempts to estimate the value of Y from the X values, this means that if the right hand side of the multiple regression equation goes up by one unit, our best estimate is that the left-hand side also goes up by around one unit. This also means that when Y goes *up* by one unit, we expect that the quantity $-50X_1 - 60X_2$ will go up by one unit. In other words, $50X_1 + 60X_2$ will go *down* by one unit. And so if we buy 50 ounces of gold and 60 ounces of silver, the value of these three investments will be $50X_1 + 60X_2$ and we will therefore have an investment that is expected to move in the opposite direction of Y.

In other words, the negatives of the multiple regression coefficients tell you how much to put in the side investments to best reduce the risk of your original investment. The R^2 value tells you approximately how much risk you can eliminate by doing this and in this case it is 40% of the total risk. Getting rid of almost half your risk is pretty good.

Exercises

7. **How much is a water view worth?** Consider the following two multiple regression equations for home values in a city: $Y = 100 + .05X_1 - 4X_2 + 30X_3$ and $Y = 125 + 75X_3$, where Y is the value of the home (in thousands of dollars), X_1 is the square footage of the home, X_2 is the age of the home (in years) and X_3 is a dummy variable which equals 1 if the home has a water view and equals 0 otherwise. Assume both regression equations fit the data reasonably well. (a) Based on these, which is a more reasonable estimate of the average value of having a water view: $30,000 or $75,000? Pick one and explain briefly. (b) Does the fact that the coefficient for the age variable (X_2) is negative mean that over time home values in this city tend to fall? Explain.

7. Kitchen Sink Regressions

We have seen that the R^2 value and the standard error both measure how well the data fit a multiple regression equation. If the R^2 is large or if the standard error is low, it means the actual Y values tend to be quite close to the forecasted Y values. This implies that when you are building a multiple regression equation for forecasting you should always be looking for ways to increase the R^2 and decrease the standard error. Mathematically it turns out that as you add X variables to a multiple regression equation the R^2 can never go down and the standard error can never go up (you could always just use coefficients of 0, so having more X variables can never "hurt"). Usually, no matter what type of X variables you add, the R^2 actually increases and the standard error decreases. Why does this happen? Even with X variables unrelated to Y, it is almost impossible for the correlation in your data set to be exactly zero—there is usually always a slight correlation one way or another. This means that as you add more X variables the resulting regression equation always fits your data better and better. So does this mean we should go on forever adding more and more variables?

Let's return to the La Quinta example for predicting profitability of a hotel. We mentioned that useful factors to consider might include the city population, the median income in the area, the number of college students within 5 miles, and the price of a room. But why stop there? Why not also add factors such as the color of the hotel, the street address, the hair color of the manager, the number of restrooms in the employee lounge, and so forth? Why not just take every possible variable you have data on and throw them all into the regression equation? This type of regression equation is often called a "kitchen sink" regression, and there is a reason why it's not so good.[5]

The truth is that your regression equation will fit your current data better and better as you add more X variables but it may actually be worse on future data. This is because the equation is specifically tailored to your data and thus is more likely to fit it better than it would fit some other set of data you might gather in the future.

> A regression equation will fit your current data better and better as you add more independent variables, but it may actually be worse on future data.

For example, suppose we look at a sample of hotels and by luck it happens to turn out that there is a small positive correlation between profitability of the hotel and the number of letters in the name of the manager (the correlation will usually be either slightly positive or slightly negative so there is not much luck involved here). So if you wanted to predict the profitability of one of these hotels, the length of the manager's name may be useful. But if you try to predict profitability for a different set of hotels, this strategy of using the length of the manager's name would be bad.

Financial analysts have long known about this problem. When a method for choosing stocks is tested on previously gathered data, they call it "back-testing" the method. It is well known that elaborate methods for picking stocks can perform very well during back-testing but can perform terribly on future data.

Ideally you should only use variables which have significant explanatory power. The principle of **parsimony** states that you should try to explain as much as you can with as few variables as possible.

---

[5]The term "kitchen sink regression" comes from the old expression "everything but the kitchen sink," meaning "practically everything possible." Surfing the Internet, I find that this expression may have come from World War II when everything possible was used to contribute to the war effort...all metal, pots, and pans, were used for the U.S. arsenal. The only objects left out were porcelain kitchen sinks.

> The principle of parsimony says to explain as much as you can with as few variables as possible.

One way to decide if a variable has significant predictive power is to see if the R^2 goes up significantly after you add it to your multiple regression equation. Even though a given X variable is highly correlated with the variable Y you are trying to predict, it may not increase R^2 by much if most of the useful information from that X variable is already contained in the other variables already in the equation (we will see an example of an extreme case of this in the next section).

Another way to decide if a variable adds significant predictive power is to look at something called the **adjusted** R^2, that essentially just equals a slightly deflated version of the R^2 value.[6] A regression equation will always fit your current data better than it will fit future data—since you actually tailored the equation to your current data. While the R^2 value measures how well a regression equation fits your current data, the adjusted R^2 is designed to measure how well the regression equation might fit similar future data, taking into account the "kitchen sink" problem. When you add a new variable to your regression equation, the adjusted R^2 will usually increase only if that variable "pulls its weight" in the equation. And if that variable doesn't "pull its weight," the adjusted R^2 usually decreases. You should use this as helpful advice in finding a good set of variables but not necessarily as an ironclad rule.

Even if adding a variable increases the adjusted R^2, you still might not want to include it for the sake of simplicity. Or you might want to specifically include that variable because it makes better physical "sense." Finding a good forecasting equation is still an art form, and you can make money as a consultant doing this. In the last chapter of this book we will spend a lot more time on methods for choosing variables to include in a multiple regression equation. But for now it is important to understand the principle of parsimony: explain as much as possible with as few variables as possible.

---

[6] *Technical note.* The R^2 value and the adjusted R^2 value are related using the following formula:

$$\text{Adjusted } R^2 \approx 1 - \left(\frac{\text{number of data points} - 1}{\text{number of data points} - \text{number of } X \text{ variables} - 1} \right) \times (1 - R^2).$$

Exercises

8. **Stock return forecasting.** The file `stockforecast.xls` contains data on the monthly return for seven different stocks (labeled Stock A through Stock G) over a period of 20 months. Using data from the first 10 months, build a regression model to predict next month's Stock A return using the current months returns for Stock B though Stock G. (a) How good are the predictions using this forecasting method? (b) Apply the forecasting method to the data from the last 10 months. Does it perform as well? Explain.

9. **Sales forecasts.** An analyst has a regression equation for forecasting sales with R^2 equal to 0.3 and adjusted R^2 equal to .2. She has two variables she is considering adding to the equation. When she adds only the first variable, the R^2 goes up to .7 and the adjusted R^2 goes up to .6. When she adds only the second variable, the R^2 goes up to .6 and the adjusted R^2 goes up to .5. When she adds both variables at the same time, the R^2 goes up to .75 and the adjusted R^2 goes up to .55. This is summarized in the table below:

Model	R^2	Adjusted R^2
With no additional variables	.2	.3
With first variable added	.7	.6
With second variable added	.6	.5
With both variables added	.75	.55

Which additional variable(s) do you suggest she use in the equation? Why?

8. Multicollinearity

We have seen how to use a multiple regression equation, how to determine if it fits the data well, and we've seen some of the problems that come up in interpreting it. We next turn to a problem called **multicollinearity** that can make the coefficients unreliable.

Usually the variables you put into a multiple regression equation are somewhat correlated with each other. This is usually no problem but it can be bad if the correlation between them is too high or if some variable is almost completely determined by some set of other variables (of course if you are trying to build an equation to predict some Y variable it is very nice if you can find X variables very highly correlated with it—the problem arises when you have X variables that are very highly correlated with each other).

To see how this causes problems, let's look at an example. Suppose a company that makes shoes would like a formula for the value of the inventory they have on hand. Suppose each pair of shoes is worth $100 and they are manufactured in pairs. Their accountant comes up with the equation

$$Y = 50X_1 + 50X_2,$$

where X_1 equals the number of left shoes they have in inventory and X_2 equals the number of right shoes they have in inventory. How good is this equation at forecasting the value of inventory? It is good, since it will essentially give $100 per pair of shoes.

Now another accountant gets the equation

$$Y = 100X_1.$$

How good is this at forecasting? This is also good, since it also essentially gives $100 per pair of shoes. This equation works just as well as the previous equation in practice since the shoes are manufactured in pairs.

A third accountant gets the equation

$$Y = 200X_1 - 100X_2$$

and is also happy because the equation gives good forecasts of $200 - $100 = $100 per pair. A fourth accountant gets the equation

$$Y = 10{,}000X_1 - 9{,}900X_2,$$

that also gives equally good forecasts. This fourth accountant then uses this equation to propose a potentially profitable proposition to management: if they keep the number of right shoes fixed and start to increase production of left shoes, they can increase inventory value by about $10,000 per left shoe! In fact, this equation suggests they should also start decreasing the number of right shoes they produce.

Obviously this is nonsense and illustrates the problem that arises when you have X variables that are too highly correlated with each other. The number of left shoes and the number of right shoes are perfectly correlated with each other here. If you let the computer fit a multiple regression equation to your inventory data, it could reasonably give you any of the four regression equations mentioned above, or switch between them as you add data, since they all would fit your data equally well. This causes the regression coefficients to become very unreliable and unpredictable.

Here is more intuition for why the coefficients can be unpredictable: one or two data points can drastically change the coefficients. Suppose that the computer gave you the fourth equation $Y = 10{,}000X_1 - 9{,}900X_2$, and you saw that X_1 had a coefficient of 10,000. Now if just a few more data points were added where the number of left and right shoes were not exactly equal, the computer would probably wake up to reality and give you $Y = 50X_1 + 50X_2$. So the coefficient might lower drastically from 10,000 to 50 with the addition of only a few data points. This means estimates of the coefficients can change drastically as you add even a very small amount of new data.

The problem of **multicollinearity**: if you have two variables that are almost perfectly correlated with each other, don't put them both into a regression equation as independent variables—it can make all the regression coefficients very unreliable.

The best solution to a multicollinearity problem is to leave out one of the two variables. In many practical situations, multicollinearity does not really cause problems unless a correlation is significantly higher than .9.

Also, it is possible to have a multicollinearity problem even if correlations between any pair of variables are weak. If one variable can be predicted perfectly from several other variables, one of the variables can be viewed as being redundant and must be left out. For example, suppose you are using several percentages in a regression equation and you know they must add up to 100 percent. If you know all of them except one, you can figure out the remaining one by subtracting the total of all the others from 100 percent. So one of the percentages is always redundant in the presence of all the others and you should leave out one of them from your multiple regression equation. Usually if you have just two X variables this is called a **collinearity** problem, and if you have several X variables that combine to give something nearly perfectly correlated with another X variable it is called a **multicollinearity** problem.

Once again, you should be extremely happy if you have an X variable that is highly correlated with the Y variable. That will make prediction of Y easier. The multicollinearity problem we are worried about above occurs when you have X variables that are too highly correlated with each other.

Exercises

10. An analyst would like to build a multiple regression equation to predict Y using the available data from X_1, X_2, X_3, and X_4. First he computes the correlation between all pairs of variables and this is shown below:

	Y	X_1	X_2	X_3	X_4
Y	1				
X_1	0.2	1			
X_2	0.95	0.3	1		
X_3	0.5	0.4	0.3	1	
X_4	0.5	0.6	0.4	0.95	1

(a) What explains the "1"s along the diagonal?

(b) Do you see a pair of variables here that could potentially have a problematic correlation? Explain.

11. Sales forecasting. Suppose a store would like to forecast monthly sales volume of a product and they have data on three variables: the number of days it was sold at the regular price during a month, the number of days it was sold at the discounted price during a month, and also the monthly sales volume. An analyst notices that the correlation between the number of days it was sold at the discount price and the number of days it was sold at the regular price equals $-.98$. What explains the very strong correlation, and is there anything wrong with including both variables in the multiple regression equation?

9. More on Dummy Variables

We explained in a previous section how you can incorporate non-numerical **categorical** variables, such as gender, into a regression equation by using

dummy variables. We did this by creating a new variable that for women equaled 1 and for men equaled 0. This was straightforward. But there are some potential multicollinearity problems that arise if you have more than just two categories.

Suppose you own a large number of stores in three types of locations: urban, rural, or suburban. You would like to forecast sales from advertising expenditures so you can maximize your investment.

At first you might have the idea to make three separate regression lines. But if you want to have more than one categorical variable in your equation, you could unfortunately quickly end up with dozens of separate lines—one for each of the different possible combinations. For example, if you had three different locations and four different store layouts, you would need three separate lines for each of the four store layouts—or in other words $3 \times 4 = 12$ equations. To avoid this mess, you should incorporate the location in a single equation.

Suppose you create the following multiple regression forecasting equation, where Y equals sales for a given store (in \$1,000s) and X_1 equals advertising expenditures at that store (in \$1,000s). For the location variable X_2, you use 0 if the store is in an urban location, 1 if the store is in a rural location, and 2 if the store is in a suburban location:

$$Y = 20 + 5X_1 + 10X_2.$$

How do we interpret the "10" coefficient? It says that increasing the location variable by one unit, with advertising expenditures the same, is associated with an extra \$10,000 in sales. And it says that going from urban to rural has the same increase in sales as going from rural to suburban. Is this reasonable?

The answer is no, because the numbering of the locations was arbitrary and there is no reason to assume that the increase in sales from Location 0 to 1 is similar to the increase from 1 to 2. If we renumbered the locations, we would get a completely different equation—and this seems like "cheating." If Location 1 just happened to have the highest average sales, there could be no linear relationship and the coefficient might turn out to be 0—even though the location is important.

The solution is to create separate dummy variables for each category. We create three new variables X_3, X_4, and X_5, as follows: let X_3 equal 1 for urban stores and let it equal 0 otherwise, let X_4 equal 1 for rural stores and let it equal 0 otherwise, and let X_5 equal 1 for suburban stores and let it equal 0 otherwise. Sometimes this technique is called "creating dummy variables from a categorical variable."

But before we replace X_2 with all these variables in the regression equation, we must always omit one of the dummy variables in order to avoid a multi-collinearity problem—the problem where some variable is almost completely determined by other variables. Since exactly one of the dummy variables here will equal 1, if you know two of the variables you automatically know the third one. But if we leave out one of the dummy variables, does this mean we will be ignoring one of the categories? Not at all. The category we leave out is indicated when all the other dummy variables are zero and so it will be the "baseline" category. We illustrate this next.

> Turn categorical variables into dummy variables for each category before you run a regression. Don't forget to leave out one of the categories, so it will be the "baseline" category.

Suppose we decide to leave out the category for suburban stores, X_5. We get the following equation:

$$Y = 40 + 7X_1 + 8X_3 - 6X_4.$$

How do we interpret the coefficients now?

We interpret the coefficient "7" as "each extra \$1,000 in advertising is associated with an extra \$7,000 in sales on average, when looking at stores in the same type of location."

We interpret the coefficient "8" as "urban stores tend to average \$8,000 higher in sales than suburban stores that spend the same on advertising." Note that since we left out the category for suburban, the dummy variable tells us the difference between urban and suburban.

We interpret the coefficient "−6" as "rural stores tend to average \$6,000 lower in sales than suburban stores that spend the same on advertising."

Leaving out the suburban variable makes it the baseline: urban is 8,000 above this, rural is 6,000 below this, and the average difference in sales between rural and urban is 14,000. Thus we could say "rural stores tend to average $14,000 lower in sales than urban stores that spend the same on advertising."

When you're faced with a lot of different variables to use in a multiple regression, make sure you identify the categorical variables and properly turn them into dummy variables. And you also might want to first group some similar categories together if there are very many of them (there are no formal rules for this—you have to use your business common sense.)

Using Excel and R to create dummy variables

Suppose you have the following data on a number of stores you own: the store size (in square feet), the annual profits, and a location variable that equals 1 if the store is in an urban location, 2 if the store is in a rural location, and 3 if the store is in a suburban location. There is one row for each store. You'd like to transform the location variable into three separate dummy variables: one for urban, one for rural, and another for suburban. The dummy variable will equal "1" if the corresponding store location is of that type.

You can look in the data file **dummy.xls** to follow along and a sample of the first few rows is shown in Figure 7.3:

Size	Profits	Location
11236	$1,223,605	1
10722	$ 853,846	3
10814	$ 746,198	1
11830	$ 810,472	2
11182	$1,046,028	2
11394	$1,112,770	1

Figure 7.3. An excerpt from the data file **dummy.xls**.

Instructions using the Excel "IF" command

Step 1: First create a new column for each category and label each in the first row with the name of the category. In this example we have three categories labeled "1," "2" and "3."

Step 2: Apply the formula "=IF($C2=F$1,1,0)" to cell F2 and drag it (using the little black dot handle at the lower right-hand corner of the cell) over to the left and down to the bottom of the sheet. The final product is shown in Figure 7.4.

	A	B	C	D	E	F
1	Size	Profits	Location	1	2	3
2	11236	$ 1,223,605	1	1	0	0 ◄¬=IF($C2=F$1,1,0)
3	10722	$ 853,846	3	0	0	1
4	10814	$ 746,198	1	1	0	0
5	11830	$ 810,472	2	0	1	0
6	11182	$ 1,046,028	2	0	1	0
7	11394	$ 1,112,770	1	1	0	0

Figure 7.4. Creating dummy variables from a categorical variable using Excel's "IF" command.

Instructions using R

Start R-Studio, import the data file "`dummy.xls`" and create a copy titled "d":

```
d=data.frame(dummy)
```

To create three new columns, one for each of the three categories, type

```
d$dummy1=ifelse(d$Location==1,1,0)
d$dummy2=ifelse(d$Location==2,1,0)
d$dummy3=ifelse(d$Location==3,1,0)
```

and we can view the results in Figure 7.5 using the command

```
View(d)
```

Exercises

12. **Home values.** An analyst creates the following multiple regression equation to predict the value of a home: $Y = 10,000 + 100X_1 + 2X_2$ where Y is the home value (in dollars), X_1 is the square footage and X_2 is the zip code for the area where the home is located. Interpret the coefficients and give some suggestions on how the equation could be improved.

	Size	Profits	Location	dummy1	dummy2	dummy3
1	11236	$ 1,223,605	1	1	0	0
2	10722	$ 853,846	3	0	0	1
3	10814	$ 746,198	1	1	0	0
4	11830	$ 810,472	2	0	1	0
5	11182	$ 1,046,028	2	0	1	0
6	11394	$ 1,112,770	1	1	0	0

Figure 7.5. Creating dummy variables from a categorical variable using R's "ifelse" command.

13. **Sales forecasting.** An analyst has a regression equation having $R^2 = .6$ for predicting retail sales of a store from the square footage. Since the store location is very important, she then adds in a variable that equals 1 if the store is located in an urban center, 2 if the store is located at an airport and 3 if the store is located in a suburban mall. This new equation has $R^2 = .61$. What does this tell you about the link between location and sales?

10. Interaction Terms

Suppose you are trying to forecast Y from X_1 along with a lot of other variables and you notice the following graph:

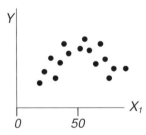

A single line doesn't describe the relationship so well but two lines may actually fit much better:

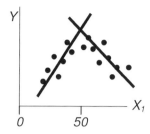

How do we get Excel to fit two lines here? We essentially want a line where the slope and intercept changes when X_1 goes from below 50 to above 50. You can see from the graph that when X_1 is below 50 the slope is positive and the Y intercept is negative, and when X_1 goes above 50 the slope is negative and the Y intercept is positive. You may be thinking we could just split the data set in half and just run two separate regression lines, but this is not always a good idea. If there are other variables involved in the forecasting equation, we may end up splitting the data set into so many pieces that we are not able to accurately estimate the coefficients anymore.

The solution is to create what are called **interaction terms**. For this, we create two new variables X_2 and X_3 as follows. Let X_2 equal 1 if X_1 is above 50 and let X_2 equal 0 otherwise. Also let X_3 be the product of X_1 and X_2 so that $X_3 = X_1 \times X_2$. We'll show you how these work together next. Suppose we get the following regression equation:

$$Y = 10 + 5X_1 + 20X_2 - 9X_3.$$

This means that when X_1 is above 50, we have $X_3 = X_1$ and $X_2 = 1$ and so we get following downward sloping regression line:

$$
\begin{aligned}
Y &= 10 + 5X_1 + 20 - 9X_1 \\
&= 30 - 4X_1.
\end{aligned}
$$

When X_1 is below 50, we have $X_3 = 0$ and $X_2 = 0$ and so we get following upward sloping line:

$$Y = 10 + 5X_1.$$

We thus have two separate lines representing the link between X_1 and Y. When X_1 is below 50, a unit increase in X_1 is associated with a 5 unit increase in Y; when X_1 is above 50, a unit increase in X_1 is associated with a 4 unit decrease in Y.

11. Fitting a Curve

Another option for non-linear data is to try fitting a curve:

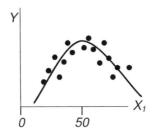

To generate a curve we create a new variable X_2 and let it equal $(X_1)^2$. This will then give you a multiple regression equation looking something like the following:

$$
\begin{aligned}
Y &= 10 + 5X_1 - 4X_2 \\
&= 10 + 5X_1 - 4(X_1)^2,
\end{aligned}
$$

the mathematical equation of a parabola. The coefficient you get may not have a straightforward managerial interpretation but the equation may end up giving better forecasts.

12. Case Study: *US News* Business School Rankings

Background

U.S. News & World Report is a popular magazine that annually ranks business schools. These rankings can influence the type of applicants a school gets in the future and also influences the prestige associated with the MBA degree from that school. People have argued that their ranking methods are not so fair but the rankings they assign are influential nonetheless.[7] Business school deans often make changes in an effort to improve their rankings.

---

[7] "Many American colleges balk at U.S. News rankings," by Janine Brady, *CNN.com*, June 20, 2007. http://www.cnn.com/2007/EDUCATION/06/20/college.rankings/index.html.

The rankings for 2005 were based on a number of factors that were combined to get an overall score from 1 to 100 for each school and then schools were ranked based on these scores. Roughly speaking, a one point increase in overall score corresponds to moving up one position in the rankings. Being one of the top 50 ranked schools nationally is considered very prestigious.

School administrators nationwide were asked to rate each program on a scale from 1 (marginal) to 5 (outstanding). Corporate recruiters were also asked to rate programs on the same scale. Also considered was the mean starting salary plus bonus and employment rates for graduates, computed at graduation and three months later. The quality of students was measured by mean GMAT scores, mean undergraduate GPA, and the proportion of applicants accepted by the school. The exact formula the magazine uses to rank the schools is not published.

Boston University's MBA program was ranked 47th in the nation in 2005 (it is currently much better); Harvard and Stanford were numbers 1 and 2 respectively. Below is how Boston University ranked nationally on each of the separate factors considered in the overall rankings. Note that lower ranks are better (except for acceptance rate—a higher rank is better for this).

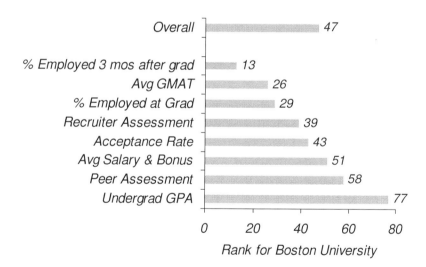

From this we see that Boston University is actually ranked much better than its national ranking on several individual factors: the percentage of graduates who were employed at graduation and three months later (ranked 13th and 29th nationally respectively), recruiter assessment (ranked 39th), and average GMAT scores (ranked 26th).

Questions

Here are a few questions we would like to answer:

- Since the exact formula for computing the rankings is not published, how can we tell which factors are most important in the rankings?
- Scholarship money can often be allocated to attract students with higher undergraduate grade point averages and GMAT scores. By how much would the overall ranking change if either of these were increased? Suppose GMAT could be increased using scholarships by 10 points. What would this do to a school's ranking?
- One of the schools seems to be very unfairly ranked given its statistics. Which school is it and how far off is its ranking? And how might this have happened?

Analysis

Below we will answer the questions but first we will show you how a naïve analysis can be very misleading. Suppose we compute the correlation between each factor and the overall score for the 80 highest ranked schools. Below is a graph of what we get. How can we interpret it?

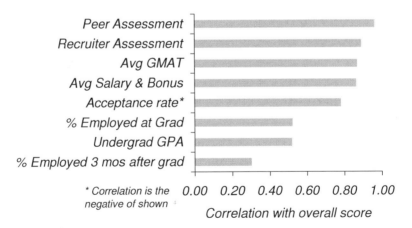

Can we say that any of these factors have a direct causal relationship with overall score? The answer is yes because the magazine article said so. It said that they base their rankings solely on these factors and so these factors alone are what cause the rankings to rise or sink. The correlation numbers themselves don't tell us there is a causal relationship but the magazine article does.

Since we now know there is a causal relationship, do the correlation coefficients above tell us the relative strength of each factor's impact on overall score? The answer is unfortunately no. It may seem, since they have the highest correlations, that the most influential factors are peer assessment, recruiter assessment and GMAT. And it may also seem as though the employment percentages and GPA are the least important. But it is impossible to tell how much of a correlation here is due to direct causation and how much is due to a correlation with some other factor which has a stronger impact on rankings. For example, suppose hypothetically that GMAT scores played very little role in the rankings but just so happened to be highly correlated with peer assessment scores—which actually were counted heavily in the rankings. This would make GMAT scores appear to have a strong influence on the rankings when in fact they didn't. We don't want to waste our money investing in scholarship money to improve the incoming class' GMAT scores if it would have no impact.

> Don't rely solely on pairwise correlation coefficients when there are many correlated variables involved—these can be very misleading.

Since we can't rely on the correlation coefficient, what can we do? We still want to know how much of an impact on overall ranking a change in any one of these individual factors would have.

Suppose we try something different. We draw the graph of GMAT versus overall score and compute the equation of the regression line. How can you interpret the coefficient 0.41?

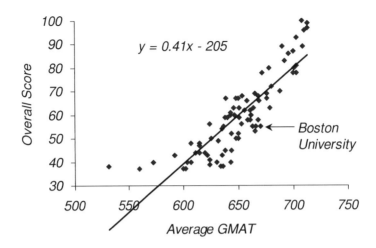

We see here that schools with higher average GMAT scores tend to be ranked higher overall. We can interpret the coefficient .41 as follows: for each extra 10 points in average GMAT score, we see an extra 4.1 points in overall score on average, or approximately an increase of 4.1 positions in the rankings.

Does this mean that, with everything else staying the same, if Boston University could increase average GMAT scores by 10 points it could expect to move up roughly 4 positions in the rankings?

The answer to this question is no; this regression line doesn't tell you what would happen if everything else stayed the same and you only changed GMAT scores. As we said before, all the different factors that go into the rankings are highly correlated with each other: schools with higher GMAT scores also tend to score higher on the other factors as well. All this regression coefficient of 4.1 tells you is that two schools that differ by 10 points in average GMAT tend to be ranked about 4.1 positions apart.

You could, however, make the following more accurate statement: if Boston University could increase average GMAT score by 10 points and also was able to increase all the other factors by amounts that are typically seen in schools having 10 additional GMAT points, then we could expect that Boston University would move up roughly 4 positions in the rankings.

This last statement is more accurate but is not so useful because it does not tell us how much of an increase we would also need in each of the other factors. Unfortunately, that's all we can really say from this regression line. Another problem is that some of these factors would be hard to influence by just changing the GMAT. We would really like to know what would happen if we increased only GMAT scores.

The solution to our problem lies in a multiple regression analysis. We will solve this in a moment, but first we will try to mislead you a couple more times.

Here is another graph showing undergraduate GPA versus overall score. How do you interpret the regression coefficient 73.4 shown?

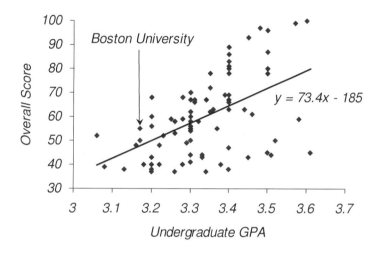

Does this mean that finding a way to raise undergraduate GPA by 0.1 grade points is expected to move a school up by about 7.34 positions in the rankings? No, again for the same reason. We could say that by raising undergraduate GPA by 0.1 grade points, and also raising all the other factors by amounts typically associated with the schools which have an additional 0.1 undergraduate grade points on average, we would expect to see the school move up by about 7.34 positions in the rankings. This is not exactly telling us how much of an impact GPA has on its own. To answer this, we need multiple regression.

Here is the last time we try to mislead you. Below is a graph showing acceptance rate versus overall score. Since the coefficient is −.78, it means that lowering the acceptance rate by 10 percentage points (and improving all the other factors to the levels of schools having this lower acceptance rate) will give an improvement of 7.8 positions in the rankings.

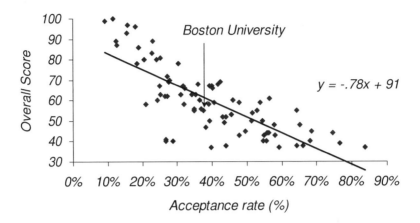

This makes it look like the acceptance rate has a large impact on rankings—but to get these 7.8 positions of rankings improvement, many other correlated factors must change as well. This does not tell us what happens when we only change the acceptance rate. We will see later that the acceptance rate plays very little role in the rankings (much less than GPA or the employment percentages), but it just so happens to be highly correlated with other factors that play a larger role. To see the impact of a single factor by itself, we need multiple regression.

Finally, we run a multiple regression where the dependent variable (the Y variable) is overall score and we use each of the above factors as the independent variables (the X variables). We get the following computer printout, which the computer says has $R^2 = .996$:

	Coefficients
Intercept	-140
Peer Assessment	11
Recruiter Assessment	5
Undergrad GPA	19
Avg GMAT	0.06
Acceptance %	-0.006
Avg Salary & Bonus	0.0002
% Employed at Grad	0.08
% Employed 3 mos after grad	0.25

This means to get overall score, you multiply each factor by its corresponding coefficient, add them all up, and add on the intercept.

Since we know that these are the only factors that go into the rankings and the R^2 value is nearly equal to 1 (it is probably slightly less than 1 due to rounding off of the numbers in the published data), we can conclude that this regression equation is essentially the formula the magazine uses to assign overall score—and hence rankings. We've now unlocked the secret of the rankings, and discovered their secret formula!

Here we see that the highest coefficient corresponds to undergraduate GPA and the smallest corresponds to average salary and bonus. Does this mean that undergraduate GPA is the most influential factor and salary the least influential factor? Absolutely not, since they are on completely different scales. Average salary figures are generally in the thousands, GPAs range from 1 to 4. This is not the proper way to compare the relative influence.

> The relative size of multiple regression coefficients does not tell you the relative importance of the different factors. The coefficients may be measured in different units and cannot be directly compared.

The proper way to compare the relative influence is to look at the interpretation of each coefficient separately. The coefficient 19 for GPA means that a 0.1 point increase in average GPA (with all the other factors remaining the same) would correspond to $19 \times 0.1 = 1.9$ extra points in overall score—or roughly 2 positions in the rankings.

From all this can we say that if Boston University were able to increase its average GPA by 0.2 grade points (with all other factors remaining the same) they could expect at improvement of approximately 4 positions in the rankings? I would say yes—this is a very sensible interpretation of the multiple regression coefficients.

The coefficient .06 for GMAT means that a 10 point increase in average GMAT scores (with all other factors staying the same) would correspond to about 0.6 points in overall score—or roughly 0.6 positions in the rankings. This is less than a one position movement in the rankings. Before when we looked at the simple regression line it appeared that we could move up 4 positions in the rankings with this much of a GMAT improvement. In reality this would only correspond to movement of less than a single ranking position. Also notice that the coefficient on the acceptance percentage is negative, meaning that a lower acceptance percentage corresponds to a higher overall score.

A very nice way to compare the influence each factor has on the overall ranking is to look at the reciprocal (1/coefficient) of the regression coefficients. These tell you how much of a change in that factor you would need to see to observe a unit change in the dependent variable on average. For example, since the recruiter assessment coefficient equals 5, that means a 1 point increase corresponds to a 5 point increase in overall score. Looking at the reciprocal, we can say that a $1/5 = 0.2$ point increase in recruiter assessment would correspond to a $5/5 = 1$ point increase in overall score. Below is a table of the reciprocals of the regression coefficients.

	Increase needed for 1 position improvement in overall ranking
Peer Assessment	0.09
Recruiter Assessment	0.2
Undergrad GPA	0.05
AVG GMAT	16
Acceptance %	−160%
Avg Salary & Bonus	$4,353
% Employed as Grad	12.5%
% Employed 3 mos after grad	4.0%

This table tells us that in order to move up one position in the ranking, a school could do any single one of the following equivalent things: increase GPA by .05 grade points, increase average salary and bonus by $4,353, increase the percentage employed at graduation by 12.5 percentage points, increase the percentage employed 3 months after graduation by 4 percentage points, or increase average GMAT by 16 points. The peer and recruiter assessments may be more difficult to influence, although they play a very large role: a 1 point increase in peer assessment by itself would move a school up 11 positions in the rankings.

Also notice that in order to move up one position in the ranking, a school would have to lower its acceptance percentage by 160 percentage points. As this is impossible, this means that the acceptance percentage plays very little role in the overall rankings; this factor by itself can't even move a school a full position in the rankings. The simple correlation analysis we did at first showed that it was more highly correlated with overall score than the percentage employed at graduation and the percentage employed three months after graduation. It seemed then as if it was a very influential factor. The correlation we saw was most likely because it was correlated with another factor that played a more important role. Further calculation shows that the correlation between acceptance percentage and GMAT score is -0.8 and its correlation with peer assessment is -0.8. Both these could explain why the acceptance percentage seemed so important at first, and it turned out later not to be so important. Relying on the correlation analysis by itself tells the wrong story and is very misleading here.

In reality if a school managed to increase GMAT scores, most likely their GPA numbers would rise as well. Trying to raise one without also affecting the other really is not realistic. If GMAT scores went up by 16 points and GPA by .05, we can add the two separate effects to then expect a 2 position improvement in the rankings.

If we rerun the multiple regression above but this time we omit the GPA variable, we get a coefficient of .083 for GMAT. This means that if we can raise GMAT by 10 points and we believe that GPA will rise by the amount typically correlated with this much of a GMAT change (with everything else staying the same), we will expect to see in increase of about .83 points in score overall—or a bit less than 1 position in the rankings. This means that we would need a $1/.083 = 12$ point rise in GMAT along with the typically correlated rise in GPA to see a 1 position improvement in the rankings.

Which school was unfairly ranked? Let's take a look at the graph of the predicted score versus the residual (the residual equals the actual score minus the predicted score). This is what we see:

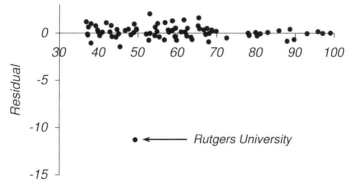

Notice that Rutgers University has a residual of about −12! This means the score assigned to Rutgers by the magazine is about 12 points—and therefore about 12 positions in the rankings—below what the formula used to score the other schools would give. This is cause for concern. As you can see, all the other schools are assigned scores very close to what the regression equation would predict. Looking at the data, we can see that Rutgers deserves to be ranked a lot higher than it was. In fact, the following year it jumped about 12 positions in the rankings, even though the numbers didn't change very much. I don't have an explanation for why this occurred, although it is a very serious thing. One possible explanation is that Rutgers was late in sending some of the statistics and these were counted as "0" (this actually happened to Boston University one time).

You may now be curious why the other schools don't line up perfectly at their predicted scores. This is because Rutgers is such an outlier that it throws off the forecasting equation for the rest of the schools. If we remove Rutgers and then rerun the multiple regression, this is what we get for a residual graph:

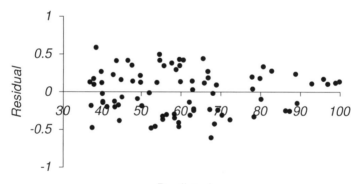

Predicted score

Now we notice that almost all the predicted scores are within half a point from the actual scores—this difference here is just due to rounding the scores. Removing Rutgers gives a slightly different forecasting equation and thus the interpretations of the coefficients above should be revised slightly.

Summary

We now summarize what we've learned here. When you have many variables and you would like to understand the effect that any one by itself has on some other variable, running a multiple regression can be very helpful. It can be very misleading to look at correlation coefficients when the different variables you are comparing are correlated with each other. Multiple regression coefficients tell you something very different from simple regression coefficients or correlation coefficients. It is important to be able to understand the difference and use the appropriate one in a given situation.

13. Exercises

14. **Hotel occupancy rates.** A large motel chain has developed a regression model to predict the occupancy rate (the average daily percentage of rooms that are occupied) of its motels. From data gathered over the past year they estimate the following regression model: $Y = 42 - .003X_1 + 2X_2$ (having $R^2 = .5$), where Y = occupancy rate (a number from 0 to 100), X_1 = total amount spent on advertising (in dollars), and X_2 is a dummy variable that equals 1 if the motel is a franchised location and equals 0 for a non-franchised location. The data fit the model reasonably well.

   (a) Give managerial interpretations for the coefficients and R^2.

   (b) What kind of occupancy rate would you forecast for a franchised location spending $2,000 in advertising?

   (c) Based on the coefficient $-.003$ can we conclude advertising does not work very well at raising occupancy rates? Explain.

   (d) Can we say that franchised locations overall on average have a higher average occupancy rate than the non-franchised locations?

15. **Gender differences in salary.** Consider the multiple regression equation $Y = 10 + 5X_1 - 3X_2 + X_3$ where X_1 is age, X_2 is gender (0 = male, 1 = female), X_3 is experience in years, and Y is salary in thousands of dollars. Assume the regression equation fits the data well.

   (a) Does this mean the overall average for men is about $3,000 higher than the overall average for women? If not, what can you say about the $3,000?

   (b) Does this mean for men with the same level of experience, older men tend to earn more than younger men?

16. **Student debt.** Data about student debt at graduation is gathered from a large number of colleges. Let Y = average student debt at graduation

(in thousands of dollars), X_1 = type of school (public = 0, private = 1), X_2 = annual average room and board expense (in thousands of dollars), X_3 = annual total tuition cost (in thousands of dollars), and X_4 = campus location (1 = urban, 2 = rural, 3 = suburban).

(a) Consider the simple linear regression equation $Y = 3X_2 + 30$ having $R^2 = .4$. Give managerial interpretations for the coefficient of X_2 and the R^2 value.

(b) Give managerial interpretations for the coefficients in the multiple regression model $Y = 30 + 4X_1 + 2X_2$ having $R^2 = .6$.

(c) Describe the extent to which the model in (b) improves the ability to predict average debt compared to the model in (a).

(d) Predict average debt for students graduating from a public university where room and board expenses are $7,000 per year.

(e) Consider the simple regression model $Y = 20 + 5X_3$ having $R^2 = .5$ and the multiple regression model $Y = 15 + 4X_2 + 3X_3$ having $R^2 = .5$. Based on just these two models, what can you say about how useful is knowing room and board expense in predicting average debt?

(f) Consider the regression model $Y = X_4 + 34$ having $R^2 = .001$. Based on just this model, what can you say about how useful knowing the location is in predicting average debt?

17. **Purchasing behavior.** A marketing department has gathered data on the purchasing behavior of its previous customers; see the file `directmarketing.xls` (an excerpt is in Table 17). They have gathered data on five potential new customers (also in the same file). Which of these five new potential customers would be the best to target? Rank the five customers in terms of how much you believe they will purchase.

18. **Production forecasting.** A factory manager would like to be able to forecast the day's production. The number of machines that are available for use tends to vary, as does the number of people available to work. Looking at the data in the file `manufacturer.xls` (an excerpt is in Table 18), how much production would you estimate for a day where there are seven machines and 18 workers available?

19. **Bank withdrawals.** A bank would like to predict the total amount of money customers withdraw from automatic teller machines (ATMs) on weekends based on the median value of homes in the area and on whether

Table 7.2. An excerpt from the file `directmarketing.xls`.

Customer #	Purchases	Distance (miles)	Gender	Income
1	$4,520.51	7.2	F	$83,931
2	$11,230.18	15.4	F	$127,998
3	$6,812.47	2.6	F	$72,127
4	$9,648.57	16.6	M	$81,437
5	$1,292.93	2	F	$88,913
6	$11,448.62	16	F	$135,822
7	$10,521.31	20.2	M	$126,091
8	$9,394.01	9.4	M	$155,241
9	$3,290.39	9	M	$91,819
10	$3,846.64	13.2	F	$48,984

Table 7.3. An excerpt from the file `manufacturer.xls`.

Day #	Units produced	Machines available	Workers available
1	1932	6	26
2	1044	2	15
3	1233	6	10
4	1201	2	19
5	1217	2	19
6	1376	2	23
7	1951	10	17
8	1372	6	13
9	2105	8	24
10	2197	10	21

the ATM is located in a shopping center or elsewhere. They developed the following multiple regression model based on historical data: $Y = 613 + 87H - 8708S$, where Y = predicted amount withdrawn, H = median home value (both measured in $1,000s), and $S = 1$ if the ATM is in a shopping center and 0 otherwise.

(a) Interpret the regression coefficients in this model.

(b) Predict the withdrawal amount for an ATM located in a shopping center in a neighborhood in which the median value of homes is $350,000.

(c) The data shows that the average amount of money withdrawn on weekends from ATMs located in shopping centers is *higher* than for ATMs not in a shopping center. Does this contradict the above regression model, or is there a possible explanation? Explain.

(d) From the information provided by the model and part (c), do you think ATMs are more likely to be located in shopping centers where the median home value in the area is high or when it is low? Explain briefly.

20. **Catalog sales.** A catalog marketing company constructs a regression equation to forecast monthly sales. The regression equation, which fits the data reasonably well, is as follows: $Y = 75 + .007X_1 - 3X_2$, having $R^2 = .4$, where Y = number of units sold (in hundreds of units), X_1 = promotional spending in dollars, and $X_2 = 1$ if the catalogs are sent out using third class postage, and equals 0 if catalogs are sent out using first-class postage.

(a) What would you forecast sales to be if \$30,000 were spent on promotions and the catalogs were sent using first-class postage?

(b) What proportion of the variability of sales is explained by variables other than promotion expenditures and postage class?

(c) Give a clear managerial interpretation of the coefficient .007.

(d) Suppose an interaction variable X_3 was added which equaled the product of X_1 and X_2. If the coefficient of X_3 turned out to equal .002, does this mean promotional expenditures are more effective or less effective when using first-class postage? Explain briefly.

(e) Suppose one analyst averages the historical sales figures for all months where the catalogs were sent out by first class mail, and another analyst does the same for the other months. Who should get the higher average? Explain briefly.

21. **Healthcare costs.** An insurance analyst would like to know if doctors at Clinic A treat patients more economically than doctors at Clinic B. He sees that the average treatment cost per patient in A is \$1,100 and is \$1,900 in B. Since \$1,100 is lower than \$1,900, he concludes that A is more economical. A second analyst notices that the average age of patients at B is higher than at A and she argues this must be taken into account. She recommends looking at the following regression equation: $Y = 10X_1 - 500X_2 + 1500$, where Y is the patient cost (in \$), X_1 is the age of the patient (in years), and X_2 equals 0 if the patient goes to Clinic A, and equals 1 if the patient goes to Clinic B. Based on the coefficient -500, she concludes that Clinic B is more economical.

(a) Which of these two analysts' reasoning seems more sensible to you? Why?

(b) A third analyst also notices that the use of less expensive generic drugs instead of expensive name-brand drugs is more frequent at Clinic B than at Clinic A, and he adds to the regression equation a variable X_3 which equals 1 if name-brand drugs were used, or equals 0 if generic drugs were used. He computes the regression equation $Y = 10X_1 + 200X_2 + 40X_3 + 1400$ and concludes A is more economical based on the coefficient $+200$ (this regression equation fits the data even better than the equation above). Which of the three analysts' reasonings seems most sensible to you? Why?

22. **Predicting sales.** The marketing department in the central office of a chain store has developed a multiple regression model to predict sales at a store. From historical data they get the following model: $Y = 20{,}000 + .4X_1 + 1{,}000X_2$, where $Y = $ sales for the year at a store (in \$1,000s), $X_1 = $ median income (in \$) of households in the area from which the store attracts customers, and X_2 is a dummy variable that equals 0 if the store is located in a highly competitive area and 1 otherwise.

(a) Give managerial interpretations for the regression coefficients .4 and 1,000.

(b) What sales would you predict for a store that attracts customers from an area with a median income of \$30,000 and is in a highly competitive area?

(c) Does the fact that .4 is a lot smaller than 1,000 mean that location is much more important than customer income? Explain briefly.

23. **Height and income.** A book discusses an interesting relationship between height and income for people in the United States.[8] The file height-income.xls contains data on height, income, and gender for a large representative sample of people. What kind of relationship do you see and what could explain it?

More Challenging Exercises

24. **Employee evaluations.** There are twenty employees in a company and there are six managers who do employee evaluations. Each employee gets

---

[8] *Teaching Statistics: A Bag of Tricks,* by A. Gelman and D. Nolan, Oxford University Press, 2002.

independent evaluations from three different managers and each manager gets randomly assigned ten employees to evaluate. Each employee is given a score from one to one hundred by each of the three managers and these are averaged together to get an average score for the employee. Here is a graph of the average scores:

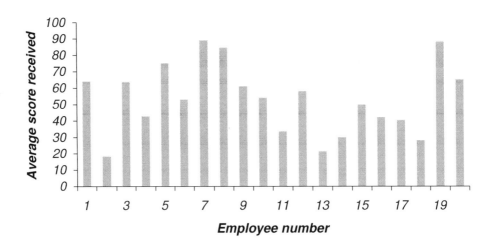

As you can see, there was a lot of variation in the scores. Some of the employees who received low scores complained that they had bad luck to be assigned to managers who used more strict criteria and tended to give lower scores to everyone they evaluated. They also constructed the below graph of the average scores given by each of the six managers:

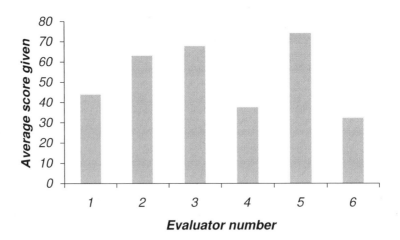

As you can see, there was a lot of variation across the different managers. Looking at the raw data (in the file `evaluation.xls`), discuss in a short paragraph how you could use multiple regression to assign scores to employees so that the differences between managers wouldn't create this type of unfairness.

25. **MBA salaries.** During the first day of an MBA statistics course, students were asked in an anonymous survey the annual salary from their most recent job, the salary they expect in their next job after graduation with an MBA, whether or not they are an international student, and their gender. The regression equation $Y = 48 + 0.9X_1 + 6X_2 - 17X_3$ is computed, where Y is the expected annual salary in the next job (in \$1,000s), X_1 is the salary from the most recent job (in \$1,000s), X_2 equals one if the student is an international student (and equals zero otherwise), and X_3 equals one if the student is female (and equals zero otherwise). The standard error is 20 and the R-squared is 30%.

   (a) Give managerial interpretations, if appropriate, for the coefficients, intercept, standard error and R-squared.

   (b) A second regression equation $Y = 102 - 20X_2 - 19X_3$ is computed. What could explain why the coefficient for X_2 in this equation is negative, while in the first equation above it is positive? Are there more stories this equation reveals about international students?

   (c) A third regression equation $Y = 61 - 28X_2 - 4X_3$ is computed, where now Y equals the annual salary in the most recent previous job (in \$1,000s). Along with the second regression equation above, what stories can you tell about gender?

   (d) Does the regression effect here say that the students who earned very high salaries in their last job will, on average, have a lower salary in their next job? Explain.

26. **High performing salespeople.** A firm having employee retention difficulties would like to identify its high performing and low performing salespeople so they can be more appropriately rewarded or retrained. Each salesperson in the company has their own territory, and the following data are available for each territory in 2009:

   S = sales (in units) of the firm's product in a given territory
   T = total sales (in units) of the firm's product plus its competitor's products in a given territory

A = dollars spent on advertising in the territory

N = number of years the salesperson in a given territory has been employed by the company

(a) The regression equation $S = 41,500 + 3.2A$ is computed. Give an interpretation for the coefficient 3.2, and explain what this says about how much an increase in sales might be expected if advertising were increased by $20,000 in a territory.

(b) A second multiple regression equation $S = 30,100 + 0.14T + 1.8A$ is computed. What could explain why the coefficient associated with advertising is lower here than in the first equation?

(c) A third equation $S = -18,952 + 0.12T + 1.2A + 380N$ is computed. For this model, $R^2 = .82$ and standard error $= 2,100$. Give the appropriate managerial interpretations of the coefficients 0.12, 1.2, and -18,952 and explain what $R^2 = .82$ means.

(d) In a particular territory, $T = 10,500$, $A = 12,000$ and $N = 12$. Last year, the salesperson in this territory sold 3,200 units. Would you conclude this salesperson is doing better than expected, about as expected or worse than expect. Briefly explain.

(e) You are considering adding the following new independent variable to the regression equation in Part (d):

C = the change in the firm's market share in the territory from 4 years ago.

Suppose after adding this new variable you see that its coefficient is positive. If you were a successful salesperson and your firm's market share in your territory had been increasing over the last 4 years, would you want this variable included in the model? Briefly explain.

27. **Business school rankings.** The file USnews 2009.xls shows the data from the 2009 *US News and World Report* business school rankings. One of the schools does not have a fair rank given its data. Can you tell which school it was and what a more fair rank for this school should have been? And why do you think this may have occurred?

28. **Hotel profit margins.** The file hoteldata.xls (an excerpt is in Table 7.4) contains data from each hotel which is part of a franchise. The hotel's current profit margin (in percent) as well as other data about

Table 7.4. An excerpt from the file `hoteldata.xls`.

Hotel #	Profit Margin	Access	Age	College
1	0.58%	2	7	386
2	13.63 %	3	13	601
3	10.53 %	1	7	331
4	6.92 %	1	12	1813
5	0%	4	3	1854
6	9.49%	3	15	3896
7	19.09%	0	11	1971
8	19.22 %	2	13	95
9	11.33 %	2	13	130
10	18.29 %	4	12	108

the hotel and the surrounding area is included. On the second page of the file is a description of the data variables and data on two potential new hotel sites under consideration: Site A and Site B.

(a) Using just the variables "College," "Price," and "State," create a multiple regression equation for forecasting profit margin. How accurate are forecasts expected to be using this equation?

(b) By how much will forecasts improve if we also use the information represented by the variable "Access"?

(c) Using the equation from Part (b), which one of the two sites do you forecast would have a higher profit margin?

29. **Retail sales.** The file `store.xls` (an excerpt is in Table 7.5) contains sales data for a given product from many months in a store, as well as the number of feet of shelf space devoted to the product, the retail price, advertising expenditures and whether or not a coupon was available that month.

(a) Create a multiple regression equation for forecasting sales using only the variables "Shelf space," and "Coupons."

(b) Based on the equation from Part (a), what impact on sales does the coupon seem to have?

30. **What makes a car beautiful?** An automobile manufacturer would like to know which features of an SUV make it most aesthetically appealing to consumers. They use detailed photos of 39 SUV models currently on the market and ask consumers to rate the overall aesthetic appeal of

Table 7.5. An excerpt from the file `store.xls`.

Month	Sales	Advertising	Shelf Space	Unit Price
1	58	$1,770	5	$4.90
2	409	$3,090	7	$4.73
3	258	$2,850	5	$5.16
4	384	$4,470	10	$4.77
5	225	$3,980	7	$5.37
6	401	$4,720	12	$5.37
7	593	$5,070	16	$5.78
8	547	$4,040	11	$5.09
9	241	$5,060	9	$4.75
10	71	$2,970	6	$5.33

Table 7.6. An excerpt from the file `suv.xls`.

Suv Model	Front	Console	Seats	Color	Rear	Size	Overall
1	5	4.5	4.8	3.6	3	5.1	5
2	4.9	2.9	3	4.2	5.2	3.9	3.8
3	5.1	6.8	4.5	4.9	4.5	6	5.3
4	4.8	5.2	3.5	4.3	4.1	4.5	4.2
5	5.1	3.1	6.3	6	6.6	4	5.6
6	4.1	3.9	6.1	7.4	5.7	6.6	4.5
7	3.9	3.3	5.7	6.4	5.2	6	6.2
8	2	3.5	4.9	3.6	3	2.9	2.6
9	3.1	7	5.8	4.8	3.4	3	3.9
10	5.9	3	4.5	5.1	4	4	3.7

each SUV on a scale from one to ten. They also ask each consumer to separately rate the aesthetic appeal of several specific features of each of the SUVs: the center console, the seats, the front view, the rear view, the color and the size. The file `suv.xls` (an excerpt is in Table 7.6) contains the average consumer ratings for each SUV model.

(a) Create a multiple regression equation for forecasting overall appeal using only the variables "Front," "Console," and "Rear."

(b) Using the equation from Part (a), give some advice to the product designers on the relative importance of these factors from Part (a) to the overall aesthetic appeal of an SUV to consumers.

Table 7.7. An excerpt from the file `customerdata.xls`.

Customer #	Purchases	Distance (miles)	Gender	Income
1	$3,681	7.2	F	$83,931
2	$9,950	15.4	F	$127,998
3	$6,091	2.6	F	$72,127
4	$8,834	16.6	M	$81,437
5	$404	2	F	$88,913
6	$10,090	16	F	$135,822
7	$9,260	20.2	M	$126,091
8	$7,842	9.4	M	$155,241
9	$2,372	9	M	$91,819
10	$3,357	13.2	F	$48,984

31. **Purchasing behavior.** The file `customerdata.xls` (an excerpt is in Table 7.7) contains information on a firm's current customers along with the total amount of purchases each customer has made.

    (a) Create a multiple regression equation for forecasting purchases using only the variables "Distance" and "Gender."

    (b) Based on the equation from Part (a), describe the type of customer who spends a lot.

32. **Fighting employee flight.** A large bank would like to know which factors affect employee performance.[9] The file `bank.xls` (an excerpt is in Table 7.8) shows data on employees and their performance evaluation ratings. Create a multiple regression equation to predict performance using only the variables "Years on the job," "Hours worked per week" and "Base pay." Interpret the coefficients.

33. **Customer satisfaction.** A manufacturer conducts a survey of the many client firms that purchase its products. The client firms are asked to rate the quality of various dimensions of the manufacturer (such as delivery speed, its website, its sales force, etc.) on a scale from 1 to 7 where 1 corresponds to poor and seven corresponds to excellent. There are

---

[9]Based on "How Fleet Bank Fought Employee Flight," by H. R. Nalbantian and A. Szostak, *Harvard Business Review*, April 2004, pages 116–126. In this article they discuss how a technique called "logistic regression" was used to identify factors affecting employee turnover. In this question we modify the example to fit the setting of multiple linear regression.

Table 7.8. An excerpt from the file `bank.xls`.

Employee #	Performance eval (100=best, 0=worst)	Years on the job	Hours worked per week	Base pay
1	39	6	78	44000
2	8	4	45	75600
3	18	7	68	68900
4	52	2	67	13900
5	3	3	76	18700
6	35	5	52	58600
7	45	4	60	50900
8	47	4	29	20300
9	19	6	18	25200
10	95	7	98	42900

also questions about overall satisfaction, the likelihood of continuing to purchase from this manufacturer, as well as other questions about the client firm. The results of the survey are in the file `customersurvey.xls`. Create a multiple regression equation to predict overall satisfaction using only the variables "Website Quality of product," "Ordering and billing," and "Salesforce Technical support." Which of these factors seem to be the most important driver of satisfaction?

34. **Cleaning products.** Market researchers are trying to determine what determines consumer satisfaction with household cleaning products. A large number of people are asked in a survey to think of a household cleaning product they frequently use and then rate the reputation of the brand, the ease of use, the price, the scent, as well as overall satisfaction the product. The data is in the file `cleaningsurvey.xls` and responses are on a scale from 1 to 5, where 5 corresponds to excellent and 1 corresponds to poor (an excerpt of the data is shown in Table 12.6). Create a multiple regression equation to predict satisfaction using only the variables "Reputation of brand," "Ease of use," and "Price." Interpret the coefficients.

35. **Evaluating drivers.** A delivery company would like to know which of its drivers are the fastest and which are the slowest. Also, it would like an easy way to forecast delivery trip times from the trip mileage, the number of delivery stops, the number of containers delivered and

Table 7.9. From the file `cleaningsurvey.xls`.

Reputation of brand	Ease of use	Price	Scent	Overall Satisfaction
4	4	4	3	4
4	4	3	3	4
3	4	2	2	3
2	2	2	2	3
2	3	3	2	3
4	4	3	3	3
4	4	3	3	4
4	4	3	3	4
4	4	3	4	4
3	3	3	3	3

the number of containers picked up. Historical data from two months, including the initials of the drivers, are in the file `delivery.xls`. Build a multiple regression equation to forecast the trip time using only the trip mileage and the total number of stops.

(a) What would this equation gives as a forecast for a 300 mile trip containing five stops?

(b) Approximately how accurate do you expect the forecast from Part (a) to be? Explain briefly.

36. **Cloud computing.** An international management consulting firm would like to know what type of people see promise in the potential business benefits of cloud computing. The file `cloud.xls` (an excerpt is in Table 7.10) contains survey responses from a large number of employees in China, Germany, Brazil and the United States. Among other things, the level of technical expertise of the employee was asked (on a scale from 1 to 9, where 9 represents the greatest technical expertise) as well as the employees opinion on the level of future cloud computing use at their company (on a scale from 1 to 9, where 9 represents the highest level of future use). Although there are differences in the average response across the different countries, there are also differences in the level of technical expertise of the respondents. Use multiple regression to uncover the differences in views of cloud computing between countries after taking into account differences in the technical expertise of the respondent.

Table 7.10. From the file `cloud.xls`.

Respondent	Country	Technical expertise	Plans for cloud computing
1	Germany	7	9
2	US	1	1
3	China	6	2
4	Brazil	6	3
5	Germany	6	7
6	Germany	8	3
7	China	5	2
8	China	2	2
9	US	3	1
10	China	3	6

37. **Catalog sales.** A large retailer sends out a special catalog of seasonal merchandise and gathers data on the purchase behavior of customers who make at least one purchase from the catalog. For these people, an analyst creates the multiple regression equation $Y = 105 + 0.0007X_1 + 0.21X_2 - 12X_3$ to predict Y, how much a customer purchases (in dollars) from the special catalog, using their income in dollars (X_1), the total amount spent for items from all catalogs over the past five years in dollars (X_2), and the total number of orders made over the past five years (X_3). The R^2 for the equation is 71%.

   (a) Give the managerial interpretation for the coefficient 0.21.

   (b) For a similar type of customer with an income of $52,000 who spent $1,200 over the last 5 years and made 5 orders, how much would you predict the person ordered from the special catalog?

   (c) Give a managerial explanation for what the R^2 value tells us about these customers.

   (d) Since the coefficient 0.21 is so much higher than the coefficient 0.0007, can we conclude that the total amount spent is more important than income in predicting spending from the special catalog?

   (e) Notice that the coefficient for the number of orders is negative. Can you explain how this could be true?

   (f) Can you quantify the uncertainty associated with regression predictions from the information given? Explain briefly.

38. **Hurricane damages.** In August, 2011, hurricane Irene threatened to

Table 7.11. From the file `hurricane.xls`.

Wind Speed (mph)	Distance to NYC (miles)	Damages (millions of $)
65	60	13,040
80	65	1,023
45	85	1
65	60	300
90	105	648
120	40	45,301
85	60	15,292
100	70	18,662
95	135	3,499
105	65	34,261

make a direct hit on New York City with wind speeds up 100 mph. A newspaper article estimates the potential damage such a hurricane could do. The article gives data on other hurricanes: how close they came to New York City (in miles), the wind speed (in miles per hour) and the total damages (in millions of dollars); an excerpt is shown in Figure 7.11. Build a multiple regression model to estimate the potential damage that could be caused. Give an estimate of the accuracy of this forecasting method and discuss briefly.[10]

---

[10] "A New York Hurricane Could Be a Multibillion-Dollar Catastrophe," by Nate Silver, *The New York Times*, August 26, 2011.

Fun Problem: Two Sided Cards

There are three cards on the table. The first card has a letter "A" written on both sides of the card. The second card has a letter "B" written on both sides of the card. The third card has a letter "A" on one side of the card and has a letter "B" on the other side.

Suppose someone shuffles these cards, picks a random card, and shows you the letter written on a random side of the card. Suppose this letter turns out to be "A." What should you guess for the other side of the card? Or does it make no difference what you guess?

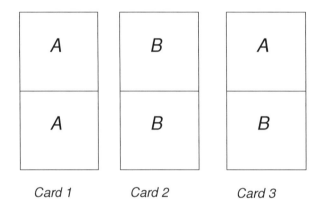

Card 1   Card 2   Card 3

Answer. It makes a big difference and you should guess "A" for the other side. On these three cards there are obviously six equally likely sides you could have been shown. And of the three sides with the letter "A" on them, two of them are on Card #1. This means you are more likely to have seen Card #1 and so you should guess "A" for the other side.

Chapter 8

Probability

1. Introduction

In 2003, Pepsi ran an unusual marketing promotion in which contestants were given a chance to win a whopping prize of one billion dollars. This prize was so large that Pepsi purchased special insurance just in case somebody actually won. Berkshire Hathaway, the company run by Warren Buffett (the world's second richest man, who recently decided to give most of his money to charity[1]), agreed to pay the prize money in exchange for an insurance premium purchased by Pepsi that was reported in the article to have cost between $1 million and $10 million.[2] Is this a reasonable price for such insurance?

In order to begin to answer this question, we must be able to calculate the likelihood of someone actually winning the prize. An important element of success in the insurance business is being able to accurately assess the likelihood of complex unpredictable events. Berkshire Hathaway makes a great deal of money by insuring automobiles and unusual items such as

---

[1]CNN.com front page story, Monday, June 26, 2006, and http://www.forbes.com/billionaires/

[2]G. Anderson, "Pepsi's billion-dollar monkey: A new contest offers a chance to win $1B. But first, a simian must pick the magic number," April 10, 2003, 1:01 PM EDT, CNN/Money. See http://money.cnn.com/2003/04/09/news/companies/pepsi_billion_game/index.htm. See also http://www.dartmouth.edu/~chance/chance_news/recent_news/chance_news_12.04.html#item3

dancers' legs and hurricanes; contestants on "don't try this at home"-type reality television shows have been insured through the famous insurance marketplace, Lloyd's of London.[3] In this chapter we will introduce some of the basic rules for calculating and combining probabilities and show you how to recognize when things get so complicated that you need to seek the advice of a statistician.

Suppose a given situation can cause some event A to happen. The **probability** of event A happening, denoted by $P(A)$, tells us the fraction of time A is expected to happen in the long run if the situation could be repeated over and over. Sometimes we also refer to this as the **chance** of A happening. If $P(A)$ equals 0, it means A is expected to happen zero percent of the time and if $P(A)$ equals 1 it means A is expected to happen one hundred percent of the time. Sometimes a probability is written as a percentage. For example, if there is a 60% chance of rain, we can say the probability of rain is 0.6 and we can write this as $P(\text{rain}) = 0.6$. For accurate weather forecasting we should expect rain on about 60% of the days when forecasters said there was a 60% chance of rain. Probabilities must be between 0 and 1, because something can't happen more than 100% or less than 0% of the time.

Probabilities are between 0 and 1:

$$0 \leq P(A) \leq 1$$

A useful property is that the probability of something happening equals one minus the probability that it doesn't happen. If the chance of rain is 60%, this means the chance of no rain is 40%. This is called the **complement rule** and two events are called **complementary** if one happens whenever the other one doesn't happen. For example, rain and no rain are complementary events. We summarize this rule as follows:

---

[3]Seehttp://www.lloyds.com/News_Centre/Features_from_Lloyds/Dont_try_this_at_home.htm

> The **complement rule**. The probability some event
> A happens equals 1 minus the probability it doesn't
> happen:
>
> $$P(A) = 1 - P(A \text{ doesn't happen})$$

Accidents.   A construction company notices that 98% of the time its projects are completed without a single accident occurring. What's the chance a project has at least one accident?

Solution.   We can compute this using the complement rule ($100\% - 98\% = 2\%$) since a project must either have no accidents or at least one accident.

And what's the chance a project has more than one accident? We cannot answer this question using the information given because the complement of having more than one accident is having either zero or one accidents. Since we don't know these probabilities, this time the complement rule doesn't help us. We lack the information to answer this question.

Exercises

1. If a stock market index goes up 70% of the time, what percent of the time does it go down?

2. An insurance company analyst sees that on 15% of days they receive at least two accident claims in the mail. Does this mean that on $100\% - 15\% = 85\%$ of the days it receives less than one accident claim? Explain.

3. Which do you think would be better news to hear: that someone calculated you have a 90% chance of making a profit, or that someone calculated you have a 110% chance of making a profit? Explain.

2. The Multiplication Rule

People are often tempted to multiply probabilities together in order to combine them. This makes sense in some situations, but not in others. In this section I will illustrate when it's appropriate to do this.

Reliability testing.   When measuring the quality of a product coming out of a manufacturing process, it is very often impossible or undesirable to test the reliability of every single unit before you sell it. With automobile crash-testing, for example, nobody would ever pay full price for a car after it was tested in a crash. The same goes for testing the lifetime of a light bulb: after it burns out and you know its lifetime, you can't sell it. Quality control inspectors usually take a small sample out of a batch of products for testing and hope that the rest of the batch is similar. For this reason quality analysts must assess the chance a sample is representative of the entire batch; we will consider the basics of this principle here.

Suppose a box contains two defective components and two good components. If quality control inspectors randomly draw two components from the box for testing, how likely is it that they will happen to get both of the good components?

Solution.   At first glance you might think the answer is 50%, but this is not correct. The real answer is much lower than 50%. Let's illustrate the solution method by starting with a diagram of the box:

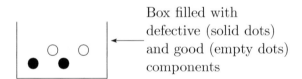

Box filled with defective (solid dots) and good (empty dots) components

The chance that the first component is good is 2/4, or 1/2. After this happens, then you are left with a box with 1 good component and 2 defective components. At this point the chance that the second draw is a good component is now 1/3 – this is shown in Figure 8.1. So if we repeated this whole process many times, half the time the first draw would be a good component and one third of that half of time the second draw would also be good component, so the answer is 1/3 of $1/2 = 1/3 \times 1/2 = 1/6$.

This is an illustration of the **multiplication rule** of probabilities. It says that to find the chance two events both happen, multiply the chance the first one happens by the chance the second one happens given that the first one previously happened.

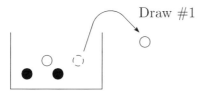 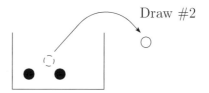

The chance of first drawing a good component is 2/4,... ...and then the chance of drawing another good component is 1/3.

Figure 8.1. Two draws from a box containing two good components and two defective components.

The **multiplication rule**: the chance two events A and B both happen equals the chance A happens multiplied by the chance B happens given we know A happened:

$$P(A \text{ and } B \text{ both happen}) = P(A) \times P(B \text{ given } A \text{ happened})$$

The probability $P(B$ given A happened) is usually written as "$P(B \mid A)$" and is read the "**conditional probability**" of B given A happened." The multiplication rule also applies to any number of events. For example in the case of four events A, B, C, and D we get the following:

$$P(A \text{ and } B \text{ and } C \text{ and } D \text{ happen})$$
$$= P(A) \times P(B \mid A) \times P(C \mid A \text{ and } B) \times P(D \mid A \text{ and } B \text{ and } C)$$

Insurance.   The profitability of an insurance company depends on its ability to accurately assess risk and set its rates according to the risk. Here we consider a simple example.

A fire insurance company insures a building that they estimate would have a 70% chance of being completely destroyed in a fire. But the company also estimates the building has only a 2% chance of experiencing a fire over the course of a year. How would you accurately estimate the chance the building will be completely destroyed by fire within a year?

Solution.   Here we can apply the multiplication rule as follows to get a 1.4% chance:

$$P(\text{fire and building completely destroyed})$$
$$= P(\text{fire}) \times P(\text{building completely destroyed} \mid \text{fire})$$
$$= .02 \times .7 = .014, \text{ or a } 1.4\% \text{ chance}$$

Warehouse operations management.   Efficient operations management is increasingly being recognized by management as a competitive edge and much of the performance of a system depends on the amount of uncertainty it faces. Analysts compute probabilities to help understand the link between demand patterns and operations performance. We consider a simple example here.

Suppose an analyst is studying the efficiency of an inventory warehouse that has five floors filled with items. Experience shows that items ordered by customers are all equally likely to be located on any of the five floors. If three items are ordered, (a) what is the chance they are all located on the top floor? (b) What is the chance the items are all located on different floors?

Solution.   (a) The chance that the first item is on the top floor is 1/5 and, given that the first item is on the top floor, the chance that the second item is also located on the top floor is 1/5. The same goes for the third item. This means the chance that all three are on the top floor equals $1/5 \times 1/5 \times 1/5 = 1/125 = .008$. (b) In this question we would like all the items to be located on different floors. This means that the first item can be on any one of the five floors, and this has a 5/5 or 100% chance of happening. Given that the first item is on a particular floor, the chance that the second item is on a different floor is 4/5, since there are four remaining floors at this point. Finally, given that the first two items are on two different floors, the chance that the third item is located on one of the three remaining floors is 3/5. This means that the chance all the items are on different floors equals $5/5 \times 4/5 \times 3/5 = 12/25 = .48$, or 48%.

Powerball.   Many states earn revenue by holding lotteries. The key to designing a profitable lottery is making sure the money you take in is more than the prizes you pay out. Understanding the chances of winning is important in forecasting the profit from such lotteries.

Powerball is a popular lottery game, with the jackpot occasionally reaching over $200 million. In addition to the jackpot there are many smaller prizes awarded and the lottery can usually accurately forecast the total amount of the prizes it will have to pay out for each drawing. But on March 30, 2005 there was a surprisingly large number of people who won smaller prizes and the lottery ended up having to pay out $20 million more than it had planned.[4] Lottery officials worried that there may have been fraud involved and spent time investigating.

Here is how Powerball works: on your lottery ticket you pick five different "white" numbers from 1 to 53 and a "red" number from 1 to 42. On the day of the drawing, five different white balls are pulled out of a drum containing 53 numbered white balls, and one red ball is pulled out of a drum containing 42 numbered red balls. You win a prize by matching all five white balls in any order and win the jackpot if you also match the red ball.

Approximately 130 people guessed all five white numbers correctly and each won between $100,000 and $500,000 for this (the amounts varied because there are other ways to win additional prizes). With around 100 million tickets sold, how many are expected to have all five white numbers correct?

Solution.   Let's start by calculating the chance someone will guess all five white numbers correctly. Suppose a person first chooses their five white numbers. To find the chance those same numbers are then drawn by the lottery, imagine that the 5 balls in the drum for these numbers are labeled as "good" and the remaining 48 balls are labeled as "bad." As in the reliability problem above, we are now interested in the chance of drawing all five "good" balls from a box containing 5 "good" and 48 "bad" balls. We can now use the multiplication rule the same way as before. The chance the first ball drawn is "good" is 5/53, and then we have 4 remaining "good" balls remaining out of 52 balls total. After drawing the second "good" ball we have three remaining out of 51 balls total. Continuing on and applying the multiplication rule, we

---

[4] "Who needs Giacomo? Bet on the fortune cookie," by Jennifer Lee, *New York Times*, May 11, 2005, National desk, Page 1.

get a chance of

$$\frac{5}{53} \times \frac{4}{52} \times \frac{3}{51} \times \frac{2}{50} \times \frac{1}{49} = \frac{1}{2,869,685}$$

This means that 1 out of every 2,869,685 tickets should win, and so out of 100 million tickets we expect around 100,000,000/2,869,685 = 35 winners.

So on March 30, 2005, Powerball officials expected 35 winners but 130 winners came forward. What could explain this big difference? It turns out that a fortune cookie company, Wonton Foods, had been printing the same five winning numbers (22, 28, 32, 33, and 39) as suggested lottery numbers in millions of fortune cookies over a long period of time, and people used those numbers when filling out their tickets. After this incident, Wonton Foods agreed to change the numbers in their cookies more often.

The jackpot. In August 2005, Powerball lottery officials increased the number of white balls from 53 to 55 to reduce the number of jackpot winners.[5] The jackpot prize rolls over to the next drawing if there is no winner and huge jackpots are well-publicized in the media and serve as very effective free advertising. By how much did this change reduce the chances of winning the jackpot?

Solution.   To win the jackpot you must match all five white numbers and then the red number. Using the multiplication rule gives us a chance of

$$\frac{5}{53} \times \frac{4}{52} \times \frac{3}{51} \times \frac{2}{50} \times \frac{1}{49} \times \frac{1}{42} = \frac{120}{14,463,212,400} = \frac{1}{120,526,770}$$

when there were 53 white balls being used. With 55 white balls being used we get a chance of

$$\frac{5}{55} \times \frac{4}{54} \times \frac{3}{53} \times \frac{2}{52} \times \frac{1}{51} \times \frac{1}{42} = \frac{120}{17,532,955,440} = \frac{1}{146,107,962}$$

Since the ratio of the chances is 120,526,770/146,107,962 ≈ .8, this means the chance of winning the jackpot was reduced by almost 20% after 2005.

Question.   Powerball numbers are drawn twice a week. If you played Powerball twice a week for 30 years, assuming there are 53 white balls, what is the chance you never win the jackpot?

---

[5]http://www.powerball.com/pb_history.asp

Solution.   Since your chance of winning on a single game would be

$$\frac{1}{120{,}526{,}770} \approx .000000008$$

the chance you lose on a single game is, using the complement rule, $1 - .000000008 = .999999992$. Twice a week for 30 years is $52 \times 2 \times 30 = 3{,}120$ games, so the multiplication rule tells us that the chance you lose all of them equals $.99999992$ multiplied by itself again and again 3,120 times, or $(.999999992)^{3,120} = .999974$, or a 99.9974% chance. It's almost guaranteed you will never win in your lifetime,. so you should definitely make other retirement plans.

Conditional versus unconditional probability

It is important to note that the "conditional" probability $P(B \mid A)$, the probability of B given that you know A already happened, is not the same as the "unconditional" probability $P(B)$, the probability of B when we don't know whether or not A previously happened. The multiplication rule uses the "conditional" probability $P(B \mid A)$.

With the box of components considered in the first example above, it is easy to see that the conditional probability

$$P(\text{second component good} \mid \text{first component good}) = 1/3,$$

but, believe it or not, if we are not allowed to see the first component drawn,

$$P(\text{second component good}) = 2/4 = 1/2.$$

Here's why these two solutions don't contradict each other. Suppose someone else is in charge of making the first draw and instead of just drawing a single component, she secretly removes three of the four initial components at random. You are then allowed to make the second draw and take the remaining component. It is easy to see that you still have an equal chance of ending up with any of the four initial components on the second draw, so the chance of choosing a good component on this draw is $2/4$, even though there is only one component remaining in the box. When you draw components out of a box and don't test them, the chance any given component is good is the same whether it's the first component or the last component you draw.

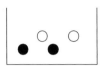

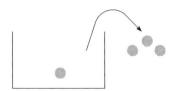

Starting with this box
of components,...

......if someone draws out
three without testing them,
the chance that the remaining
one is good is 2/4.

Scratch lottery controversy. Confusing conditional probabilities with unconditional probabilities also has created controversy among lottery customers and public-relations difficulties for government officials. Here's an example

In a scratch ticket lottery, you scratch the ticket to reveal whether or not you've won a prize. Suppose there are 1,000 prize-winning tickets out of a total of 100,000 tickets. Clearly, if you buy a single ticket your chances of winning are $1000/100,000 = 1/100$, or a 1% chance. Now suppose before you buy your ticket someone steals a box containing 20,000 random tickets and they are all lost. If you now buy one of the remaining 80,000 tickets, what is your chance of winning a prize?

Solution. Your chances of winning are still 1%. Imagine if someone stole all the tickets except a single ticket. That last remaining ticket would have the same chance of being a prize-winning ticket as any other ticket—a 1% chance.

There was a recent controversy surrounding scratch tickets for the Massachusetts State Lottery.[6] The state misplaced a large number of lottery tickets and people believed that this would affect their chances of winning. A lottery spokesperson had to explain on television to the general public that the probability of winning is not affected when random lottery tickets are lost. The spokesperson was not so convincing and there were many complaints and protests. In the end, the lost tickets were found.

---

[6] "Case of Lost Lottery Tickets is Solved, They Weren't Lost After All," by John McElhenny, Associated Press, August 16, 2001.

Exercises

4. **Quality inspection.** A quality inspector selects five random components, removing them one at a time from a box containing five defective components and twenty working components. (a) What is the chance of selecting all the defective components? (b) What is the chance none selected are defective? (c) What is the chance the fourth component selected is defective, given that the first three components selected have not yet been tested?

5. **Lottery chances.** In a lottery you choose three different numbers out of a list of twenty possible numbers. The lottery then selects three random numbers to be the winners. (a) What is the chance all your numbers are selected? (b) What is the chance at least one of your numbers is selected?

6. **Health insurance.** A health insurer notices that 70% of its customers have no health expenses in a year. In a pool of five random patients, what's the chance none of them have any health expenses in a year?

3. Independent Events

When the chance of an event changes after you learn some other event has occurred, the events are called **dependent**. If the chance of one event doesn't change regardless of what happens with the other event, these are called **independent**. For example, earthquake insurance claims from different houses in a city would be dependent events, since the whole city is influenced by the same earthquake. In contrast, fire insurance claims from different houses in a city are more likely to be independent events, since most fires are isolated incidents (unless of course a wildfire engulfs the whole city, which has happened before in California). Insurance companies generally prefer to have independence between claims so they won't be suddenly swamped by a large number of claims at the same time. And this, incidentally, is one of the reasons earthquake insurance is more risky than fire insurance for insurance companies to offer.

> Two events are **independent** if the chance for one of them occurring is not affected by knowing whether or not the other one occurred.

The multiplication rule says that to find the probability several events all happen, you take the probability of the first event multiplied by the conditional probability of the second event given the first happened multiplied by the conditional probability of the third event given the first two events happened, and continuing on like this with the rest of the events. However, with independent events the conditional probabilities are the same as the unconditional probabilities, so the probability all the events happen will be equal to the product of their individual probabilities. This applies to any number of events, but we summarize it for two events as follows:

> If A, B are independent events, then
>
> $$P(A \text{ and } B \text{ happens}) = P(A) \times P(B)$$

Insurance.   An insurance company insures three offshore oil platforms in different parts of the country against severe weather damage. They estimate each platform has a 10% chance of being damaged. (a) What's the chance that all three platforms will be damaged? (b) How would the answer to (a) change if the three platforms are located near each other?

Solution.   (a) If the platforms are in different parts of the country, it seems reasonable that damage would be an independent event: damage at one of the platforms would not increase the chance of damage at another platform. This means the chance of damage at all three would be $0.1 \times 0.1 \times 0.1 = .001$, or a .1% chance.

The answer to (b) would be higher than the answer in (a). When the platforms are near each other, they all are influenced by the same weather so the events would be dependent. The probability the first platform is damaged will still be 0.1, but the probability the second platform is damaged given the first platform is damaged would be larger than 0.1—since if we know the first platform is damaged it tells us there was bad weather in the area.

The same reasoning applies to the third platform, so multiplying these all together would then give us a higher probability of damage.

DNA crime evidence.   Prosecutors often use DNA evidence during a trial to link a defendant to the crime scene and the strength of the evidence is usually presented in terms of probability. Investigators in a laboratory can examine several locations along the DNA strands gathered from both the crime scene and from the defendant, and can determine if they both have the same genes in the same locations. Then they try to estimate the chance that a random person would match in the same way.

For example, suppose the defendant's DNA has Gene A in Location 1, Gene B in Location 2, and Gene C in Location 3. Also, suppose that DNA from the crime scene has these same genes in these same locations. This looks pretty bad so far for the defendant, but how strong is the evidence against him? Suppose in the general population 4% of the people have Gene A in Location 1, 5% have Gene B in Location 2, and 6% have Gene C in Location 3. Investigators typically multiply these to estimate the probability a random person matches all three positions: $.04 \times .06 \times .07 = .00012$, or about 1 out of 8000. In legal jargon, multiplying the numbers together is called the "product rule." This 1 out of 8000 chance is usually presented to the jury as the chance that any given random person, other than the defendant, matches the crime scene's DNA. Does the "product rule" make sense? What assumption is behind this rule?

Solution.   Multiplying the individual probabilities together only makes sense if we assume that the appearance of genes in one location is completely independent of the appearance of genes in another location. In other words, knowing that a person has one gene in one location doesn't change the chances that he or she also has another gene in another location. This is a controversial assumption about human genetics and numerous convictions have been appealed based on criticisms of this assumption. This assumption does not hold for many other characteristics of people. For example, people with black hair are more likely to have brown eyes than are people with blond hair. The moral is to be cautious when multiplying probabilities for events that are not completely independent. Also, if you are about to be convicted for a crime you didn't commit based on DNA evidence, make sure you mention this to the jury.

> If you have several probabilities and would like to combine them into a final answer, don't just blindly multiply them—that is often the wrong approach.

The random walk model. It's widely recognized that the direction the stock market moves often appears to be completely random and independent from day to day. This is usually described as the **random walk model**, and this model is used by most options traders on Wall Street and is studied by Nobel prize-winning economists.[7] The simplest random walk model assumes that during any given day a stock's price has a 50% chance of going up and a 50% chance of going down, regardless of what happened in the past. Under this random walk model, (a) what is the chance a stock's price goes up the next two days in a row? (b) What is the chance it goes in the same direction (either up or down) for two days in a row, and then changes direction on the third day? (c) What is the chance it goes in the same direction for three days, and then changes direction on the fourth day?

Solution. (a) Because the directions each day are assumed to be independent, we can use the multiplication rule to get $.5 \times .5 = .25$, or a 25% chance that the stock market will go up two days in a row. (b) This time we don't care which direction it goes on the first day, so the chance of success on this day is 100%. But on the second day it must go in the same direction, which has a 50% chance of happening. Then on the third day it must go in the opposite direction and this also has a 50% chance of happening. We then can use the multiplication rule to get $1 \times .5 \times .5 = .25$, or a 25% chance. (c) This time we get $1 \times .5 \times .5 \times .5 = .125$, or a 12.5% chance. Notice that the chance here is half the chance in (b), and the answer would be cut in half again if we wanted to see yet another day in the same direction.

Does the stock market follow the random walk model? Below are three sequences of letters and each represents the period from January 3, 2005 to June 16, 2006. Each letter represents a single day of the stock market in chronological order from left to right: a "U" means the stock market went up that day (as measured by the S&P 500 stock market index), and a "D" means it went down that day. The upper-left letter corresponds to January

---

[7] http://nobelprize.org/economics/laureates/1997/

3, 2005 and the lower-right letter corresponds to June 16, 2006. The twist is that only one of these sequences is the real stock market and the other two are completely faked. Can you tell which one is the stock market index and which two are the fakes?

Sequence A

UDDUDUDUDUUDDDDUUUDUUUDUDUDUUUUUDUDUUUDUDUUUDDUDUDDDUDDDUDUDUDD
UUUUDUUDDDUUDUDUDUDUUDUDDUDUDDUUUUDUUDUUDUUDUDDUDUUUUUDDUDDDUD
DUUDUUUUUUUDUUDUDUUUDUUUDDDUDUDUDUDUUDDUDUDUUDUUDUDDDUUDDDUUUU
UUUDDDDUDDDDDUUDUDUUDDDUUDUUUUDUUUDDUUUUUUUUDUDUUDUDDUUUUDDDDUUUU
DUDDUUUUUDUDUDDUDUUDUUUUDUDDUDUDUDUUUDDUDUUDUDDDDUDUUUUUUDDUDUD
DUDDUUUDDUDUUDUUUDDDUUUDUDUUDUDDDUDDDUDDUUUDUUUDDDUDDDUU

Sequence B

UDUUDDUUDUUDUDDDDUUDUUUDUUDUUDUUUDUDDUDUUDUDUUDDDUDDDUUDUUDUDD
UDUDDUDDUUDDDUDDUUDUDDUDUDUUDDUUDUDUDUUUDUDUDDUDUDUDDUDUUDUUD
UUDUUDDUDUDUDDUUDUUDDUUDDUDDUUDDUUDUUUDUDDDUDDUDDUUDUDUDDUDDDU
UDUUDUUDUDUUDUDDUDUDUDDUDDDUDUDUUDUDUDUDDUUDUUDDUDDUDUDUDUUUUD
UUDDUUDDUDDUUUDDUDUUDUUDDUUUUDUUUDDUDUDDUUDUUDDUUDUDDUUDUUDUU
DUUDDUUDUDDDUDDUDUUDUUDUDUUDUDUUDDDDUUDDUUDDUUDDDUDUUDDUUU

Sequence C

UUUUUUUUUUUUUUUDUUDDUUUUUUUUUDDDDDUUUDDUUUDDDDDDDUUDDUUDDUUUU
UDUUUUDDDDDDUUUUUUUUDDUUUUUUUDDDUUUUDDDDDDDDDDDDDDDDUUUDDDDU
UDUUUUUUUUUUUUUUUUUUUUUUUUUUUUUUUUUUUUUUUUUUUUDDDUUUUUUUUUUUU
UUUUUUUUUUUUUUUDUUUUUUDDDDDDDDDDDUUUUUUUUUUUUDUUUUDDDUUUUUUUUU
UUUUUUUUUUUUUUUUUDDDDDUUUUUUDUUDDDDDDDDUUDDDDDUDDDUUDUUUUUUDDD
DDDDDDDDDDDDDUUUDDDDDUUUUDDDDDDDDUUDUDUUUUUUUUUUUUUUUUUD

Solution.   Well, each sequence looks as if it has about the same percentage of ups and downs, so this doesn't give us any clues. But it seems as though Sequence C has lots of ups and downs all together in a row. For example there is a run of 15 'U's at the very beginning of Sequence C. This looks somewhat suspicious. To investigate this, we will count the number of times the stock price goes in the same direction for two days in a row and then changes direction on the third day. This is called a 2-day run. We will also count the number of 3-day runs: the number of times the stock price goes in the same direction for three days in a row and then changes direction on the fourth day. When we do this separately for 2-day, 3-day, and all the way up to 7-day runs, and draw histograms for each sequence, we get the following graphs:

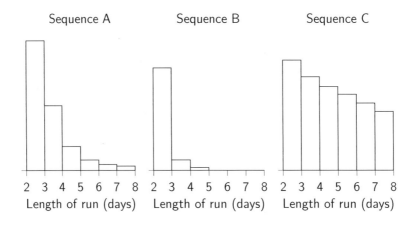

Which one is the real stock market? In Part (c) of the previous question we learned that, under the random walk model, the chance of a three-day run is half the chance of a two-day run. And the chance of a four-day run is half the chance of a three-day run, and so on. This means we should see half as many three-day runs as we do two-day runs, and half as many four-day runs as we do three-day runs. And this then means the bars should drop down by half each time as we go from left to right. The graph for Sequence A most closely fits this description:

Sequence A

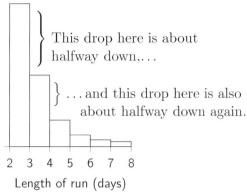

And, in fact, Sequence A is the real stock market—the other two are fakes! In Sequence B there are not enough long runs, and in Sequence C there are too many. So it looks as if the stock market actually is consistent with the random walk model.

The **efficient market hypothesis** of economics suggests that the random walk model is reasonable because any information from the past should already be incorporated into the stock's current price, and so future price movements should be caused by events independent of the past.[8] This hypothesis also suggests that you can't consistently make money in the stock market. Though this theory is accepted by many economists, it is very controversial and reportedly inspired the multibillionaire Warren Buffett to say "I'd be a bum on the street with a tin cup if the markets were always efficient."[9]

Incidentally, most people if asked to write down a random sequence of heads and tails tend to write down something similar to Sequence B, in which the runs are too short. In fact, a completely random walk would frequently have longer runs than most people think.

Exercises

7. Power grid. A city has two power generators that each break down about 2% of the time. But whenever one breaks down, it puts extra strain on

[8]B. Malkiel, "The Efficient Market Hypothesis and Its Critics," *The Journal of Economic Perspectives*, Volume 17, Number 1, March 1, 2003, pp. 59–82.

[9]http://www.global-investor.com/quote/2710/Warren-Buffett

the other and increases its chance of breaking down. (a) Can we estimate the fraction of time both of them are down as $.02 \times .02 = .0004$? If not, is $.0004$ an underestimate or overestimate? (b) Can we estimate the fraction of time at least one of them is down as $1 - .98 \times .98 = .0396$? If not, is $.0396$ an underestimate or overestimate?

4. The Addition Rule

It's sometimes tempting to add percentages or chances together in order to combine them, but this often does not make sense. In this section I will show you when this does make sense and will give you some other ways of calculating chances when adding does not make sense.

Market research.   A market research survey of consumer traveling habits finds that, in the last year, 60% of the people have taken a vacation trip, 70% have taken a work related trip and 10% have taken neither. Does this mean that $60\% + 70\% = 130\%$ have taken at least one of the two types of trips (vacation or work related) within the past year?

It's pretty easy to see that this can't be correct because it's impossible to have more than 100% of the people doing anything. Because many people have taken both a vacation and work related trip, simply adding up the percentages ends up counting these people twice—and therefore gives the wrong answer. To answer the question correctly, we can look at the complementary event: notice that 10% have taken neither type of trip and so $100\% - 10\% = 90\%$ must have taken at least one of the two types of trips.

We can illustrate this in a diagram. Below, the square represents 100% of the people and each oval represents a subgroup of people. To find the percentage in the shaded area (people who have taken at least one of the two types of trips), if you add the area of each oval you will double count the portion in which people have taken both types of trips:

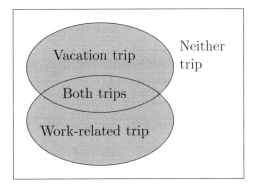

The moral here is that if adding percentages ends up counting something more than once, it doesn't make sense to add. In this situation it is often helpful to find the chance of the complementary event and subtract from one hundred percent.

> If you want to find the chance that at least one of several of events occurs and adding the percentages would double count some of the chance, instead try finding the chance of the complementary event and subtract from 100%.

Suppose the market research survey above also shows that 20% have not taken a vacation trip but are planning to take one later in the year. Does this mean that 60%+20% = 80% have either taken a vacation or are planning to take one later in the year? This time the percentages add up to less than 100%, but that doesn't prove it makes sense to add them together. Does it really make sense to add up the percentages this time?

This time adding them makes sense because we are not counting anyone twice: someone among the 60% who have taken a vacation can't also be among the 20% who have not taken one. The moral of the story here is that it's safe to add percentages if you don't end up double counting anything.

> It's safe to add percentages as long as nothing gets double counted.

The same rule applies to probabilities, and we will look at an example of this next.

Call centers. A company studying its call center efficiency finds there is a 40% chance an incoming call requires service from a live operator and otherwise it can be handled by an automated system. If two calls arrive in a row, does this mean there is a 40% + 40% = 80% chance at least one of them requires a live operator?

Answer. Here adding does not make sense because it's possible both calls require a live operator and so we would end up double counting this chance.

Let's instead look at the chance of the complementary event, in which neither call requires an operator, and subtract from 100%. There is a 60% chance that each call can be handled by the automated system, and so, if we assume the calls have independent requirements, this means that there is a $.6 \times .6 = .36$ or 36% chance that neither call requires an operator. This shows us that the chance at least one requires an operator is 100% − 36% = 64%. The assumption of independent call requirements is reasonable because the callers are different people.

The rules for adding probabilities are the same as those for adding percentages and we summarize these next. A commonly-used technical term related to the addition rule is that events are called **mutually exclusive** if not more than one of them can happen. For example, a heat wave and a snowstorm in the same place are generally mutually exclusive events, because they don't happen together. A heat wave and a rainstorm, on the other hand, are not mutually exclusive because they could happen together. A heat wave on one day and a snowstorm on another day are not mutually exclusive events, because they could both happen.

> Events are called **mutually exclusive** if not more than one of them can happen.

The addition rule applies to any pre-determined number of events, but we summarize it for two events as follows:

The **addition rule**: if events A, B are mutually exclusive, then

$$P(A \text{ or } B) = P(A) + P(B)$$

As with percentages, looking at the complementary event can be helpful if you have events that are not mutually exclusive:

If A, B are not mutually exclusive, try using

$$P(A \text{ or } B \text{ or both happen}) = 1 - P(\text{neither } A \text{ nor } B \text{ happen})$$

Fire insurance.   An insurance company insures 10 independent buildings against fire damage. It estimates a 5% chance of fire damage at each building. What's the chance at least one of the buildings has fire damage?

Solution.   The first thing to notice is that we can see damage at more than one building so the events are not mutually exclusive. This means we can't simply add up the chances. We should try looking at the complementary event, in which none of the buildings have damage, and then subtract it from 100%. Since the chance of damage is .05, the chance of no damage is .95. With the assumption all buildings are independent, which is reasonable for fire damage in separate buildings, the chance of no damage at any of the buildings is $.95 \times .95 \times \cdots \times .95 = (.95)^{10} \approx .6$, or a 60% chance. This means the chance of damage to at least one building is $100\% - 60\% = 40\%$.

Lottery with five prizes.   A lottery has a hundred tickets total and five of them have a prize. If you buy ten tickets, is the chance of getting at least one prize equal to $5\% + \cdots + 5\% = 50\%$? Explain why adding makes sense or explain why it does not make sense.

Solution.   Since it's possible to win with more than one ticket, the events are not mutually exclusive and adding the chances does not make sense. Let's instead compute the chance by looking at the complementary event, in which all 10 tickets are losers, and then subtract from 100%.

The chance the first ticket is a loser equals 95/100. Given the first ticket loses, the chance the second ticket also loses is 94/99, since there are 94 losing tickets remaining out of 99 tickets remaining. The chance of losing for the next ticket is then 93/98, and then 92/97, and finally 86/91 for the last ticket. Multiplying these together, the probability all the tickets are losers equals

$$\frac{95}{100} \times \frac{94}{99} \times \frac{93}{98} \times \frac{92}{97} \times \frac{91}{96} \times \frac{90}{95} \times \frac{89}{94} \times \frac{88}{93} \times \frac{87}{92} \times \frac{86}{91} \approx .58$$

This means the chance at least one ticket wins is $100\% - 58\% = 42\%$.

Lottery with one prize.   A lottery has a hundred tickets and one prize. If you buy 10 tickets, is the chance of getting at least one prize equal to $1\% + \cdots + 1\% = 10\%$? Explain why adding makes sense, or explain why it does not make sense.

Solution.   This time adding makes sense, since it's not possible to win with more than one ticket—the events are mutually exclusive and the addition rule applies.

Many lotteries.   Consider a weekly raffle having 100 tickets, one of which wins a prize each week. If you want to maximize your chance of winning something, is it better to buy 10 tickets in one raffle or one ticket in 10 different raffles? Or is it the same?

Solution.   With any single ticket the chance you win is 1/100, or a 1% chance. So if you have ten tickets in one raffle your chance of winning is $1\% + 1\% + \cdots + 1\% = 10\%$. You are allowed to add up these chances since you cannot win with more than one ticket at the same time—they are mutually exclusive.

When you have one ticket in each of 10 raffles, you are not allowed to just add the chances. Since it is possible to win with more than one ticket, they are not mutually exclusive. To compute this chance correctly, find the chance you don't win with any of the 10 tickets and then use the complement rule. The chance you lose with any single ticket is 99/100, and so if you play 10 raffles your chance of losing all of them is $.99^{10} = .904$. Your chance of winning at least once is then $1 - .904 = .096$, or a 9.6% chance. This is slightly lower than the 10% chance from before, so it's better to buy all your tickets in one raffle.

Airline safety violations. Suppose each time an airplane flies there is a 5% chance it experiences some minor safety violation. Does this mean that after 20 flights we can use the addition rule to show there's a $5\% + 5\% + \cdots + 5\% = 100\%$ chance of a minor safety violation?

Solution. No. Since it's possible to see a violation on more than one flight, the events are not mutually exclusive. This means we can't simply add the chances. Looking at the complementary event in which there are no violations and subtracting from one, we see the answer is $1 - (.95)^{20} = .64$, or a 64% chance (this is under the assumption that each flight outcome is independent).

Airline crashes. Suppose each time an airplane flies there is a 5% chance it crashes and is destroyed. Since you can't have a crash on more than one flight, does this mean we can now use the addition rule to show that after 20 flights there's a $5\% + 5\% + \cdots + 5\% = 100\%$ chance of a crash?

Solution. No, this is a tricky issue and I will try to explain it here. It's true that it's not possible to crash on more than one flight, but the addition rule can only be used if the total number of events is pre-determined. Here the actual number of flights is not determined in advance—we could have up to a total of 20 flights, but the actual number may be less if there is a crash. This means we can't apply the addition rule, and so to get the correct answer we must instead use the approach of the previous question to get the same answer: a 64% chance. The moral here is that to apply the addition rule, the number of events must be fixed and pre-determined in advance.

> The addition rule can only be used if the number of events is fixed and determined in advance.

Pepsi-Cola's Billion-Dollar Contest. At the beginning of this chapter we briefly discussed Pepsi-Cola's billion-dollar contest. Here we will finish the discussion.

On September 14, 2003 the final round of Pepsi-Cola's billion-dollar contest was aired on television. On this show a contestant was given a chance to win one billion dollars, an amount so large that Pepsi had to purchase insurance to cover the prize money just in case somebody actually won.[10] At the beginning of the contest, a thousand contestants each picked a different number from a list of a million possible numbers. After several rounds of elimination, one final contestant was given a chance to win the billion dollars. This contestant would win if any of the 1,000 numbers chosen by all the contestants matched a number chosen in advance by a complex process that at one point even involved a monkey choosing numbered billiard balls from a bag. (a) What's the chance the final contestant wins the billion dollars? (b) Berkshire Hathaway promised to pay the prize money in exchange for an insurance premium from Pepsi reported to be between $1 million and $10 million. Is this a reasonable amount to charge for this insurance?

Solution. (a) Since each number has a one out of a million chance of being the winning number and there will be only one winning number, we can use the addition rule to get $1{,}000/1{,}000{,}000 = 1/1{,}000$, or a 1 out of 1000 chance of winning. (b) If we imagine the contest were held many times, we would expect one winner out of every one thousand games. Thus in order for the insurance company to make money from its premiums in the long run, one thousand premiums should be able to cover the $1 billion prize they would expect to pay out. This means they should charge at least $1 billion$/1{,}000 = \$1$ million for the insurance. Since they charged somewhere between $1 million and $10 million, this seems like the right ballpark. The $1 million is often called the **expected value** of the game. Unfortunately for Pepsi-Cola, and fortunately for Berkshire Hathaway, nobody ended up winning the billion dollars.

Incidentally, most of the profit in the insurance business comes not from the insurance premiums themselves, but from profit made by investing these premiums for a while before having to pay out claims to customers. This money is called "**float**."[11]

---

[10]G. Anderson, "Pepsi's billion-dollar monkey: A new contest offers a chance to win $1B. But first, a simian must pick the magic number," April 10, 2003, 1:01 PM, *CNN/Money*. See http://money.cnn.com/2003/04/09/news/companies/pepsi_billion_game/index.htm. See also http://www.dartmouth.edu/~chance/chance_news/recent_news/chance_news_12.04.html#item3

[11]Warren Buffett, "2005 letter to shareholders of Berkshire Hathaway," page 5, http://www.berkshirehathaway.com/letters/2005ltr.pdf

5. Binomial Probabilities

Airlines routinely sell more tickets than they have seats because they know not everyone will show up. But then there is a risk that too many people will show up and an airline must carefully assess this risk in deciding its sales policy. This approach is often called **yield management**.

Yield management.   Suppose an airline sees that 10% of the customers who reserve seats don't show up. If it sells four tickets, (a) what's the chance everyone shows up? (b) What's the chance only three people show up? You can assume customers behave independently of each other.

Solution.   We know there is a 90% chance a given customer shows up. For the answer to Part (a) we use the multiplication rule to get $.9 \times .9 \times .9 \times .9 = (.9)^4 \approx .66$, or a 66% chance. For Part (b) imagine that the tickets are sold one after another. You might then think the answer should be $.9 \times .9 \times .9 \times .1$, but this is the chance the first three people to buy tickets show up and the last person fails to show up. To get the chance any three people show up you need to add this to the chance that only the second person fails to show up, which is $.9 \times .1 \times .9 \times .9$, and so on with the third person and then the fourth person. The reason it makes sense to add these products together is because they each represent mutually exclusive outcomes: if the second person is the only person who fails to show up, it's impossible that the third person is also the only person who fails to show up. Therefore, we have

$$
\begin{aligned}
\text{P(only three people show up)} \\
= (.1 \times .9 \times .9 \times .9) + (.9 \times .1 \times .9 \times .9) \\
+ (.9 \times .9 \times .1 \times .9) + (.9 \times .9 \times .9 \times .1) \\
= .29, \text{ or a 29\% chance}
\end{aligned}
$$

If you want to figure out the chance only two people show up, the problem becomes more complicated and you need to add up six different possibilities for who the two people could be: the first person and the second person, the first and third, the first and fourth, the second and third, the second and fourth, and the third and fourth. This process can get messy pretty quickly as the number of people gets larger, and so there is a formula built into Excel that computes these types of probabilities for you. These are called **binomial** probabilities, and I will describe the Excel function for them next.

Suppose you have n independent trials of an experiment (such as a person deciding whether or not he or she will show up for a flight), the probability of "success" on any given trial is p and you would like to find the chance of seeing exactly k successes anywhere among these n trials, then the Excel formula for the answer is

$$= \texttt{BINOMDIST}(k, n, p, 0)$$

If you would like to know the chance of getting k or fewer successes, just plug in the number "1" instead of the number "0" at the end above. It's important to note that this formula only works if the trials are independent, otherwise it will give you the wrong answer. Here are a couple of examples.

Example.   An insurance company insures ten projects, which each independently have a 5% chance of requiring a claim to be paid. What is the chance the insurance company must pay for exactly three claims? Here we have $n = 10$, $k = 3$, $p = .05$, and so the formula for Excel is "`=BINOMDIST(3,10,.05,0)`" and the answer Excel gives us is .0105.

Example.   In the previous example what is the chance the insurance company must pay for two or fewer claims? Here we have $n = 10$, $k = 2$, $p = .05$, and plugging in a "1" at the end to indicate "two or fewer," we get the Excel formula "`=BINOMDIST(2,10,.05,1)`" and the answer Excel gives us is .988.

Example.   Suppose a raffle with 100 tickets has 10 winning tickets. If you buy five tickets, what's the chance you buy exactly 3 winning tickets? Can we use the binomial formula? Unfortunately we can't because the trials are not independent. Incidentally, there is another Excel formula for this type of sampling problem: the formula "`=HYPGEOMDIST(k, n, K, N)`" gives you the chance of drawing exactly k "good" items out of n draws without replacement from a population having K "good" items out of a total of N items. This is called the **hypergeometric** formula. In this case we have $k = 3$, $n = 5$, $K = 10$, $N = 100$, and the Excel formula "`=HYPGEOMDIST(3,5,10,100)`" gives us the answer .0064.

Exercises

8. **Birthday coincidences.** At a White House press conference President George W. Bush held on his 60th birthday, a reporter in the audience revealed it was also his birthday.[12] The president invited him up to the podium for a photograph and then asked if there were any more in the audience having the same birthday. Two more people then said they shared the same birthday as the president, prompting him to exclaim "amazing, everybody's birthday today!" If there were approximately one hundred people in the audience, would it really be "amazing" to find at least three people with the same birthday as the president?

6. Gambling

Understanding probabilities is also useful when gambling, or when designing new casino games. In this section we will look at some examples of probabilities associated with cards, dice, and other common gambling games. While I don't condone gambling here, such games do serve to illustrate the laws of probability.

Cards. A regular deck of playing cards contains 52 cards. There are four suits: spades (♠), clubs (♣), diamonds (♢), and hearts (♡). Clubs and spades are black cards; hearts and diamonds are red cards. In each suit there are thirteen cards: cards numbered 2 through 10, along with a Jack (J), Queen (Q), King (K), and Ace (A).

Example. If you deal a single card from a shuffled deck, P(get a queen) = 4/52, because there are 4 queens and you are equally likely to get any of the 52 cards.

Example. If you deal three cards from a regular deck, find the chance they are all spades.

---

[12] "Media leak spoils Bush birthday gift. President marks 60th with calls, surprises and a spilled secret," Thursday, July 6, 2006, CNN.com. `http://www.cnn.com/2006/POLITICS/07/06/bush.bday.ap/index.html`

Solution. The multiplication rule says P(all spades) = P(first is a spade) × P(second is a spade given the first was a spade) × P(third is a spade given the first two are spades). The chance the first card is a spade is 13/52 and, given one spade has been removed, the chance the second card is also spade is 12/51 and, given two spades have been removed at this point, the chance the third card is also a spade is 11/50. So, applying the multiplication rule, the chance they are all spades is $13/52 \times 12/51 \times 11/50 = .013$.

Example. Suppose you have three shuffled decks of regular cards and you turn over the top card from each deck. What is the chance they are all spades?

Solution. The same multiplication rule applies, but the numbers turn out differently. The chance that the first card is a spade is 13/52. Given that the first card turned over is a spade, the chance the second card is also spade remains at 13/52—since the two decks are separate. Given two spades have been turned over at this point, the chance the third card is also spade is still 13/52. So, using the multiplication rule, the chance they are all spades is $13/52 \times 13/52 \times 13/52 = .016$.

Example. If you deal three cards from a single shuffled deck, what is the chance of getting no spades?

Solution. By the same reasoning as before, the chance that the first card is not a spade is 39/52 and, given that a non-spade has been removed (and so 38 of the remaining 51 cards are not spades), the chance that the next card is also not a spade is 38/51 and, given the first two cards removed are not spades, the chance that the third card is also not a spade is 37/50. So, using the multiplication rule, the chance of getting no spades is $39/52 \times 38/51 \times 37/50 = .41$.

Example. If you deal three cards from a shuffled deck, show that

(a) P(all the same suit) $= 52/52 \times 12/51 \times 11/50 = 132/2550 = .05176$

(b) P(all different suits) $= 52/52 \times 39/51 \times 26/50 = 1014/2550 = .3976$

(c) P(not all spades) $= 1 - $ P(all spades) $= 1 - 13/52 \times 12/51 \times 11/50 = 1 - .016 = .994$

Example.   Suppose you roll two dice, what is the chance you roll a "4" at least once?

Solution.   Here we look at the chance of getting no "4"s and subtract from 1. The chance the first roll is not a "4" equals 5/6, and the chance the second roll is not a "4" also equals 5/6. Thus the chance neither rolls are "4" equals $5/6 \times 5/6 = 25/36$ by the multiplication rule. So this means the chance at least one of the two rolls is a "4" equals $1 - 25/36 = 11/36$.

Example.   Suppose you deal two cards from a shuffled deck without replacing them. What is the chance at least one of them is red?

Solution.   Here we look at the chance of getting no reds and subtract from 1. The chance the first card is not red equals $26/52 = 1/2$. The chance that the second card is not red given the first card was not red equals $25/51$, and so the chance neither are red equals $26/52 \times 25/51$. So the answer is $1 - 26/52 \times 25/51 \approx .755$.

Example.   Suppose you are dealt two cards from a shuffled deck without replacing them. What is the chance you get the ace of spades?

Solution.   We can apply the addition rule here. The chance the first card is the ace of spades equals 1/52. The chance that the second card is the ace of spades also equals 1/52. Note that the addition rule uses unconditional probabilities and so the denominator is 52 even though there are 51 cards remaining. And since the chance both cards are the ace of spades equals zero, the events are mutually exclusive. This means that the chance you get the ace of spades equals $1/52 + 1/52 = 2/52$.

Exercises

9. **Cards.** Two full decks of playing cards are placed near each other on a table. The top two cards from each deck will be turned over, for a total of four cards turned over. (a) Find the chance no diamonds are turned over. (b) Find the chance the cards turned over cover all the four suits.

10. **Dice.** A six-sided die is rolled four times. What is the chance that (a) you roll a 3 or more on each roll? (b) you never roll a 3 or more? (c) on at least one roll you roll less than a 3?

11. **Perfect deal.** A mini-deck of twelve cards has three cards from each suit. The cards are shuffled and dealt out to four people so that each person gets three cards. A "perfect deal" is when each person gets all of their cards in a single suit. What is the chance of a perfect deal?

12. **Roulette.** A roulette wheel has 18 red numbers, 18 black numbers, and two green numbers that are all equally likely to come up when you spin the wheel. If you spin the wheel five times, (a) what's the chance you get the same number each time? (b) What's the chance you never get the same number twice? (c) What's the chance you get no odd numbers (there are 18 odd numbers on the wheel) (d) What's the chance you get no red numbers until the fourth spin?

13. **Poker.** Poker is popular card game in which each player is dealt five cards. If all of your cards are in the same suit, it is called a "flush." This is an excellent hand and you will probably win if you ever get this. (a) If you are dealt five cards from a regular deck, what's the chance you will get a flush? (b) Suppose two people are playing poker. What is the chance they both have flushes in different suits?

14. **Dice.** You roll a six-sided die 10 times. What is the chance you never roll the number "6"?

15. **Cards.** Two cards are dealt from a shuffled deck. Find the chance that (a) the first card is an ace (b) the second card is an ace given the first is an ace (c) the second card is an ace (d) at least one of the two cards is an ace (e) both cards are aces.

7. Exercises

16. **Hurricane insurance.** An insurance executive estimates that for a single hurricane season, there is an 80% chance that a major hurricane will hit at least one of the two hurricane-prone coasts: the Gulf coast or the East Coast.[13] (a) If we assume both coasts are equally likely to experience hurricanes, does this mean each coast individually would have roughly a 40% chance of experiencing a major hurricane? Explain. (b) Does this

---

[13]See http://www.lloyds.com/News_Centre/Briefings_and_speeches/Insuring_against_the_next_big_one.htm

mean that over the course of two similar hurricane seasons there would be $80\% + 80\% = 160\%$ chance of a major hurricane on at least one of the two coasts? Explain.

17. **Quality inspection.** A quality inspector selects three random components, removing them one at a time, from a box containing five defective components and fifteen working components. (a) What is the chance of selecting no defective components? (b) What is the chance the third component selected is defective given that the first two selected are tested and found to be defective? (c) What is the chance the third component selected is defective given that the first two components selected have not yet been tested?

18. **Server reliability.** An important data server breaks down for 40% of the time and is operational the other 60% of the time. How many independent servers should be running so that there is a 99% chance at least one is operational? Assume the servers break down independently.

19. **Insurance claims.** An insurance company estimates that over one year there is a 2% chance a given house will experience hurricane damage and a 1% chance the house will experience fire damage. (a) For two similar houses in a town, is it reasonable to estimate the chance both experience hurricane damage as $0.02 \times 0.02 = .0004$? What assumption would this calculation make, and is it reasonable? (b) Is it reasonable to estimate the chance both houses experience fire damage as around $0.01 \times 0.01 = .0001$? What assumption would this calculation make, and is it reasonable?

20. **Security weaknesses.** A company's computer system has three potential weaknesses criminals could exploit to gain unauthorized access. Suppose we know criminals have a 2% chance of discovering the first weakness, a 3% chance of discovering the second weakness and a 4% chance of discovering the third weakness. (a) If we assume independence, what's the chance a given criminal discovers at least one of the weaknesses? (b) Independence is not such a good assumption here, since a criminal smart enough to find one weakness is more likely to find other weaknesses as well – and a criminal who can't find one weakness is unlikely to be able to find the other weaknesses. Would the real answer to (a) be larger or smaller than the number you got for (a) above?

21. **Credit default.** A bank extends credit to five firms and assesses the risk of default to be around a 5% chance for each firm. (a) If the firms each do business in independently operating sectors of the economy, estimate the chance at least one defaults. (b) If the firms each do business in the same sector of the economy, would the chance at least one defaults be larger than or smaller than your answer in (a)? Explain.

22. **Pharmaceutical product launch.** A large pharmaceutical firm has developed a new drug and is scheduled to release it to the market at the end of next year. The profit it earns depends on whether or not it is the first firm to bring this type of drug to the market. The company is concerned because three other smaller competitors are independently working on similar drugs: the firm estimates that the first competitor has a 10% chance of bringing its drug to the market sometime before the firm's market release, the second competitor has a 15% chance, and the third competitor has a 5% chance. (a) What is the chance the large firm is the first to bring its drug to the market? (b) What is the chance the large firm is the last to bring its drug to the market?

23. **Market research.** Market researchers choose a panel of five people at random from a pool of fifty volunteers and let these people test a new product. If the pool consists of 10 men and 40 women, (a) what's the chance at least one man is on the panel? (b) What's the chance the panel is all men?

24. **Scratch lottery.** In a scratch lottery game, tickets are printed with various prizes written on them. Players scratch the ticket to reveal the prize and only one ticket overall has the grand prize on it. If one city sells twice as many tickets as another city, is the chance of finding the grand prize winner somewhere in the first city twice the chance of finding the grand prize winner somewhere in the second city? Explain.

25. **Powerball lottery.** People playing the Powerball lottery game select a number and win the grand prize if their number matches the one that lottery officials draw at random. Tickets for each game are sold across many states, but they all use the same winning number. If more than one person picks the winning number, they share the prize. If one state sells twice as many tickets as another state, is the first state twice as likely to have a winner compared with the second state? Explain.

26. **Powerball lottery.** In the Powerball lottery between 1997 and 2002, five different numbers between 1 and 49 were chosen in each drawing. (a) What is the chance all the numbers chosen are even? (b) Over 515 games, all five numbers came out even in 10 of these games.[14] Is this the expected number of times this should happen, or is something suspicious here?

27. **Consumer survey.** In a large survey of consumers, 60% have used your company's product, 30% have used your competitor's product, and 25% say they haven't yet used your competitor's product but they may in the future. (a) Can we say 30% + 25% = 55% have either used your competitor's product or say they may use it in the future? Explain. (b) Can we say 60% + 30% = 90% have used either your product or your competitor's product? Explain.

28. **Human resources.** Human resources records at a company show that 10% of its staff are executive level managers, 20% are outside contractors, and 5% of the staff overall has been working there less than one year. (a) Can we say that 10% + 20% = 30% of the staff are either executive level managers or outside contractors? Explain. (b) Can we say that 20% + 5% = 25% of the staff are either outside contractors or have been there less than one year? Explain.

29. **Health insurer.** A health insurer notices that for a given year only 5% of its customers have any health expenses over \$100. In a pool of 10 random patients, what's the chance none of them have any health expenses over \$100?

30. **Quality control.** A quality control department uses random inspection sampling to help find defective components. Suppose that there happen to be three defective components and twelve good components in a batch produced. An inspector decides to randomly choose four of these fifteen components and test them one at a time. (a) What is the chance the inspector finds at least one defective component? (b) What is the chance he doesn't find any of the defective components? (c) What is the chance the inspector only finds one defective component, and it also happens to be the last one he tests?

---

[14]http://www.powerball.com/powerball/winnums-text.txt

31. **Market testing.** Three of the fifty states will be chosen at random for a market test of a new product. (a) What is the chance none of the six New England states get chosen? (b) What is the chance that Massachusetts gets chosen?

32. **Call center.** A company studying its call center efficiency notices that 25% of all customer telephone calls require a live operator, and 75% of the calls can be handled by an automated system. Does this mean that if four calls arrive in a row, there is a $25\% + 25\% + 25\% + 25\% = 100\%$ chance at least one of them will require a live operator? If not, compute the chance of this.

33. **Product test volunteers.** Twelve people are chosen at random for a product test from a pool of volunteers that contains the same number of men and women. (a) Is it possible to compute the chance no men are chosen? Explain briefly. (b) If we assume the pool of volunteers is very large, approximately what's the chance no men are chosen? (c) If we assume the pool of volunteers is very large, approximately what's the chance exactly two men are chosen?

34. **Online retailer.** A retail web site finds that 60% of customers purchase one item, 30% purchase two items, and 10% purchase more than two items. With four customers what is the chance that (a) they all purchase more than one item? (b) None of them purchase more than two items? (c) None of them purchase less than two items?

35. **Airline delays.** An airline notices when they schedule a daily departure for 8 AM there is a 10% chance the flight is delayed by the control tower, but if they schedule the departure for 7:45 AM the chance is only 5%. (a) What's the chance at least one flight gets delayed over the course of seven days with an 8 AM departure? (b) If you repeated part (a) with a 7:45 AM departure instead, would you get an answer equal to half the answer for part (a)?

36. **Order fulfillment.** An operations manager would like to have 95% of customer orders filled without any errors. An analysis shows that orders for any single item the company produces are currently filled with this level of accuracy. The problem is that customers typically ask for approximately 50 separate items in a single order and these items must often originate from several different manufacturing divisions. If any item in

the order is incorrect, the customer is unhappy. An analyst says that since the accuracy for each item ordered is 95%, it means that regardless of the size of the order we should expect 95% of customers to have their orders filled without any errors. (a) Do you agree with this reasoning? Explain. (b) Approximately how high does the accuracy for orders of a single item need to be to get 95% of customer orders error free? Explain. What are the assumptions behind your calculations and are they reasonable?

37. **Call center.** An analyst is studying the efficiency of a call center. Experience has shown that each customer who calls in is equally likely to require service from one of six different departments. If four calls arrive in a row, (a) what is the chance they all require service from the same department? (b) What is the chance they all require service from different departments?

38. **Drug development.** A pharmaceutical company has a lengthy drug development process in which a drug must successfully complete a large number of phases before it can be brought to market. Only about 10% of the new drugs considered actually make it into Phase I of development. Only about 20% of drugs that make it into Phase I actually go on to start Phase II. If the company considers ten independent drugs for development, how likely is it that at least one of them makes it into Phase II?

39. **Growth and profitability.** A *Harvard Business Review* article says companies often damage profitability by trying to achieve more growth and, on the other hand, often slow growth by trying to achieve profitability.[15] The article says that companies that can successfully do both simultaneously in the same year more than half the time will become top industry performers. The article mentions that General Motors (GM) did both in about one-third of the past 20 years. If we took the fraction of the 20 years where GM successfully achieved profitability and multiplied this by the fraction of the 20 years where GM had successful growth, would we get a number higher than 1/3, lower than 1/3, or about 1/3? Explain briefly.

---

[15] "Managing the Right Tension," by Dominic Dodd and Ken Favaro, *Harvard Business Review*, December 2006, page 62.

40. **Stock prices.** Stock A has a 40 percent chance of going up, and Stock B has a 30 percent chance of going up. (a) An analyst uses this to estimate that there is a $30\% + 40\% = 70\%$ chance at least one of the two stocks goes up. Is this sensible reasoning? (b) A second analyst estimates the chance both stocks go up as $.30 \times .40 = .12$, or a 12% chance. Is this sensible reasoning?

41. **Lottery coincidence.** In the late 1980s both MA and NH state lotteries had as the winning number the randomly selected 4 digit number 7923. National news programs proclaimed this to be a rare event. According to a statistical expert, the probability both the MA and NH numbers were 7923 equals $(1/10,000)^2 = 1/100,000,000$. Given the lotteries are held each weekday, this should happen once every 400,000 years. Does this mean it was an extraordinarily rare event, perhaps worthy of investigation?

42. **Product launch.** A firm plans to release two related new products (Product A and Product B) to the market. Market research shows that Product A has a 60% chance of success, and Product B has a 50% chance of success. If Product A turns out to be a success, then Product B has an 80% chance of success. (a) What is the chance they are both successful? (b) What is the chance at least one fails? (c) Are successes for these two products independent events? Explain. (d) Suppose B turns out to be a success. Knowing this information, what is the chance A is also a success?

43. **CFA exams.** There are several levels of exams that must be passed over a period of several years to be certified as a financial analyst by the CFA Institute.[16] Part of the prestige of the designation is the difficulty of the exams. The percentage of exam takers who passed the Level I, Level II, and Level III exams in 2004 was 34%, 32% and 64% respectively. Since $.34 \times .32 \times .64 = .07$, does this mean that about 7% of test takers pass all three levels of exams? Explain.

44. **Extreme jobs.** A *Harvard Business Review* article discusses a study concluding that jobs with high performance requirements (fast-paced with tight deadlines, 24/7 client demands, and so on), are much more

---

[16]see http://www.cfainstitute.org/cfaprogram/cfaprofile/cfa_exam.html

common among men than among women.[17] The study also shows that women who work a high number of hours per week (at least 60 hours) are more likely to be in high performance jobs compared with men who work a high number of hours. The conclusion from the article is that "women are not afraid of the pressure or responsibility of extreme jobs—they just can't pony up the hours." Below is a table showing the percentage of the study's subjects falling into each of eight different categories. (a) Describe the relationship between hours worked and job performance requirements. (b) Use this data to numerically illustrate the study's two conclusions described above.

	Low hours worked		High hours worked	
	Men	Women	Men	Women
High performance requirements	15%	4%	17%	4%
Low performance requirements	33%	12%	12%	2%

45. System reliability. You are evaluating two systems to detect when your process is no longer working correctly. For each of these systems, there is a chance that the system will fail and give misleading readings. System A consists of two components: Component 1 has a .72 chance of working correctly, Component 2 has a .83 chance of working correctly and the components are independent. This system will give correct readings if at least one of the components works correctly. (a) What is the chance System A will give a correct reading? (b) System B consists of two components: Component 1 has a .72 chance of working correctly and Component 2 has a .96 chance of working correctly. However, the components are not independent. If Component 1 is not working correctly, there is a .10 chance Component 2 will not work correctly. Again, this system will give a correct reading if at least one of the components works correctly. What is the chance System B will give a correct reading?

46. Ask Marilyn. A recent letter to the *Parade Magazine* newspaper column, "Ask Marilyn" by Marilyn vos Savant (who in her biography writes she

---

[17] "Extreme Jobs: the Dangerous Allure of the 70 Hour Work Week," by Sylvia Ann Hewlett and Carolyn Buck Luce, *Harvard Business Review*, December 2006, page 49.

is listed in the Guinness Book of World Records under "Highest IQ"[18]), reads "My wife and I won entry into a drawing for a $10,000 prize. In all, 13 families who'd won entry arrived at the drawing. So 13 keys—one of which would open the gate to the grand prize—were placed in a basket, and the families lined up. We believe the last family in line had the best chance, and we were in 11th position, yet the prize was won by the family in 10th position. We never had a chance to draw a key. Which position do you think had the best chance?"[19] What do you think Marilyn wrote back as her answer?

47. **Marbles.** A box contains four marbles. One marble is red, and each of the other three marbles is either yellow or green—but you have no idea exactly how many of each color there are, or even if the other three marbles are all the same color or not. (a) Someone chooses one marble at random from the box and if you can correctly guess the color you will win $1,000. What color would you guess? Explain. (b) If this game is going to be played four times using the same box of marbles (replacing the marble drawn each time), but you have to make your guesses ahead of time, what four guesses would you make? Explain.

48. **Ask Marilyn.** Marilyn vos Savant discusses the following question in her *Parade Magazine* column "Ask Marilyn:" "Suppose I am taking a multiple-choice test. One question has three choices. I randomly choose A. Then the instructor tells us that C is incorrect. Should I switch to B before turning in my paper?"[20] What is the answer here?

49. **Cargo inspection.** There are five shipping containers and two of them contain suspicious cargo. A homeland security port inspector chooses three out of the five containers randomly for inspection and then inspects them one at a time.

   (a) What's the chance that among the three chosen she finds some suspicious cargo?

   (b) What's the chance she finds no suspicious cargo?

   (c) What's the chance she finds both containers with suspicious cargo?

---

[18]http://www.marilynvossavant.com/bio.html

[19]Marilyn vos Savant, "Ask Marilyn," *Parade Magazine*, October 29, 2006. See http://www.parade.com/articles/editions/2006/edition_10-29-2006/Ask_Marilyn

[20]"Ask Marilyn," *Parade Magazine*, May 18, 2003.

50. 25% of the employees in a company have MBA degrees and 15% of employees in a company have managerial positions. Also, 60% of managers have MBA degrees.

    (a) If an employee is selected at random, what's the chance that employee is a manager and has an MBA degree?

    (b) Is having an MBA degree independent of being a manager? Briefly explain.

51. Experimental design. There are two primary clinics and six satellite clinics affiliated with the Boston Medical Center. Researchers involved in a study of a new IT system will randomly choose four of these clinics to be considered as the treatment group (where the new system will be made available), and the remaining four clinics will form the control group.

    (a) What is the chance at least one of the primary clinics is in the treatment group?

    (b) What is the chance none of the primary clinics are in the control group?

    (c) What is the chance both primary clinics are in the treatment group?

    (d) Suppose a second group of researchers is conducting a separate study where they also randomly choose four of these eight clinics to be considered as a treatment group – with the remaining four clinics as the control group. What is the chance that both of these studies end up with the same treatment assignments (that is, the same clinics in the treatment group and the same clinics in the control group)?

Fun Problem: Bayes' Rule

Question. Suppose there is a rare disease that generally only infects around 1 out of 1,000 people. There is also a test for the disease that is 95% accurate. This accuracy means that if you have the disease there is a 95% chance you test "positive," and if you don't have the disease there is a 95% chance you test "negative." Suppose you take the test and it comes up positive (so the test says you have the disease). What is the chance you actually do have the disease?

Solution. Even though the test is 95 percent accurate, you will only have about a 2 percent chance of actually having the disease. This may seem a little hard to believe so let's illustrate it below.

Out of a group of about 1,000 people who go in to get tested, we would expect roughly one person to have the disease, and 999 to be disease-free. We can diagram this as follows:

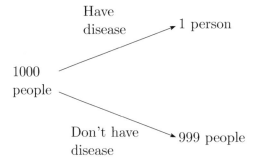

Next, we expect about 95% of each of these two groups of people to have accurate test results. This means 95% of 999 people (approximately 949 people) should test negative in the group that doesn't have the disease, and 95% of 1 person (approximately 1 person) should test positive in the group that has the disease. We can diagram this as follows:

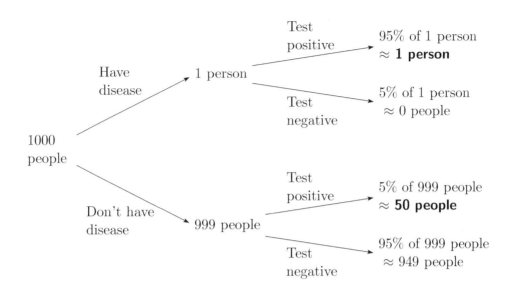

Notice that out of the approximately 51 people who test positive (the ends of the two branches with bold-face type respectively contain 1 person and 50 people), only one of them actually has the disease. So if you were one of those 51 people who tested positive, your chance of actually being the unfortunate one with the disease is about 1/51 or about a 2% chance.

This surprising result is really due to the fact that the disease was very rare to begin with. The vast majority of the people who get tested don't have the disease, so doctors get swamped with lots of "false positives." This is one of the problems with screening everybody for a rare disease—there will be so many "false positives" that the test results will be almost meaningless. One of the first things a doctor does when you test positive for rare disease is to give you a second or even a third test. After testing positive two or three times, your chance of actually having the disease is much higher. The issue of "false positives" is becoming a major concern with the increasing and largely unregulated market for at-home genetic tests.[21]

---

[21] "Too Much Information: Results of Home DNA Tests Can Shock, Misinform Some Users," by Sandra G. Boodman, *Washington Post*, Tuesday, June 13, 2006.

Technical notation

If A stands for "testing positive" and B stands for "having the disease," we know that P(test positive given you have the disease) = P(A given B) = .95 and we use this to compute P(have the disease given you test positive) = P(B given A) ≈ .02. The general technique used above to compute P(B given A) from P(A given B) is called **Bayes' Rule**.

Chapter 9

Sampling Variability and Standard Error

1. Introduction

USA Today reported that approximately 60% of all TV sets in the country were tuned to the 2006 Super Bowl and that this was an increase of around five percentage points from 2005.[1] This turns out to be about 90 million viewers which makes it one of the most-watched television shows in history. Furthermore, it was reported that women made up approximately 44% of the audience.[2] Because of the size of the audience, the network was selling commercial time for the outrageous sum of almost $6 million per minute—almost $100,000 per second.

But how could anybody possibly know how many TV sets were tuned in to the Super Bowl? And how could they possibly know if viewers were men or women? Nobody from the newspaper came around to my house looking to see who was watching and I doubt anyone came around to your house either. Since the value of television advertising is directly linked to being able to measure the size of the audience, is there any accuracy to these figures? Or are they just pulled out of thin air?

---

[1] http://www.usatoday.com/life/television/news/2006-02-06-super-bowl-rating_x.htm

[2] http://www.forbes.com/2006/01/31/advertising-super-bowl-cz_af_0201ads_super06.html

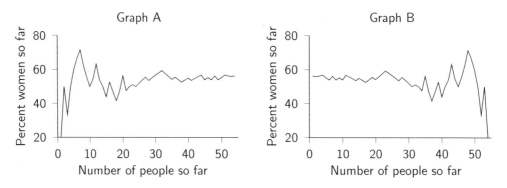

Figure 9.1. Which graph would we see?

ACNielsen, the company that computes these figures, is able to very accurately estimate these percentages simply by monitoring a very small sample of households—usually only around a thousand households spread across the entire country. The same principle applies with political polls: even if the pollster doesn't speak to every American directly, a small sample can be very representative. In this chapter we will discuss how this process works and exactly how accurate such small samples tend to be. There is a simple but strong theoretical underpinning to estimating the accuracy of such samples and the purpose of this chapter to describe it and apply it to different situations.

2. The Law of Averages

Let's start by introducing, in an intuitive way, the simple principle upon which this entire chapter rests. Suppose we stand on a street in downtown Boston and watch the people walk by. As each person walks by, we graph the percentage of people we've seen so far who are women. Which graph in Figure 9.1 would most likely represent what we would see—Graph A or Graph B?

If you said Graph A, you would be correct. Graph A starts out with large fluctuations and then starts to stabilize at around 50%. Graph B is the mirror image; it starts out with small fluctuations and then has larger fluctuations later on. The principle in action here is the idea that as the sample gets larger, the percentage of women in the sample tends to stabilize very close to the true population percentage (somewhere around 50% here in Boston).

In a small sample, there is more likely to be a lot of random variation. This is because in a small sample any single person can change the overall percentage quite a bit, but as the sample gets larger each individual person has very little impact on his or her own. This also means larger samples become more representative of the population. This principle is called the **Law of Averages**, because the same thing also applies to a sample average fluctuating around the population average.

> The **law of averages** says that with a large sample, a percentage observed in your sample will tend to be very close to the true percentage in the population. With a small sample, you are more likely to observe a large fluctuation away from the true population percentage.

Another thing to notice about Graph A in Figure 9.1 is that the percentage quickly stabilizes even after only 50 people. This means that a sample as small as 50 people from a city like Boston with millions of residents can be reasonably representative. I will discuss later in this chapter and in the next chapter how to determine the sample size you should use.

Example.   In the example above, would you be more likely to see more than 70% women in a random sample of 20 people or 200 people?

Answer.   Twenty people. With 200 people you are very likely to see close to 50 percent women; in a small sample of only 20 people it is more likely you might see a fluctuation away from this number.

Example.   In the example above, would you be more likely to see more than 40% women in a random sample of 20 people or 200 people?

Answer.   This time the answer is 200 people. In a large sample you are very likely to see close to 50% and thus more than 40%.

Example.   Which gives you a larger chance of coming out ahead financially: playing a slot machine two times or 50 times?

Answer. Unfortunately, you are more likely to come out ahead if you play the slot machine two times (and probably even more likely to come out ahead if you play it only once). If you play the machine a very large number of times you are likely to lose money, since the odds are against you on all slot machines. If you are hoping for a fluctuation away from these unfortunate odds (as all gamblers of course are), you should play as few times as possible.

In this section we developed an understanding of how larger samples tend to be more representative of the population compared with smaller samples, due to the law of averages. In the next section we will examine how this can be quantified and how this can help you estimate how large of a sample you need in a given business situation.

Exercises

1. **Medical errors.** Medical errors are a problem of increasing public concern and hospital executives are eager to understand which management practices could reduce them. Analysts studying one particular medical condition looked at hospital records to see if hospital caregivers actually correctly gave patients all the treatments their doctors prescribed for them. For this medical condition, overall only about 80% of patients nationwide correctly received all the treatments prescribed by their doctors. This percentage varied widely from one hospital to another and analysts noticed that smaller hospitals tended to show much more variability in this percentage compared with the larger hospitals. This is illustrated by the funnel shape in Figure 9.2, in which each dot corresponds to one hospital. An analyst says the funnel shape may indicate that large hospitals have better standardization procedures in place that reduce variability. Can you think of another explanation for the funnel shape?

2. **Website hits.** Since the amount of money a web site can charge for advertising is closely linked to the daily number of hits, most sites keep track of such numbers. To look for trends, one web site graphed the daily number of hits averaged over each week, as well as the daily number of hits averaged over each month. One graph in Figure 9.3 shows a point for each of 26 months, and the other graph shows a point for each of 26 weeks. Which graph is which? Justify your answer.

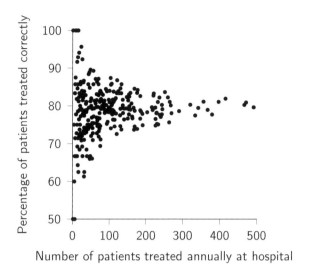

Figure 9.2. The percentage of patients treated correctly versus the number of patients treated annually at a selection of hospitals. What could explain the funnel shape?

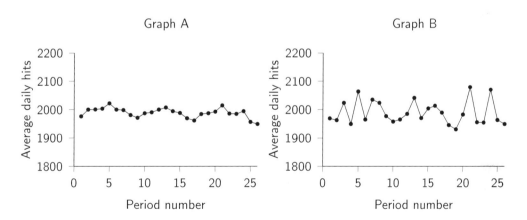

Figure 9.3. Average number of web page hits over time.

3. **Airline overbooking.** An airline has a policy of selling 5% more tickets than it has seats on each airplane, because experience has shown that about 10% of people who buy tickets never actually show up for their flights. In the event too many people show up, the airline must 'bump' customers to other flights and pay them compensation. Assuming the flights are sold-out, is it more likely the airline will have to 'bump' people from a large airplane or a small airplane? Answer and explain briefly how the law of averages applies.

4. **Index funds.** A stock index fund invests in a very large number of different stocks, so the return you get from it is roughly the average return for the stocks in the index. (a) Which of these two choices would give you a better chance of tripling your money within a month: investing all your money in the index fund, or investing all your money in a few randomly chosen stocks that make up the index? (b) Which of these two choices would give you a better chance of losing almost all of your money within a month: investing all your money in the index fund, or investing all your money in a few randomly chosen stocks that make up the index?

5. **People watching.** Suppose you stand in downtown Boston and count people who walk by with red hair. Would you be more likely to see that more than half the people have red hair in a random sample of 20 people or 200 people? Explain.

3. The Standard Error for a Sample Percentage

We have seen that if we stand on a corner in downtown Boston and take a sample of the population, as our sample gets larger the sample percentage will approach the population percentage because of the law of averages. Applying this reasoning to a political poll, we can imagine that a poll will come close to reflecting the opinions of all Americans as the sample becomes larger. In other words, all U.S. citizens do not have to be contacted for the poll to be accurate. But the law of averages only suggests a direction: use large samples. It does not tell us how large the sample needs to be. And it does not tell us how confident we can be that various sample sizes will give us an accurate reading. Answering this question is our next step.

Suppose you want to estimate the percentage of people in a population who support a given political candidate. Since there are always fluctuations in a sample percentage, how accurately can we confidently estimate the percentage based on a sample size of, say, only 100 people? And how would our accuracy change if we increased the sample size to 1,000 people?

To answer this question, we must use a statistical formula that quantifies the amount of variation, or error, you would expect to see in a sample percentage. This expected amount of error is called the **standard error for the sample percentage** (SE for %), and calculating it depends on the sample size and roughly what you expect the true percentage to be. For now I will just show you the formula and how to use it, and at the end of the next section we will see why the formula is reasonable. The formula is

$$\text{SE for \%} = \sqrt{\frac{\text{true percentage} \times (100\% - \text{true percentage})}{\text{size of sample}}}$$

where the "true percentage" is the true percentage in the population that you're sampling from—or your best rough estimate of it so far. This formula applies only to what is called a **simple random sample**, where every possible group of people in population has the same chance of getting selected to be in the sample. In other words, this type of random sampling is just as if you pulled random names on slips of paper out of a hat. (Most large surveys and polls use a slightly different way of choosing the sample, and they must therefore use a slightly different formula for the error. There are entire books written on the many different standard error formulas for each common type of sampling method.)

It may seem like circular reasoning here to use a rough estimate of the "true percentage" in order to compute the accuracy of that estimate itself, but it turns out to work well in practice. If you absolutely have no idea at all about the true percentage you can plug in 50% to get a conservative overestimate of the standard error (this will give the largest possible standard error in the formula—plug in a few numbers and see it for yourself). This type of circular reasoning is called a **bootstrap estimate** due to the old saying that you can help yourself by "pulling yourself up by your bootstraps."

But before you start using this formula, I must emphasize that one standard error does not give a realistic picture of the range of possible errors you might see from a sample—it only gives the typical error. Just as the standard deviation only gives the typical distance of numbers to the average, you get a more realistic picture of the range of a data set if you go two or three standard deviations in each direction from the average. For this reason, people traditionally double the standard error to get what is called the **margin of error**, and this can be viewed as a rough estimate of the largest error you would reasonably expect to see. Of course it is still possible for the error to come out even larger than the margin of error, but this only happens rarely. In the next chapter I will discuss how rarely this happens. To be extra conservative you could triple the standard error, but the common convention is to double it.

The **standard error for a sample percentage** (SE for %) estimates the typical error you expect to see in a sample percentage:

$$\text{SE for \%} = \sqrt{\frac{\text{true percentage} \times (100\% - \text{true percentage})}{\text{size of sample}}}$$

The **margin of error** is twice the standard error, and estimates the largest error you reasonably expect to see:

$$\text{ME for \%} = 2 \times (\text{SE for \%})$$

Let's use the formulas above and try them with sample sizes of 100, 1,000, and 5,000, just to see what happens. Let's say we have no idea what the true population percentage is, so we use 50% to give us the largest possible standard error:

- With sample size 100, the square root of 2,500/100 is 5%, so the margin of error is 10%.
- With sample size 1,000, the square root of 2,500/1,000 is approximately 1.6%, so the margin of error is approximately 3.2%.
- With sample size 5,000, the square root of 2,500/5,000 is approximately 0.7%, so the margin of error is approximately 1.4%.

Note first that as we increase our sample size, the standard error and margin of error fall. Second, for a small sample size of 100, the margin of error is 10%, which is not so bad. It is a manageable 3.2% for a sample size of 1,000 and it drops, but not precipitously, for a sample size of 5,000. This means political polls with a sample size of only 1,000 people usually are accurate to within a few percentage points. No wonder most Americans have never been called when these presidential polls are done.

There are obviously many other possible sources of bias and error in a political poll, and the standard error formula only measures the error due to simple random sampling. If the pollsters only sampled one demographic group hoping this would be representative of the whole country, this could introduce a serious bias beyond the scope of the formula. And if questions in the poll are worded in a politically- or emotionally-charged way, this may be another source of bias beyond the scope of the formula. When polls to predict an election include too many people who don't vote, this can also introduce a bias beyond the scope of the formula. This, incidentally, is one reason pollsters go to great lengths to only count the opinions of people who they believe are very likely to actually vote.[3]

Example.   Suppose that based on a random sample of 100 people, 80 percent approve of the president. Attach a margin of error to this percentage and interpret it.

Solution.   Since the size of our sample is 100 and the best rough estimate we have so far of the true population percentage is 80%, we apply the formula above for the standard error and get

$$\text{SE for \%} = \sqrt{\frac{80 \times (100 - 80)}{100}} = 4\%$$

This tells us that we expect there generally to be about four percentage points of error in this survey. The approximate outer limit of error we expect to see, the margin of error, equals twice 4%, or 8%. This means we are quite confident that 80% plus or minus 8% (somewhere between 72% and 88%) of people overall approve of the president. We would quote 8% as our margin of error for this survey.

---

[3]In political polls, often the published numbers change from poll to poll even if public opinion doesn't change at all. Pollsters usually only include "likely voters" in their polls, and the changing numbers may reflect the changing likelihoods of voting by one group or another. See "Public opinion polls diverge because they are still partly an art," by Sharon Begley, *The Wall Street Journal*, November 24th, 2004, p. B7.

Example. Suppose we now have a new candidate whom we know has an approval rating of 85%. Can we write a newspaper headline saying this competing candidate has a higher approval rating than the president?

Discussion. We cannot say whether this new candidate's approval rating is higher or lower with any real confidence here. Although the 85% rating may seem higher than the president's 80% rating, the 8% margin of error tells us that, as far as we know, the president's rating in reality could reasonably be anywhere between 72% and 88%. Since this range does not rule out an approval rating over 85%, we can't say with any real confidence that the new candidate actually has a higher approval rating. Sometimes you might hear this situation called a "statistical dead heat," in which two candidates are within the margin of error of each other. However, if we knew with certainty the new candidate had a 90% approval rating (or any rating above 88%), then we could in fact say with confidence that the president's rating is below it.

> In general it is not honest to make any statements about the results of a survey that require an accuracy sharper than the margin of error to make them.

Suppose we want to reduce the margin of error. How much larger a sample would be needed to cut the margin of error in half? By examining the formula for the standard error, we notice that to cut the margin of error in half we must quadruple the sample size and to divide the margin of error by three, we must multiply the sample size by $3^2 = 9$. The general rule is the following:

> To divide the margin of error by some number, you must multiply the sample size by that number squared.

In our example above, the margin of error was 8% based on a sample size of 100. To get the margin of error down to around 4% (in other words, cut it in half), a sample size of 400 (quadruple the original sample size) would be needed. To get the margin of error down to about 1% (in other words,

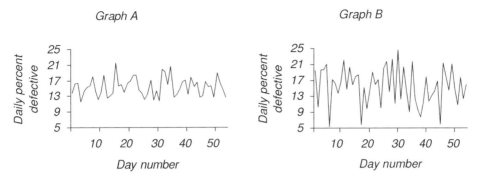

Figure 9.4. Day-to-day fluctuation in the percentage of defective items produced. Which graph would we more likely see?

divide it by 8), a sample size of 6,400 (i.e., $8^2 = 64$ times the original sample size) would be needed.

Example.   A manufacturing process produces 15 percent defective items on average. For 55 days, we manufacture 200 items each day, and we graph the daily percentage of defective items we get. Which graph in Figure 9.4 would we be more likely to see? Why?

Solution.   To estimate the amount of day-to-day fluctuation we expect to see in the daily percentage of defectives, we can imagine that each day we are taking a sample of 200 items from a population that has a true percentage of 15% defective items. We can then apply the standard error formula to get

$$\text{SE for } \% = \sqrt{\frac{15 \times (100 - 15)}{200}} \approx 2.5\%$$

We then compute the margin of error by doubling the 2.5% to get 5%. This 5% margin of error means that the day-to-day percentage of defectives should generally stay in the range 15% plus or minus 5% or, in other words, inside the range from 10% to 20%. If you look at Figure 9.5, Graph A tends to just barely stay in this range and Graph B goes way outside this range quite often. For this reason Graph A is probably what we would see. Notice that we did not need to put the 55 days into the formula anywhere; this is not the size of the sample but rather is the number of different individual samples of size 200 we are observing overall. (Incidentally, these graphs are one way quality control engineers can know when their machines need re-calibration.

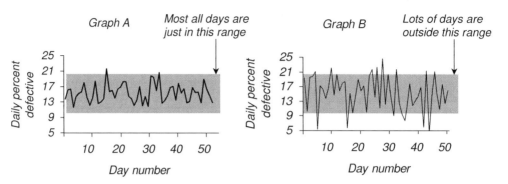

Figure 9.5. The margin of error depicted for day-to-day fluctuation in the percentage of defective items produced.

If the graph usually looks like A and then all of a sudden starts looking like B, they know there's a problem!)

Tying it all together

Here is a way to tie together the concept of standard error with the concept of standard deviation. Recall that the standard deviation tells you how far, in general, values tend to be from their average. The standard error tells you how far, in general, a sample percentage tends to be from the population percentage. This comparison means that if you took many different samples and wrote down the sample percentage for each one, the SD of that list of percentages would come out to be very close to what the standard error formula gives us. In the previous manufacturing example, we showed this in Graph A, so the standard deviation of the data graphed there should equal about 2.5%—the standard error for the sample percentage computed from the formula.

> The standard error for a sample percentage (SE for %) estimates the standard deviation of a list of sample percentages taken from repeated samples.

Example. Four different surveys are based on the same sample size and each estimates the percentage of people who support the president. The four survey results are: 50%, 60%, 50% and 60%. (a) Give a very rough estimate of the margin of error here. (b) Would you estimate the sample size for each poll was closer to 100 people or 1,000 people?

Solution.   (a) The standard deviation of the list 50, 60, 50, and 60 equals five, since the average is 55 and each number is five points away from the average. This means the standard error for the sample percentage should be around 5% and the margin of error should be around 10%. (b) Above we showed that the margin of error for samples with 1,000 people is usually less than a few percentage points. With 100 people, it's as high as 10 percentage points. This means that the sample size was probably closer to 100 people.

Example.   Suppose you want to estimate the percentage of people who smoke in New York State as well as the percentage of people who smoke in Maine. If you take a simple random sample of 1,000 people from each state and ask them if they smoke or not, in which state will the survey be more accurate?

Solution.   New York State has about 20 million residents, while Maine has only about one million residents. You might think that the accuracy of the survey should be a lot better in Maine, since there are 1/20 times as many people there. But, believe it or not, the accuracy of both surveys would be just about the same. To see why this is true, I will convince you in two different ways. First, if you look at the formula for the standard error above you see that it only depends on the size of the sample—not the size of the population. Since both states are using the same sample size of 1,000 people, the error, and hence the accuracy, should be about the same. (This is assuming that the true percentage of smokers is roughly in the same ballpark for both states.)

> The accuracy of a sample does not depend on the size of the population, but really only depends on the size of the sample. This rule holds provided the population is quite large compared with the size of the sample.

To convince you a second way, imagine that you have a new job as a professional soup taster. Your job is to test if the soup is properly seasoned. Now suppose you have two pots of soup to taste: one pot is very small and the other pot is gigantic. Do you need to take a bigger taste from the bigger pot? Or can you take the same size spoonful from each pot? If you stir up the soups, you should really only need the same size spoonful from each pot to get the flavor—one small spoonful should be equally representative of both pots of soup.

This is exactly what is happening with our samples from New York and Maine. If a sample is randomly selected from a population, having a bigger population does not mean you need a bigger sample. And the same size sample should give the same accuracy—regardless of the size of the population (provided it is large when compared with the size of the sample: if, for example, you sample the entire population you don't get any error). And, incidentally, random samples are usually more difficult and expensive to take out of a large population than out of a small population.

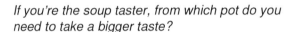

If you're the soup taster, from which pot do you need to take a bigger taste?

Little pot of soup

Big pot of soup

Technical Note: The Finite Population Correction Factor

Powerball. When the Powerball lottery originally started, five different winning numbers from the numbers 1 to 49 were chosen at random twice a week (some additional numbers were added after 2002). If we compute the percentage of winning numbers in each drawing that were odd numbers, we see that this varies from 0% on 7/11/01, when the five even numbers 16, 24, 26, 10, and 18 were drawn, up to 100% on Valentine's Day in 1998, when the five odd numbers 41, 23, 5, 39, and 7 were drawn.[4] How variable do we expect

---

[4]http://www.powerball.com/powerball/winnums-text.txt

this percentage to be overall? Since there are 24 even numbers and 25 odd numbers, the percentage of odd numbers in the population is $25/49 = .51$, or 51%. This means the standard error for the sample percentage should be:

$$\text{SE for } \% = \sqrt{\frac{51 \times (100 - 51)}{5}} \approx 22.4\%$$

But if we compute the standard deviation of this list of percentages over the roughly 500 drawings made during the period 1997 to 2002, we only get 21%. The difference between 22.4% and 21% may seem like a small difference, but the stakes are very high with the lottery. Officials at the lottery commission are always on the lookout for fraud or sources of bias. What could explain why the actual variation is smaller than what the standard error formula tells us it should be?

Discussion. Here the population consists of 49 numbers, and the sample size each time consists of five numbers. Since $5/49 = .1$, it means the sample is almost 10% of the whole population. The standard error formulas given in this chapter are most accurate when the sample size is less than a few percent of the population size, which is almost always the case in practice. But if the sample is a lot larger than this, the formulas given above tend to be overestimates. If we imagine an extreme version of the lottery in which all 49 numbers are drawn each time (instead of just 5 numbers), there would be no variation in the percentage of odd numbers drawn—we would see 51% odd numbers each time.

To account for this phenomenon, when your sample is a noticeable fraction of the population, you should take the number given by the standard error formula above and multiply it by the **finite population correction factor**

$$\sqrt{\frac{\text{population size} - \text{sample size}}{\text{population size} - 1}}$$

which statisticians have mathematically justified to be exactly the right correction to use. In this case the correction factor equals $\sqrt{(49 - 5)/(49 - 1)} = .957$, and so the actual variation should be about 95.7% of the 22.4% that the standard error formula tells us, or $.957 \times 22.4 = 21.4\%$. This is much closer to the 21% we actually saw in the data.

This correction factor is essentially negligible as long as the sample size is small compared with the population size. For example if you take a sample of 1,000 out of a population of 1 million the correction factor equals

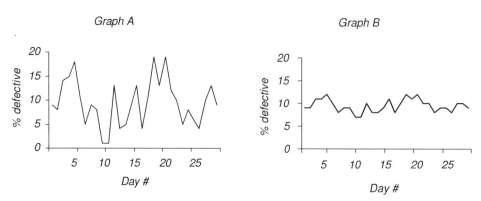

Figure 9.6. The percentage of defective items produced over time.

$\sqrt{999,000/999,999} = .999$. This means the actual standard error will be .999 times what the formula above gives and thus this factor can usually be safely ignored.

Exercises

6. **Television ratings.** A company that publishes television ratings uses a simple random sample of 2,000 households and finds that 10% were watching a particular show. Estimate the margin of error and interpret it.

7. **Manufacturing quality.** A manufacturing process produces 10 percent defective items on average. Every day a random sample of 500 items is tested and the percentage of defective items is graphed. Over a period of 30 days, which graph in Figure 9.6 would we be more likely to see: Graph A or Graph B? Justify your answer with relevant calculations.

8. **Market research.** Market researchers conduct a poll of consumers and compute the margin of error to be plus or minus 12%. Since this is a fairly large margin of error, how many times larger a sample would be necessary in order to get the margin of error down to plus or minus 3%?

9. **Job satisfaction.** Psychologists conducted a survey ten years ago and found the percentage of people satisfied with their jobs to be 65%. This year, they conducted an identical survey and found the percentage has decreased to 55%. The margin of error for the survey is quoted as 10%.

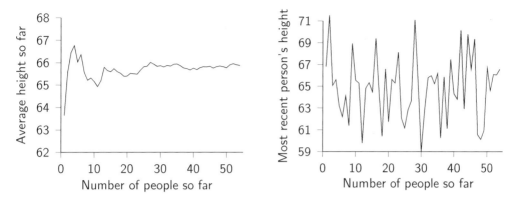

Figure 9.7. Graphs of the average height (on the left) and height of each person passing by (on the right) over time.

A newspaper headline reads, "Job satisfaction down!" Is this headline appropriate? Explain the relevance of the margin of error.

4. The Standard Error for a Sample Average

So far we have talked about the variability of a percentage computed from a sample. If your data are not categorical (for example, yes or no, true or false, male or female, etc.) but are instead numerical values (for example, income, age, profit, etc.) you may not want to summarize the sample in terms of a percentage. In this case you might summarize your sample using an average. We now turn to the issue of how to attach a margin of error to a sample average and how to estimate the standard error.

Suppose we are back people-watching on the street in downtown Boston. This time we measure everyone's height as they walk by and make two graphs: we graph the average height we've observed so far (shown in Figure 9.7 on the left) and also graph each person's height individually (shown in Figure 9.7 on the right). Notice that the law of averages is operating here in the left graph: even though there is a lot of variability in an individual's height (in the right graph), we can see that the average tends to stabilize (in the left graph). The picture on the left, and the law of averages, tells us that as the sample gets larger there tends to be less and less fluctuation (or error) in the sample average.

The **standard error for a sample average** tells us how far off in general the sample average tends to be from the true population average. It's a measure of the typical amount of error we would expect to see if we used the sample average to estimate the population average. To get the margin of error, the reasonable outer limit of the amount of error we would expect to see, we can double the standard error as before. The concept is analogous to the standard error for the sample percentage described previously, but this time it applies to the sample average.

You can compute the standard error for a sample average by taking the standard deviation of the population data (or your best estimate of it so far) and dividing by the square root of the size of the sample. This is sometimes called the **square-root law**. For now I will just show you the formula and demonstrate how to use it, and afterwards we will explain where it comes from.

The **standard error for a sample average** (SE for avg) tells you how far the sample average typically tends to be away from the true population average:

$$\text{SE for avg} = \frac{\text{SD of population}}{\sqrt{\text{size of sample}}}$$

The **margin of error** is twice the standard error, and estimates the largest distance you would reasonably expect to see between the sample average and the population average:

$$\text{ME for avg} = 2 \times (\text{SE for avg})$$

In the previous example where we measured the heights of people walking past us in downtown Boston, from Figure 9.7 we see that with 50 people, the average height is 66 inches. How far away is this number from the true population average? To apply the formula for the margin of error we first need to look at the height data to compute the standard deviation. This works out to be around three inches. The standard error formula above then gives $3/\sqrt{50} \approx .42$ inches, and this means the margin of error will be approximately .84 inches. This means we expect our sample average to be within around .84 inches from the population average.

Example.   If we survey a random sample of 100 people and record their incomes, we get an average of $50,000 with an SD of $10,000. Attach a margin of error to the $50,000.

Solution.   We can calculate the SE for the average as $10,000/\sqrt{100} = \$1,000$, and the margin of error is twice this, or $2,000. This means that we can be quite confident that the average income in the population from which we took this sample equals $50,000 plus or minus $2,000 or, equivalently, in the range from $48,000 to $52,000.

Rent-a-car class action lawsuit.   A rental car company illegally over-charged millions of customers by a few dollars here and there. Someone noticed and filed a giant class action lawsuit. The company knew it was guilty and decided to settle out of court. During the period of overcharges there were over 10 million rentals spread over thousands of offices and many records were on paper only. The company decided to estimate the average overcharge using a random sample. Both sides agreed they wanted the margin of error for the overall average to be less than one penny, since they were going to multiply this by the millions of suspected records to get the total lawsuit settlement amount. How large a sample should they take?

Solution.   A look at a small sample of customer records showed that the standard deviation in the overcharges from record to record was somewhere around $1 (this is an estimate of the "SD of the population"). Some records had no overcharge while some records had a $2 or $3 overcharge. Since they wanted the margin of error to be $.01, this means that the standard error should equal half of this, or $.005 (this is the desired "SE for avg" in the formula). We can solve for the sample size using the standard error formula as follows:

$$\$.005 = \frac{\$1}{\sqrt{\text{size of sample}}} \Rightarrow \sqrt{\text{size of sample}} = \$1/\$.005 = 200$$

and therefore

$$\text{size of sample} = (\$1/\$.005)^2 = 200^2 = 40{,}000$$

They would need to sample 40,000 random records to get a margin of error of around $.01.

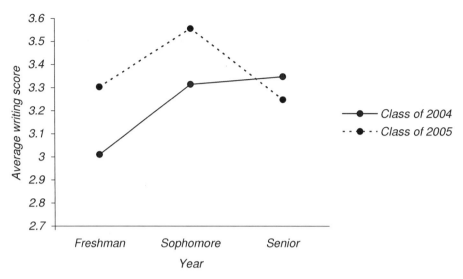

Figure 9.8. Average writing scores over time for two college graduating classes.

Training effectiveness.   A business school decides to have a number of carefully graded writing assignments throughout its curriculum in order to improve the business writing skills of its graduates. To monitor the effectiveness of this approach, the school has student papers graded by the same team of outside experts during the freshman, sophomore and senior years for two recent graduating classes of about 100 students each. Each paper is given a score from 1 to 6 and the class averages are graphed in Figure 9.8.

For the Class of 2004, average writing scores appear to steadily increase with each year. This is the good news that the school was hoping to see. But something seemed to go terribly wrong during the senior year for the Class of 2005 when the average score dropped sharply. Is this a bad case of "senior-itis"? Or does the program only work during the first year?

Solution.   Before we try to come up with an explanation for the graph, we should ask about the margin of error for each of the points in the graph. If it turns out that all the points lie within the margin of error from each other, it's not honest to claim anything more than sampling variability to explain any of the trends here.

Looking at the raw data we see that the SD of the scores within each class was around 2 points. This means for any batch of 100 random scores, we would expect to see an error (the standard error for the sample average) of around $2/\sqrt{100} = .2$ points with a margin of error twice this, or $.2 \times 2 = .4$ points. Since all the movements of these averages are all smaller than .4 points each time (the big senior year drop for the class of 2005 was from around 3.5 points to around 3.2 points: a drop of .3 points), the margin of error is not sharp enough to honestly attribute any of the movements to anything other than natural variability. Although it seems as though the school should congratulate itself for the rises and be concerned about the falls, the margin of error is not sharp enough to warrant either. The only story we can honestly tell is that there is a lot of variability from year to year.

Because the changes in the averages are less than the margin of error, we say that the changes are not **statistically significant**. This is a concept we will take up in the next two chapters.

Why the formula for standard error works

Why is the formula

$$\text{SE for avg} = \frac{\text{SD of population}}{\sqrt{\text{size of sample}}}$$

correct and why is there a square root in the denominator? Here we will look at a very simple numerical example to illustrate why the square root is correct.

Suppose we have a large population of people half of whom earn $50,000 and the other half earn $30,000. This gives an average salary of $40,000 with a standard deviation of $10,000. Suppose we then pick random pairs of people, so we have a lot of samples of size two people each, and then write down the average salary in each pair. The standard error formula tells us that the standard deviation of this list of pair averages will be around

$$\text{SE for avg} = \frac{\text{SD of population}}{\sqrt{\text{size of sample}}} = \$10,000/\sqrt{2} \approx \$7,000$$

To see why this is a reasonable figure, notice that when we randomly pick pairs of people there are four equally likely possibilities for the pair: (1) the first person picked earns $50k and second earns $30k—an average of $40k, (2) the first person picked earns $50k and the second earns $50k—an average of $50k, (3) the first person picked earns $30k and the second earns

$50k—an average of $40k, and (4) the first person picked earns $30k and the second earns $30k—an average of $30k. When we write down the list of pair averages, we thus get the numbers (in thousands) 40, 50, 40, 30. This list has an average of 40 and an SD of

$$\text{SD} = \sqrt{\frac{(40-40)^2 + (50-40)^2 + (40-40)^2 + (30-40)^2}{4}} = 10/\sqrt{2} \approx 7$$

and means the average salaries for pairs of people will have a standard deviation of around $7,000.

Why the sample percentage standard error formula works.   Where does the formula

$$\text{SE for } \% = \sqrt{\frac{\text{true percentage} \times (100\% - \text{true percentage})}{\text{size of sample}}}$$

come from? A sample percentage can be also viewed as a sample average if you look at it the right way. Suppose you want to estimate the percentage of people who have some characteristic. The usual way would be to count up the number of people who have the characteristic in your sample, divide by the sample size, and then multiply by 100%. An alternative way would be to assign everybody in your sample a number: people who have the characteristic get assigned a 100, and people who don't have the characteristic get assigned a zero. Then if you average together everybody's numbers, you will get the percentage of people who have the characteristic. This means the sample percentage is just another type of average. The numerator in the formula above is just a short-cut formula for computing the SD of the population.

For example, consider a population with five members where two of them have some characteristic you're interested in. If the people with the characteristic are assigned the number "100" and the people without the characteristic are assigned the number "0" then the average of the numbers will equal

$$\text{Average} = \frac{100 + 100 + 0 + 0 + 0}{5} = 40$$

and the standard deviation of the numbers equals

$$\begin{aligned}
\text{SD} &= \sqrt{\frac{(100-40)^2 + (100-40)^2 + (0-40)^2 + (0-40)^2 + (0-40)^2}{5}} \\
&= \sqrt{2400} \approx 49.
\end{aligned}$$

We also get the same number if we use the shortcut formula

$$SD = \sqrt{40 \times (100 - 40)} = \sqrt{2400} \approx 49$$

5. Exercises

10. **Account auditing.** An accounting firm is hired to check the sales entries for a large corporation and their goal is to certify that less than 5% of the entries have errors. There are around 1 million entries for them to check and it usually takes around 10 minutes to do the research required to check a single entry. (a) How long would it take to compute the percentage of entries having errors if they check all one million entries? (b) How long would it take to estimate the percentage of entries having errors using a random sample big enough to get a margin of error of less than 2%?

11. **Mutual funds.** On average, the stocks listed on the New York Stock Exchange (NYSE) went up last year. Which of the following two mutual funds would have had a bigger chance of going up over the year: Fund A which invested equally in 5 randomly chosen NYSE stocks, or Fund B which invested equally in 50 randomly chosen NYSE stocks? Explain briefly.

12. **Medical benefits.** The majority of businesses nationwide do not offer medical benefits. One researcher selects a simple random sample of 50 businesses nationwide, and another researcher will do the same for a sample of 200 businesses nationwide. Which researcher is more likely to have a sample where the majority of the businesses offer medical benefits? Or are both researchers approximately equally likely to get this? Explain your answer.

13. **People's heights.** People in a large population average 60 inches tall. You will take a random sample and will be given a dollar for each person in your sample who is over 65 inches tall. For example if you sample 100 people and 20 turn out to be over 65 inches tall, you get $20. Which is better: a sample of size 100 or a sample of size 1,000? Choose one and explain. Does the law of averages relate to the answer you give?

14. **Inspection sampling.** A production process fills boxes with 20 ounces of product with a standard deviation of 2 ounces, and the amount of product in each box tends to follow a normal distribution. (a) An inspector samples a random box and finds it contains 22 ounces of product. Is this unusually high? Explain. (b) Another inspector takes a random sample of one hundred boxes and finds an average of 22 ounces per box. Is this unusually high? Explain.

15. **Web site transactions.** A study of a web site finds that over the course of the whole year 10% of people who visit actually initiate a transaction. The researchers then compute this percentage for each of four days when around 1 million people per day on average visited. Which of the following would be a more reasonable guess for these four percentages if you round them to the nearest percentage point? Pick either choice (i) or (ii), and justify your answer briefly. (i) 15%, 5%, 12%, 8% (ii) 10%, 10%, 10%, 10%.

16. **GMAT scores** Incoming full time Boston University MBA students are divided into three cohorts. The average GMAT score for these students was 640 with an SD of 20. Based on this, which of the following would be the most reasonable guess for the average GMAT score in each of the three cohorts? Pick either choice (i), (ii), or (iii) and justify your answer with an appropriate numerical calculation, mentioning any assumptions you may need to make: (i) 620, 640, 660 (ii) 630, 640, 650 (iii) 637, 640, 643.

17. **Public war sentiment.** A June 2004 *USA Today* article states, "Most Americans now say that sending U.S. troops to Iraq was a mistake, a USA Today/ CNN/ Gallup Poll finds… It is the first time since Vietnam that a majority of Americans has called a major deployment of U.S. forces a mistake… 54% say it was a mistake."[5] The article reports that the margin of error for this poll is plus or minus 5%. Then the article goes on to claim "Among likely voters nationwide, Bush leads Kerry 48% to 47%, with independent candidate Ralph Nader at 3%." The margin of error for this second poll is given to be plus or minus 3%. Comment on the relevance of the margin of error to each of the statements the article makes.

---

[5]From "Poll: Sending troops to Iraq a mistake," by Susan Page, *USA Today*, June 24, 2004.

18. **Hourly wages.** Based on a simple random sample of one hundred, an analyst estimates the average hourly wage earned by workers in a city to be $30 and computes the margin of error to be $5. Can we conclude from this that most workers there earn between $25 and $35 per hour? Is this the right interpretation for the margin of error?

19. **Pharmaceutical simulation.** To estimate the expected profit from a pharmaceutical business deal, a computer generates a random sample from a very large population of possible simulated profit scenarios, computes the average profit for these sampled scenarios, and prints out the answer. The longer the program is run the larger the sample it will take and the smaller the margin of error will be for the answer it gives. The difficulty is that when the program is run for one hour (and therefore takes a very large sample), there still appear to be large fluctuations in the answers it gives. When the program is run on four separate occasions for one hour each time the four answers it gives (in billions of dollars) turn out to be (rounded to the nearest billion) 1, 3, 3, and 1. For how much total time do you estimate the firm needs to run the program to get the margin of error down to about 100 million dollars? Explain briefly. (Recall that 1 billion = 1,000,000,000).

20. **Smoking survey.** Government regulators want to estimate the percentage of people who smoke in New York State as well as the percentage of people who smoke in Maine. They decide to survey a random sample of people amounting to one-tenth of 1 percent of the population of each state. In which state will the survey be more accurate? Or will the accuracy of the survey be about the same in both states?

21. **Market research.** Two market researchers studying alcoholic beverage brand-name recognition each plan to take simple random samples of registered undergraduates at Boston University. One plans to randomly select 50 students and the other plans to randomly select 200 students. Which researcher is more likely to get a sample in which the average age is above 22 years? What reasonable assumption do you need to make to answer this?

22. **Election polling.** Twenty-five independent polls of one thousand random people each are taken during the course of an extremely close election campaign. People in the poll are asked which of the two candidates they favor and the percentage who favor the incumbent candidate is

computed for each poll. (a) Estimate the standard error and the margin of error for each of these polls. Justify your answer, and briefly explain the difference between them. (b) How large a poll sample would be required if a margin of error of only 10% was desired? (c) One of the other candidate's approval ratings averaged around 48% over all the twenty five polls. If you computed the standard deviation of the list of all twenty five poll percentages for this candidate, approximately what number would you expect to get? Explain.

23. **Presidential election.** Polls showed the two main candidates in the 2004 presidential election were nearly tied on the day before the election. To predict the winner a newspaper would like to conduct a poll that has a margin of error of less than 1%. Roughly how large a sample would be needed for such a poll?

24. **Hotel reservations.** A hotel finds that only 80% of people who reserve rooms actually show up. If it has 85 rooms available and takes 100 reservations, is it very likely there will be enough rooms for the people who show up? Justify your answer.

25. **Deposit forecasting.** For planning purposes an investment firm would like to have a forecast of the average amount its current customers will deposit over the coming year. It would like to estimate this average with a margin of error of less than $2,000. So as not to disturb too many customers, it decides to telephone a small random selection of 25 of its customers to ask each how much they plan to deposit. The average amount revealed is $25,000 with a standard deviation of $30,000. Is this a large enough sample to give the margin of error the firm would like? If not, how much larger a sample does it need to take?

26. **Health insurance premiums.** A recent *Wall Street Journal* article discusses an online survey and writes, "the online survey of 2,325 U.S. adults found that 53% of Americans think it is fair to ask people with unhealthy lifestyles to pay higher insurance premiums than people with healthy lifestyles."[6] Instead of attaching a margin of error, the article later says "with pure probability samples of 2,325 adults, one could

---

[6]"Many Americans Back Higher Costs For People With Unhealthy Lifestyles," July 19, 2006, *The Wall Street Journal*, http://online.wsj.com/public/article/ SB115324313567509976-80SD01MJZEFBPNjvlePlec962fE_20070719.html?mod=tff_main_ tff_top

say with a ninety-five percent probability that the results have a sampling error of $+/-$ 3 percentage points. However that does not take other sources of error into account. This online survey is not based on a probability sample and therefore no theoretical sampling error can be calculated." What do they mean by all this and are they making sense here?

27. **Asbestos lawsuits.** A recent *Wall Street Journal* article describes how many firms face bankruptcy due to the large number of asbestos related lawsuits filed against them.[7] It goes on to describe how legal experts believe up to 30% of these claims do not meet minimum medical requirements to be a valid claim and it takes considerable time by medical experts to determine if any given claim is valid or not. Creditors for one bankrupt company asked that a random sample of the 23,000 claims filed against it be taken because they believed the 30% figure was too low. When a random sample of 1,691 claims was chosen, an expert determined that 87% were not valid. Even though they sampled less than 10% of all the claims, is this enough to confidently conclude that the 30% figure is too low?

28. **Government bonds.** Bond investors often like to buy shares in mutual funds that give returns that closely follow commonly quoted indices, such as the Lehman Brothers Government Index. One popular index is computed from several thousand different bond types. Roughly how many bonds do you think would be needed in a portfolio that could reliably give one year portfolio returns within half a percentage point of the index? You can assume bond returns typically vary in range from -5% to $+5\%$ with a standard deviation of about 2.5%, but state any other assumptions you make for this.

29. **Disability claims.** In 2005 the Veterans Administration (VA) reported that the $25 billion in annual disability claims it pays had been rising steadily with post-traumatic stress disorder (PTSD) claims making up the bulk of the increase. The VA planned to review 72,000 such claims submitted by veterans between 1999 and 2004 because of concerns a large number may have been fraudulent.[8] This would be an enormous task

---

[7] "An Asbestos Exit," *The Wall Street Journal*, Wednesday, November 17th, 2004, p. A16.

[8] "Veterans Get Reprieve from Pentagon PTSD Review," by Brendan Coyne, Nov. 11, 2005. http://newstandardnews.net/content/index.cfm/items/2586/printmode/true

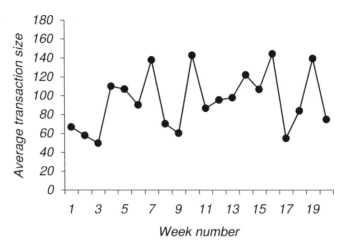

Figure 9.9. Average transaction size (in thousands of dollars) for each week.

because the VA would have to carefully examine the medical evidence associated with each claim. A decision was made to sample 2,100 claims and determine for each claim whether there was evidence of fraud or not. But advocates for the veterans expressed concerns that this sample size was not large enough. How could this be assessed in advance of taking the sample? Discuss briefly and give any relevant calculations.

30. **Commodity trading.** A commodity trader who usually makes around 100 transactions per week graphs the average transaction size for each week – this is shown in Figure 9.9. (a) Would you estimate the overall average transaction size (in thousands of dollars) to be closer to 50, 100, or 150? (b) Would you estimate the standard deviation for transaction sizes (in thousands of dollars) to be closer to 2, 200, or 2,000?

31. **Manufacturing defects.** A manufacturing process produces on average 20% defective items. Every day for 14 days a batch of 25 items is produced and the percentage of defective items is graphed. Which of the two graphs in Figure 9.10 is most likely to be what we would see? Explain with any relevant calculations.

32. **Cancer rates.** Figure 9.11 shows a map of the United States where counties are shaded having the highest 10% of kidney cancer rates and another map showing the counties with the lowest 10% kidney cancer rates

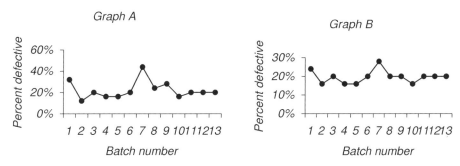

Figure 9.10. Percentage of defective items per batch over time.

(after an adjustment has been made for age differences). What could explain why most of the counties in both maps are scattered throughout rural areas and in the West?[9]

33. **A statistical tie.** A news article reports on the results of a political poll of 1,212 adults that included 485 Democrats and 374 Republicans.[10] The article reports "A majority of Democrats favor Clinton, whereas fewer than a third of Republicans favor their front-runner, Giuliani....but when the two front-runners are pitted against each other, Clinton leads Giuliani by just two percentage points, 49 percent to 47 percent, a statistical tie." The article also says that the margin of error for the Democratic poll is 4.5 percentage points, for the Republican poll it is 5 percentage points, and for general election questions it is 3 percentage points.

   (a) Why are there three different margins of error?

   (b) What do you think they mean by "a statistical tie"?

   (c) Based on the information given, do the quoted margins of error seem correct?

34. **National mood polling.** A news article reports on the results of a political poll of 991 people: "The number of Americans who believe the country is on the wrong track jumped four points to 66 percent....Bush's job approval rating fell to 24 percent from last month's record low....of 29 percent," and also the percentage of Americans who rate their personal

---

[9]From *Teaching Statistics: A Bag of Tricks,* by A. Gelman and D. Nolan, Oxford University Press, 2002.

[10]"Poll: As Thompson's star fades, Clinton's on the rise," by Bill Schneider, *CNN,* October 16, 2007, http://www.cnn.com/2007/POLITICS/10/16/schneider.poll/index.html

Highest kidney cancer death rates

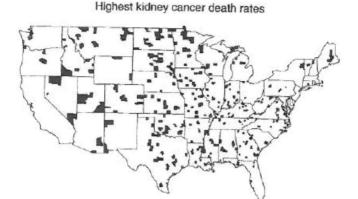

Lowest kidney cancer death rates

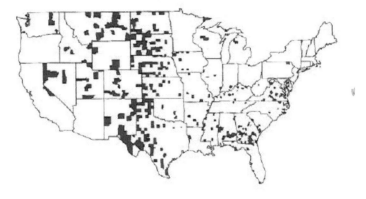

Figure 9.11. Maps of the United States where counties are shaded having the highest 10% of kidney cancer rates (at the top) and the lowest 10% of kidney cancer rates (at the bottom).

financial situation as excellent or good dipped to 54 percent from 56 percent."[11] Could these changes be reasonably attributed to sampling variation, or are they indicative of real changes in the national mood?

35. Soldier desertions. A news article states, "Soldiers strained by six years at war are deserting their posts at the highest rate since 1980...While the

---

[11] "Voters unhappy with Bush and Congress," by John Whitesides, Reuters, Wed Oct 17, 2007, http://www.reuters.com/article/politicsNews/idUSN1624620720071017.

totals are still far lower than they were during the Vietnam War, when the draft was in effect, they show a steady increase over the past four years and a 42 percent jump since last year...According to the Army, about nine in every 1,000 soldiers deserted in fiscal year 2007, which ended Sept. 30, compared to nearly seven per 1,000 a year earlier. Overall, 4,698 soldiers deserted this year, compared to 3,301 last year."[12] How variable do you expect the desertion rate to be from year to year? What does this tell you about the difference between this year and last year?

36. **Profit simulation.** A computer simulation of a risky business deal creates a large random sample of potential outcome scenarios from a (very large) population of all possible outcome scenarios. The computer reports that the average profit from these (sampled) scenarios is $1.5 billion with a standard deviation of $6 billion. The standard error is reported as $0.1 billion.

    (a) In a couple of brief sentences, give a managerial interpretation for the $6 billion and the $0.1 billion.

    (b) If the sample size were doubled, would we expect the standard deviation to increase, decrease, or stay about the same? Explain briefly.

37. **Consumer survey.** Market researchers are preparing an online consumer survey to estimate the percentage of consumers nationwide who recognize a particular product. They would like to have a margin of error of approximately 5 percentage points.

    (a) Approximately how large a sample size do they need? Explain briefly.

    (b) After receiving the responses, a critic says that responses were primarily solicited from university students and are not very representative of all consumers. A researcher involved with the study argues that this limitation is fairly expressed by the 5% margin of error. Do you agree with the critics or the researcher? Explain briefly.

---

[12] "Army Desertion Rate Highest Since 1980," *Associated Press*, Nov 16, 2007, by Lolita C. Baldor. http://apnews.myway.com/article/20071116/D8SV1IU01.html

Chapter 10

Confidence Intervals

1. The Central Limit Theorem

In the last chapter you learned how to calculate the margin of error for a sample average or percentage. Since this margin of error is not an ironclad guarantee, how likely is it that the truth actually lies somewhere within the margin of error? In other words, how likely is it that a sample percentage is actually close enough to the true population percentage to be within the margin of error? A famous statistical result called the **central limit theorem** can help quantify this chance in most common situations. The central limit theorem says that with a reasonably large random sample you can use the normal curve to estimate the chance a sample average is within a desired number of standard errors from the true population average — even if your data don't follow a normal curve. This principle for averages applies to percentages too. (How large is a "reasonably large" sample? It depends on how skewed the histogram is for the data in the population, but any sample a manager might consider using in a real business situation will usually be large enough to apply the central limit theorem. Some people say a sample of size 30 or more is necessary, but this is certainly not a magic number you should memorize.)

Let's illustrate the central limit theorem using data from the Powerball lottery, a very popular lottery game with jackpots of over $200 million. Twice a week, five winning numbers are drawn from the numbers 1 through 55 and

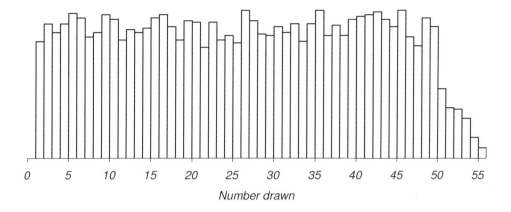

Figure 10.1. Histogram for winning Powerball numbers.

the winning numbers over the past 10 years are posted online.[1] If we look at the histogram in Figure 10.1 for these winning numbers, it looks fairly level until around number 50 when it drops off sharply. What explains the sharp drop-off?

The fairly level part of this graph, from 1 to 49, shows us that the numbers 1 through 49 have each been chosen roughly the same number of times over the past 10 years. The drop-off above 49 reveals that numbers above 49 have been chosen much less frequently. And number 55 has been chosen least of all. What explains all this?

It turns out that up until October 2002 only numbers up to 49 were used. Lottery officials then increased this to 53 to reduce the chance of winning. The prize money rolls over to the next drawing when there is no winner and huge jackpots get plenty of media attention.[2] After 2005 the numbers were increased again so that now numbers up to 55 are used. The sharp drop-off in the histogram is because the higher numbers have only been used for the last few years.

The important thing to notice about this histogram to illustrate our main point is that it does not look anything like a normal curve. Now let's investigate these lottery numbers from another angle by computing the average

---

[1] http://www.powerball.com/powerball/winnums-text.txt

[2] "Too many players winning Powerball, lottery says," *Chicago Sun-Times*, Mar 22, 2005, by Joe Mandak. Also see http://www.palottery.state.pa.us/lottery/cwp/view. asp?A=11&Q=468345

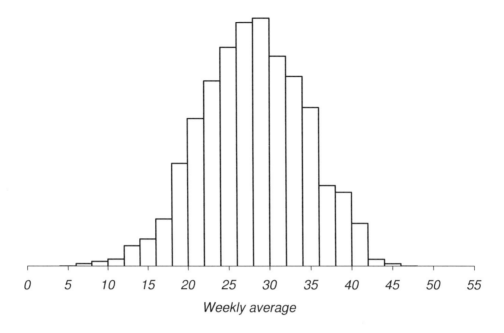

Figure 10.2. Histogram for the weekly average of the winning Powerball numbers.

of the five winning numbers from each drawing. We see that this average ranged from a low of 5.2 on June 19, 2002, when the winning numbers were 4, 7, 3, 11, 1, to a high of 42.4 on October 13, 2001, when the winning numbers were 37, 40, 41, 46, 48. If we draw a histogram for all these averages, what shape do we expect to see? Since there were never any drawings with an average below 5 or above 43, this histogram is more concentrated than the previous one. In fact, Figure 10.2 shows that it looks very much like a normal curve.

The reason a normal curve magically appears, even though the previous histogram looked nothing like a normal curve, is because of the central limit theorem. The central limit theorem says that, even if the histogram for the numbers drawn does not follow a normal curve (recall that the first histogram was fairly flat, with a sharp drop above 49), the histogram for the average number from each drawing will tend to follow a normal curve. Usually we only have a single sample, rather than repeated drawings like we do here, and this central limit theorem can help us understand the likelihood that our sample average varies from the true population average.

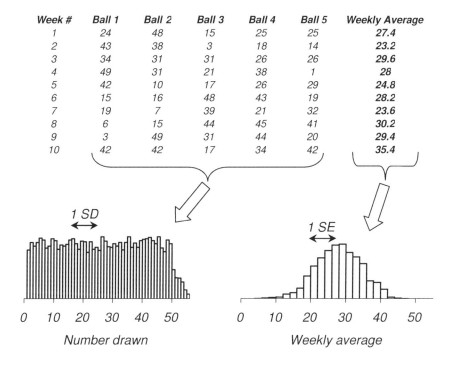

Week #	Ball 1	Ball 2	Ball 3	Ball 4	Ball 5	Weekly Average
1	24	48	15	25	25	27.4
2	43	38	3	18	14	23.2
3	34	31	31	26	26	29.6
4	49	31	21	38	1	28
5	42	10	17	26	29	24.8
6	15	16	48	43	19	28.2
7	19	7	39	21	32	23.6
8	6	15	44	45	41	30.2
9	3	49	31	44	20	29.4
10	42	42	17	34	42	35.4

Figure 10.3. An illustration of the central limit theorem. Notice that the histogram for the winning numbers (on the left) looks nothing like a normal curve but the histogram for the average of the winning numbers (on the right) does look like a normal curve.

We can summarize the central limit theorem with the diagram in Figure 10.3. The table only shows data from the first 10 weeks, whereas the histograms show all the data. Notice how we've indicated the size of one standard deviation on the histogram for the numbers drawn and the size of one standard error on the histogram for the weekly averages.

To see a business application of this, suppose we use a random sample of tax forms to estimate the average income of a population. Even though incomes do not follow a normal curve (income histograms usually have a long right-hand tail) we can use the normal curve to estimate the chance that our sample average is within a desired amount from the average of the population. If the standard error for the average turns out to be estimated at $1,000, we can say that there is a 68% chance that the average income in our sample is within \$1,000 (one standard error) of the true average income in the population we are studying and a 95% chance the average income in our sample is within \$2,000 (two standard errors) of the true average income. The 68% and 95% figures come from areas under the normal curve: 68% of the area under the normal curve is between +1 and −1 and 95% of the area is between +2 and −2. Even though incomes do not follow the normal curve at all, you can use the normal curve to estimate the chance that the average income in the sample falls in one range or another.

Again note that the central limit theorem does not say that the incomes in your sample will follow a normal curve or that the chance individual people in your sample are in one income range or another can be estimated with a normal curve (if the data histogram is skewed you can't use the normal curve for this)—it only says that the chance the sample average is in one income range or another can be estimated with the normal curve. The theorem talks about fluctuations in the sample average—not the individual data values themselves. The theorem is true because any odd quirks in the data histogram's shape end up influencing the sample average very little when you have a large sample to average over.

> The **central limit theorem** says that with a reasonably large sample you can use the normal curve to estimate the chance a sample average is within some desired number of standard errors from the population average—even if the data values themselves don't follow a normal curve. The same principle applies to sample percentages.

As another example, suppose a political poll finds 65 percent of the population favors your candidate but the poll has a margin of error of 4 percentage points. Does this mean we can guarantee exactly 65% of the population favors your candidate? Of course not because there is the margin of error to consider. But does this margin of error mean we can guarantee that somewhere between 61% and 69% of the population (65% plus and minus 4%) favors your candidate? The answer is no again because the margin of error is not an ironclad guarantee. We can, however, quantify our uncertainty as follows. We can say that out of 100 similar polls we expect 95 of them to be accurate to within the margin of error. This means we can say we are 95%

confident that somewhere between 61% and 69% of the population favors your candidate. The number 95 comes from the fact that the margin of error equals 2 standard errors and the area under the normal curve between +2 and -2 is 95%.

This is why the margin of error range is also called a "**95% confidence interval**"—because we are 95% confident that it's correct. Those are pretty good odds for the uncertain world we live in. If that's not good enough for you, you can calculate your own customized margin of error by tripling the standard error to get 99.7% (3 standard errors) confidence instead of 95% confidence (the 99.7% figure comes from the fact that 99.7% of the area under the normal curve is between -3 and $+3$). You can similarly calculate confidence intervals for averages as well as percentages. (Although it's not so common in practice, you can find an interval for any level of confidence you like. For example, for 80% confidence you use the sample average plus or minus 1.3 times the standard error. This 1.3 comes from the fact that 80% of the area under the normal curve is between -1.3 and 1.3.)

> A **95% confidence interval** is the sample average plus or minus the margin of error (which equals twice the standard error).
>
> A **99.7% confidence interval** is the sample average plus or minus three times the standard error.

Example. Suppose we survey a random sample of 100 people, record their incomes and get an average of $50,000 with an SD of $10,000. Give a 95% confidence interval for the average income in the population. Also give a 99.7% confidence interval.

Solution. Earlier we calculated the standard error to be $1,000 and so the margin of error is $2,000. The 95% confidence interval is thus $48,000 to $52,000. The 99.7% interval will be $50,000 plus or minus 3 standard errors, or $47,000 to $53,000.

Question. Does this mean 95% of people have incomes in the range $48,000 to $52,000?

Answer. No, it means we are 95% confident that the average of all people in the population falls in that range. There may actually be no single person who falls in that narrow income range but we are 95% confident the overall average does!

A 95% confidence interval **does not** mean 95% of the population falls in that interval or that we are 95% confident someone from the population falls in that interval.

A 95% confidence interval **does** mean we are 95% confident the population average or population percentage falls in that interval.

Rent-A-Car lawsuit. In the rent-a-car lawsuit example from the previous chapter, suppose they took a random sample of 100 records and saw that 25% were overcharged. Give a 95% confidence interval for the overall percentage of the 10 million records that were overcharged.

Solution. Since

$$\text{SE for } \% = \sqrt{\frac{25 \times (100 - 25)}{100}} \approx 4.3\%,$$

the margin of error is twice this, or 8.6%. This means the 95% confidence interval for the percentage of records overcharged in the population of 10 million is 25% plus or minus 8.6%. This means we can confidently say (with 95% confidence) that somewhere between 16% to 34% of the records were actually overcharged.

Monte Carlo Simulation. To estimate the expected profit from a potential business deal, a computer can be used to generate random samples from a very large population of possible simulated profit scenarios. The computer can then record the profit for each of these sampled scenarios and the average of these is often used as an estimate of expected profitability. This approach to estimating something using random scenarios is called the "Monte Carlo simulation" approach, named after the famously wealthy gambling city in

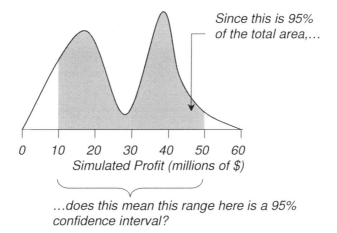

Figure 10.4. A histogram of simulated profits for a business deal. Is the shaded region a 95% confidence interval?

Monaco off the coast of France (random scenarios are used every day at the gambling tables). Pharmaceutical companies, for example, are starting to use the Monte Carlo simulation to evaluate the many uncertainties surrounding development of a new drug. Many other industries are using this as well.[3]

A sample of ten thousand profit scenarios is generated and the histogram in Figure 10.4 is drawn for the simulated profits. The region covering 95% of the total area is shaded—it goes from $10 million to $50 million. Is this a 95% confidence interval?

Discussion.   No. The shaded range tells us that 95% of the simulated profits were between $10 million and $50 million. We could also say that we believe there is a 95% chance that the profit from this deal, if we decided to go ahead with it, would be between $10 million and $50 million.

A 95% confidence interval is supposed to be the range we are 95% confident contains the true mean of the profit distribution—the mean of the population from which the samples were drawn. The shaded range above tells us where we are 95% sure the actual profit will lie. Even though the actual profit can be very uncertain in most deals, the mean profit (or expected profit) can be quite sharply estimated. It is common that even with highly risky deals

---

[3]See "The Flaw of Averages," by Sam Savage, *Harvard Business Review.* November 2002, pages 20–21.

the mean profit for one deal is compared with the mean for another deal in order to see which deal might be better.

We compute the 95% confidence interval as follows. We first need more information from the data. We compute the average profit to be $28 million and the SD to be $10 million. Since our sample size was 10,000, we can then compute:

$$\text{SE for avg} = \$10 \text{ million}/\sqrt{10{,}000} = \$.1 \text{ million} = \$100{,}000$$

Doubling this to get the margin of error, we get $200,000. This tells us that the 95% confidence interval for the expected profit of this deal is $28 million plus or minus $200,000.

Once again, we are not 95 percent confident that the actual profit will be $28 million plus or minus $200,000—the correct range for this statement would be $10 million to $50 million. We can say that we are 95 percent confident that the mean of the profit distribution (or the expected profit from this deal) is between $27,800,000 and $28,200,000.

Question. Why doesn't the histogram for profit look more like a normal curve? Doesn't the central limit theorem say it should?

Solution. The central limit theorem does not say the data in your sample will follow a normal curve. It says that the chance your sample average is in one range or another can be estimated with a normal curve. The central limit theorem talks about variations in the sample average—not about variations of the data values themselves.

> The central limit theorem does not say the data in a large sample will follow a normal curve. The central limit theorem talks about variations in the sample average—not about the data values themselves.

Household sizes. Figure 10.5 is a histogram for the number of people living in each household across the United States based on data from the US Census Bureau. This histogram shows an average of around 2.6 and census data confirms that this is the average household size in the United States. This

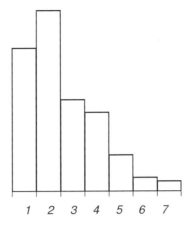

Number of people in household

Figure 10.5. Histogram for the number of people living in households across the United States.

histogram does not look at all like a normal curve because it has a long right hand tail.

If we compute the average household size in each city across the country, we see it ranges from a low of 2.1 people per household in Cincinnati and 2.2 people in Boston to a high of 4.3 people per household in Santa Ana. But when we look at the histogram in Figure 10.6 for these averages it does not look very much like a normal curve. Doesn't the central limit theorem tell us that it should?

Solution. It's true that if each city was a random sample of people and each city was approximately the same size, the histogram would look a lot more like a normal curve. But even if the cities were all the same size, cities are not random samples of people: different cities tend to have significantly different demographic characteristics which can explain the different household sizes. The central limit theorem only applies to random samples and can be misleading when applied to such non-random samples.

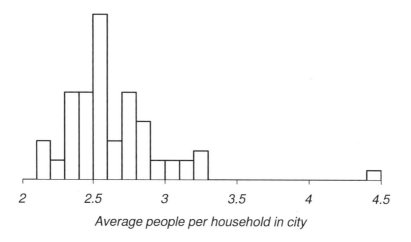

Figure 10.6. Histogram for average household size in cities across the United States.

Technical Notes

The usual mathematical notation for a confidence interval for a population average is

$$\left[\overline{X} \pm z_{\alpha/2} \frac{s}{\sqrt{n}} \right]$$

where n is the sample size, $\overline{X}$ is the sample average, s is the estimate of the SD based on the sample data, $z_{\alpha/2}$ is the $(1 - \alpha/2) \times 100$th percentile of the normal distribution, and $(1 - \alpha) \times 100\%$ is the level of confidence.

Exercises

1. **Television watching.** A market research survey of 1,000 randomly selected people shows that the average number of hours of television they
   watch per week is 4 hours with an SD of 5 hours. And 10% of the people
   watch no television at all.

   (a) Give a 95% confidence interval for the percentage in the population
   who don't watch television.

   (b) Give a 95% confidence interval for the average number of hours
   watched per week in the population.

   (c) Does this mean approximately 68% of people watch in the range of
   4 hours plus or minus 5 hours per week? Explain.

2. **Hourly wages.** Based on a simple random sample of one hundred, re-searchers calculate a 95% confidence interval for the average hourly wage earned by construction workers in the city of Boston. This interval turns out to be from $16 to $22.

   (a) Does this mean we can say that roughly 95% of construction workers in the city earn hourly wages in the range $16 to $22? Explain briefly.

   (b) In order to have an estimate with a margin of error of $1, how much bigger a sample would researchers need?

3. **Estimating salaries.** An accountant takes a simple random sample of 100 people to estimate the average salary in a large population of workers. She computes a 95% confidence interval and gets $85,000 to $95,000.

   (a) True or False and explain: We can estimate that approximately 95% of the people in the population earn in the range $85,000 to $95,000.

   (b) True or False and explain: We are 95% confident that a random person in the population earns in the range $85,000 to $95,000.

   (c) True or False and explain: We are 95% confident that the average person in the population earns in the range $85,000 to $95,000.

   (d) Based on this confidence interval, what is your best guess for the standard deviation of the salaries? Pick the best of the following choices and explain: $2,500, $5,000, $25,000, $50,000.

4. **How large a sample do you need?** How large a sample do you need to estimate an average with a margin of error of $1 when the preliminary data you've gathered has a standard deviation of around $5?

2. Creating a Monte Carlo simulation

To estimate the range of profits possible from a potential business deal, a Monte Carlo simulation can be created in Excel to generate random samples from the population of possible outcomes. Here we give a simplified example, though the same idea can be used to study real problems. The basic idea is to incorporate random numbers into an Excel spreadsheet to reflect uncertainty about various quantities, and then replicating this many times to see how this variation impacts the bottom line.

New product development. Suppose demand for a new product is ex-pected to be around 600 units, the profit from each unit sold is expected

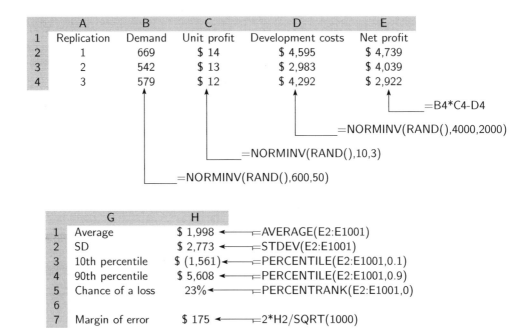

Figure 10.7. The file profitsimulation.xls.

to be around $10 and the development costs in total are expected to be around $4,000. It looks like net profit is expected to be $2,000, but how much risk is involved? What's the chance of losing money? There is uncertainty about each of the three quantities, and one way to handle this is to assume that they will be generated from a normal distribution with some given mean and standard deviation. Suppose that market experts judge that there is good chance that demand will be within 50 units of the expected 600 units and it is almost certain to be within 100 units of the expected 600; this means we might model it as following a normal distribution with a mean of 600 and a standard deviation of 50 units (though this may allow demand to be negative, it's unlikely and won't impact the simulation much). And suppose for unit profit the standard deviation is $3 and for development costs it is $2,000.

The file profitsimulation.xls, shown in two pieces in Figure 10.7, illustrates how to set this up in Excel. You can use the Excel command =NORMINV(RAND(),600,50) to generate random numbers following a normal distribution with a mean of 600 and a standard deviation of 50.

The command =NORMINV(RAND(),10,3) can be used for unit profits and =NORMINV(RAND(),4000,2000) for development costs. The net profit is simply the demand multiplied by the unit profit and then development costs are subtracted. And suppose we decide to take a sample of 1,000 possible outcome scenarios: each row numbered from 1 to 1,000 is a different replication of the simulation (the figure shows only the first three out of the 1,000 replications). You may notice that the random numbers shown above are different than the ones in the spreadsheet and they keep changing every time you make any changes; don't be alarmed but it's just the way random numbers behave in Excel.

In Column H we have computed the average, standard deviation, percentiles, the chance of a loss and the margin of error for the net profit. How are each of these quantities to be interpreted? The average of $1,998 tells us what we expect the net profit to be. The standard deviation of $2,773 tells us about how much uncertainty we expect in the profit outcome. More specifically, the 10th and 90th percentiles tell us that there is about a 10% chance that profit is below negative $1,561 and another 10% chance that profit is above $5,608. We also see that the chance of a negative net profit is approximately 23%. The margin of error of $175 tells us how accurately we are estimating the average of $1,998. In particular, it says that if we repeated the simulation many times, we should see that the average profit rarely varies by more than $175. This tells us something about the accuracy of the simulation; if we had taken a larger sample this margin of error would have been smaller.

So does this mean we expect profit to be within $175 from approximately $1,998? The answer is no: we expect profit to be most likely between negative $1,561 and positive $5,608 and the average of the profit distribution is likely to be within $175 from approximately $1,998.

European call options.   The price of a stock is expected to increase by 1% with a standard deviation of 1% during each week, and is normally distributed. Currently the stock price is $40. Consider a stock option that gives you the option to purchase one share of stock at the end of three weeks for $41. This is referred to as a European call option. What is the expected payoff from this option?

Solution.   The file `calloption.xls`, shown in two pieces in Figure 10.8, illustrates how to set this up in Excel. You can use the Excel com-

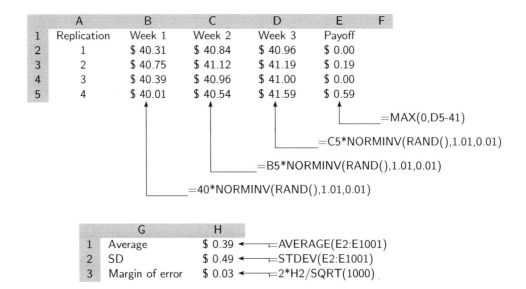

Figure 10.8. The file `calloption.xls`.

mand `=NORMINV(RAND(),1.01,.01)` to generate random numbers follow-ing a normal distribution with a mean of 1.01 and a standard devia-tion of .01. This means the price at the end of the first week can be `=40*NORMINV(RAND(),1.01,.01)`, and subsequent weeks are based on the previous week's price. The payoff at the end is the maximum of zero and the final price minus 41; this means if the final price is less than 41 then the payoff will be 0 and otherwise it will equal the final price minus 41. We decide to take a sample of 1,000 replications, and the figure shows only the first four replications. We see from Column H that the average payoff is $0.39, and our estimate of this average is accurate to within the margin of error of $0.03. Note that this does not mean we expect the actual payoff to be between $0.36 and $0.42; it simply means we expect the average payoff over a large number of such options to be in that interval.

Generating other distributions. Suppose that development costs for a product do not follow a normal distribution. In the simplest case suppose is believed that there is a 20% chance that the cost will be $90 million and an 80% chance that the cost will be $70 million. We can simulate this in Excel as follows. The Excel command `=RAND()` generates a random number that is uniformly between 0 and 1 and so the Excel command

	A	B	C
1	Chance	Cutpoints	Value
2	0.2	0	200
3	0.5	0.2	500
4	0.3	0.7	700
5			=B3+A3
6	Demand	500	=VLOOKUP(RAND(),$B$2:$C$4,2)

Figure 10.9. The file `general.xls`.

`=IF(RAND()<0.2,90,70)` will generate a random number that is either 70 or 90 and with a 20% chance of being 90.

For a more general setting, suppose demand for a product is not expected to follow a normal distribution but researchers believe there is a 20% chance that demand will be for 200 units, a 50% chance that it will be for 500 units and a 30% chance it will be for 700 units. We can simulate this in Excel as shown in Figure 10.9 (also in the file `general.xls`). First, set up a column of "cutpoints" for each value to divide up the interval from 0 to 1 in a way such that the first cutpoint begins at 0 and each subsequent cutpoint begins at the previous one plus the chance of that value. Then the command `=VLOOKUP(RAND(),$B$2:$C$4,2)` generates the desired random demand.

Exercises

5. **Inventory management.** A clothing retailer expects demand for a particular item to be 500 units with a standard deviation of 200 units and demand typically follows a normal distribution. The item costs $100 and can be sold for $200. Unsold items can be returned to the manufacturer for a refund of only $25 per item.

   (a) What is the expected profit if the retailer orders 600 units? Attach a margin of error and interpret it.

   (b) Would the retailer be expected to make more profit if it ordered 400 units instead of 600 units?

6. **Product performance.** A firm manufacturing expensive filter cartridges must be able to guarantee that its performance is within a particular range specified by the customer. The cartridges contain two layers of filtering material and each layer is manufactured by a process to give it a performance index having an average of 25 with a standard deviation of 5. When the two layers are combined, the performance index for the

cartridge will equal $1/(1/x + 1/y)$, where x and y are the respective performance indices for the two layers.

(a) What range of performance indices will 80% of cartridges fall within?

(b) What standard deviation for each layer would be required so that 80% of cartridge performance indices falls in the range from 12 to 13?

7. **Exotic call options.** The price of a stock is expected to increase by 1% with a standard deviation of 1% during each week, and currently the stock price is $40. Consider a stock option that gives you the option to purchase one share of stock at the end of three weeks for the lowest end-of-week price of the stock over the three weeks. What is the expected payoff from this option?

8. **Long-term investing.** An investor decides to deposit $1,000 each month into an investment account. The monthly percentage return is expected to follow a normal distribution with mean 1% and standard deviation of 2%. How likely is it that the investor has at least $30,000 after 25 months?

9. **Investing in stocks and bonds.** An investor starts with $1,000 invested 20% in bonds and 80% in stocks and would like to know how much this will be worth at the end of 20 years with no further deposits. Each year bonds are expected to yield a return of 3% with a standard deviation of 1%; stocks are expected to yield a return of 5% with a standard deviation of 8%. At the beginning of the 10th year if the investor does not have at least $1,500 then she will put all her money into stocks for the remainder of the horizon.

(a) What is the expected final balance?

(b) What is the chance she has more than $3000 at the end?

3. Exercises

10. **Firm sizes.** A random sample of 100 small firms is chosen from some population and a histogram for the number of employees at each firm is graphed. The data values are spread fairly evenly inside each of the bars below.

(a) Give an estimate for the 95th percentile of this data.

(b) Which of the following is most likely the 95% confidence interval for the average number of employees per firm in the population? Choose the best one of the following intervals and explain. (i) 20 to 70 (ii) 30 to 60 (iii) 40 to 45.

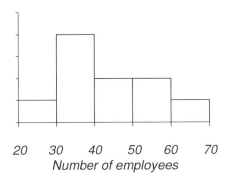

Number of employees

11. **Overdue bills.** A large population of overdue bills has balances that follow a normal curve. When we take a sample of 100 of these, the average is $500 and the SD is $100.

    (a) What statement can you make about the range $300 to $700?

    (b) What statement can you make about the range $480 to $520?

12. **Population income.** A simple random sample of 400 people is taken to estimate the average income in a large population. It turns out that the average income in the sample is $30,000 with an SD of $10,000. Answer true or false and explain briefly:

    (a) We are 95% confident the average income in the population is between $29,000 and $31,000.

    (b) Roughly 95% of the people earn between $29,000 and $31,000.

13. **Web surfing at work.** Reports from a new system set up to monitor Internet use by company employees indicate that 145 out of a random sample of 1100 employees at a corporation accessed Internet sites that were clearly not related to their jobs over the last week. On average, employees who accessed non-job related sites spent 4.8 hours per week at such sites with a standard deviation of 2.1 hours.

    (a) Give a 95% confidence interval for the percentage of employees at the corporation who access non-job related sites.

    (b) Briefly explain in one short sentence how to interpret this interval.

(c) Statistics show that about 10% of people nationwide access non-job related sites at work. Is this company really any different from the national statistics?

(d) Can we say that roughly 95% of people at this company spent in the range of 4.8 hours plus or minus 4.2 hours on non job-related web sites?

14. **Election polling.** Twenty five polls of one thousand random people each are taken during the course of an extremely close nine month election campaign. People in each poll are asked if they approve of each candidate or not, and each of the two candidate's approval ratings averages out to right around 50% over all the twenty five polls.

(a) Estimate the margin of error for each of these.

(b) In approximately how many of these polls would you expect the poll to be inaccurate by more than the margin of error you gave in (a)? Explain.

(c) If you computed the standard deviation of the twenty five approval rating percentages for each candidate, approximately what number would you expect to get? Explain.

(d) How large a poll sample would be required if only a margin of error of 10% was desired?

15. **Determining likely voters.** Pollsters try to determine whether or not a person is a "likely voter" before they count their opinion in a poll. If we assume 40% of the registered voters will actually vote, in a random sample of 100 registered voters we can be 95% confident that somewhere between _____ and _____ of them will actually vote. Fill in the blanks with a number.

More challenging exercises

16. **Airline yield management.** When an airline sells tickets for a flight, only about 90% of the people who buy a ticket actually show up.

(a) For a flight with 400 seats, the airline can be 95% confident that between _____ and _____ percent of the people will actually show up. Fill in the blanks with a percentage.

(b) Approximately how many tickets should they sell so that there is less than a 10% chance they run out of seats?

17. **Coin flipping.** Someone will flip a coin 50 times every day and will compute the percentage of flips that day that come up heads. You must guess a range within which you think this percentage will fall and you must use the same range each day. You get one point each day you are correct but two points are deducted each day you are incorrect. You start with 100 points and your object is to end up after many days with as close to 100 points as possible. What range should you use?

18. **Hotel booking.** A hotel finds that only 80% of people who reserve rooms actually show up. If they have 90 rooms available and take 100 reservations, what is the chance the hotel will have enough rooms for the people who show up? Justify your answer.

19. **Hurricane aid fraud.** In 2005, hurricanes Katrina and Rita flooded 80% of New Orleans with water more than 20 feet deep in some places.[4] About federal aid for victims, CNN wrote, "investigators estimated that 16 percent of the FEMA's (Federal Emergency Management Agency's) disaster relief payments were made to people who submitted invalid registrations, to the tune of about $1 billion. Because the figures were calculated using a statistical sample, however, the agency said the amount could range from $600 million to as much as $1.4 billion."[5] Why did the agency use a statistical sample and what was the standard error the investigators computed?

20. **Product ratings.** A market researcher asks 100 consumers to rate a product from 1 to 10. She gets answers that average 7.2 with a standard deviation of 2. She calculates a 95% confidence interval to be the range 7.2 plus or minus .4 or, equivalently, the range from 6.8 to 7.6.

    (a) Does this mean that about 95% of the answers given were in the range 7.2 plus or minus .4?

    (b) Does this mean we are confident that 95% of consumers in the population would rate the product in the range 7.2 plus or minus .4?

    (c) Does this mean we are 95% confident that the average of all consumers in the population would be in the range 7.2 plus or minus

---

[4] "New Orleans shelters to be evacuated. Floodwaters rising, devastation widespread in Katrina's wake," CNN.com, Wednesday, August 31, 2005. See http://www.cnn.com/2005/WEATHER/08/30/katrina/index.html

[5] "FEMA hurricane cards bought jewelry, erotica: Federal audit finds $1 billion in potential fraud," CNN.com, Wednesday, June 14, 2006. See http://www.cnn.com/2006/US/06/14/fema.audit/index.html

.4?

21. **Bank deposit forecasting.** For planning purposes an investment firm would like to forecast the average dollar amount their current customers will deposit over the coming year. They would like to accurately estimate this average with a margin of error of less than $2,000. They decide to telephone a random selection of 25 of their customers to ask each how much they plan to deposit. The average amount reported is $25,000 with a standard deviation of $30,000. Is this a large enough sample to give the margin of error they would like? If not, how much larger a sample do they need to take?

22. **Asbestos lawsuits.** A recent *Wall Street Journal* article describes how many firms face bankruptcy due to the large number of asbestos related lawsuits filed against them.[6] Unfortunately some percentage of these claims do not meet minimum medical requirements to be a valid claim, but it takes considerable time by experts looking at medical records to determine if a claim is valid or not. Creditors for one bankrupt company asked that a random sample of the 23,000 claims filed against it be taken to estimate the percentage that is valid. In a random sample of 1,691 claims it was determined that 87% did not satisfy the minimum medical requirements to be valid.

    (a) Attach a margin of error to the 87%.

    (b) Since it was previously believed this percentage was around 30%, a lawyer claims this sample is too small to prove the 30% figure was wrong and that sampling error could easily be responsible for why the sample came out the way it did. Is this a sensible criticism? Explain briefly.

23. **Government bonds.** Financial analysts would like to design a mutual fund to have a value that closely follows the Lehman Brothers Municipal Bond Index, a commonly quoted bond market index. There are many thousands of different bonds comprising this index, and these averaged in 2005 a return of 0.9% with a standard deviation of around 2%. Instead of buying all of the thousands of different bonds for the mutual fund, analysts decide to buy a random selection. How many randomly selected

---

[6]"An Asbestos Exit," *The Wall Street Journal*, Wednesday, November 17th, 2004, p. A16.

bonds would be needed in a portfolio that would confidently yield a return within plus or minus 1/10 of a percentage point of the index?

24. **Job time study.** From historical data you know that the time it takes to do a certain job is normally distributed with a mean of 130 minutes and a standard deviation of 20 minutes. To monitor performance, each week a sample of 25 jobs is selected and the average time it takes to do the job is recorded. In a particular week the average time to do the job was 128 minutes with a standard deviation of 16 minutes. The distribution of times in the sample comes out bell-shaped. What can you say, if anything, about each of the following intervals: (a) $128 \pm 2 \times 16$, (b) $130 \pm 2 \times (20/\sqrt{25})$, (c) $128 \pm 2 \times (16/\sqrt{25})$, (d) $128 \pm 2 \times (20/\sqrt{25})$.

25. **Manufacturing quality.** From historical data you know that the diameter of a rod you manufacture has a mean of 10 millimeters (mm) and a standard deviation of 0.2 mm. Actual diameters are normally distributed. To monitor performance of the production process, each day you take a sample of 25 rods and calculate the average diameter.

    (a) You want to establish control limits so that as long as the sample mean is between these limits you will assume the process is working as expected. When the sample mean is outside the limits you will stop the process and look for a cause of the large deviation. What control limits would you recommend? Explain briefly and justify your answer with relevant calculations.

    (b) One day the machine is inadvertently set to produce rods with average diameter 10.1 mm with a standard deviation of 0.2mm. What is the chance the sample mean for that day would fall within the control limits calculated in Part (a)?

    (c) A new customer will place a large order for rods if your process meets the following specification: at least 95% of the rods produced will be between 9.7 mm and 10.3 mm. Does your current manufacturing process meet these specifications? Briefly explain.

26. **Workflow planning.** In order to plan work flow in a manufacturing process it is important to know how long it takes to perform various manufacturing tasks. Analysts would like to estimate the average time one particular task takes as well as the percentage of time the task takes longer than 20 minutes. The analysts plan to observe a random sample of times the task is performed.

(a) To estimate this percentage with a margin of error of less than 1%, roughly how many replications do you advise they should observe? Justify your answer by showing relevant calculations.

(b) After taking a random sample the analysts estimate the percentage of time the task takes longer than 20 minutes to be approximately 30%. Then they separately compute a 95% confidence interval for the average time and get a range from 18 minutes to 20 minutes. A manager raises a concern about the calculations, pointing out that it's not possible at the same time for 95% to be between 18 to 20 minutes and 30% to be over 20 minutes. Is this a valid concern? Explain briefly.

27. **Deposit forecasting.** An investment firm with 10,000 clients would like to accurately forecast the average dollar amount their current customers will deposit over the coming year. They decide to telephone a random selection of 25 of their customers to ask how much they plan to deposit, and they would like to keep this sample as small as possible so the calls do not annoy too many customers. Since they will be multiplying this average by the total number of customers to get an overall forecast, they would like to accurately estimate this average with a margin of error of less than $4,000. Last year the average deposit for all 10,000 clients was $25,000 with a standard deviation of $30,000. Do you think a sample of 25 is enough to give them the margin of error they want? If not, how large a sample do you suggest they need to take? Justify your answer with relevant calculations.

28. **Profit simulation.** A computer is used to simulate many thousands of possible profit scenarios for a business deal. The average profit is estimated to be $10 million and the standard deviation is $5 million. The 95% confidence interval is computed to be the range from $9 million to $11 million. Complete the conversation below between a person who does understand and a person who does not understand the difference between the standard deviation, the standard error and the confidence interval. Use the following as a beginning and then write a few more sentences afterwards:

Person 1: Does this mean there is a 95% chance our profit will be between $9 million and $11 million?

Person 2: No, actually,...

Person 1: Aha, you mean...

Person 2: Yes,...

THE END

29. **Suspicious audit delays.** A study investigates audit delay, the length of time from a company's financial year-end to the date of the auditor report. There are suspicions that bad news becomes delayed more than good news. A random sample of 100 companies is taken, the 95% confidence interval is calculated to be the range from 75 days to 85 days, and the delay times follow a normal distribution.

   (a) Give a one sentence interpretation of the confidence interval.

   (b) The percentage of companies with delay times in the range from 70 to 90 days is closest to 99%, 90%, 60%, 30%, or 1%. Explain briefly.

   (c) If the delay times did not follow a normal distribution but instead had a tail to the right (but had the same average and standard deviation), how would you expect the 95% confidence interval to change? Choose one and explain briefly: (i) The upper limit would increase. (ii) The upper limit would decrease. (iii) It wouldn't change.

   (d) In the sample there were 40 companies reporting primarily good news and 40 companies reporting primarily bad news (the remainder reported mixed news). The confidence interval for the companies reporting good news was from 70 to 73 days and for the companies reporting bad news was from 80 to 95 days. What can you conclude from these intervals?

30. **Television repairs.** The manufacturer of a television set claims 90% of its sets last at least 5 years without needing a single repair. However, some customers whose television sets needed repairs in the first 5 years are threatening to take the manufacturer to court for false advertising. Consumer advocates find that in a random sample of 50 units, half of them required repairs within the first five years. In its defense, the manufacturer argues that a sample of 50 units is way too small to cast any blame since it sold over 100,000 units total and there have been very few other complaints. Does the manufacturer raise a legitimate point here? Explain briefly.

31. **Food processing.** A flour bag filling machine is set to fill bags to 5 pounds. The standard deviation in bag weights is 0.1 pounds and the

distribution of weights is normal. Every week the company takes a random sample of 10 bags of flour and measures the weight of each bag.

(a) About what percent of bags are expected to weigh over 4.75 pounds?

(b) The average weight of the 10 bags will be in the range _____ to _____ pounds in around 90 out of every 100 weeks. Fill in the blanks.

32. **Evaluating investments.** A computer generates a large sample of simulated outcome scenarios for a complex investment, computes the profit for each outcome and arranges the list of profits into a dataset. An analyst computes the standard error and a 95% confidence interval from this, and also computes the 2.5th and 97.5th percentiles of the dataset. Which do you think would be of more interest to top management: the confidence interval or the percentiles? Briefly explain.

33. **Toxic mortgages.** A newspaper article discusses how Bank of America, faced with a lawsuit on allegedly toxic mortgages, attempted to delay the trial by demanding that files for each of 368,000 or more disputed loans be evaluated individually by the plaintiff. It was estimated that this would have cost $75 million and it would have taken a team of 24 people more than four years. Instead, the judge ruled that a random sample of 6,000 disputed loans could be used to keep the trial on schedule. To estimate the percentage of loans with irregularities, approximately what margin of error would be expected using this approach? How long would it take and how much would it be expected to cost?[7]

34. **Identifying high-valued customers.** After an analysis of a large commercial database, market researchers have identified characteristics of potential customers likely to consistently purchase large amounts of your product over time; these are called high-valued customers. In the large database, 20% of potential customers have these characteristics. Market researchers plan to take a random sample of customers and hope to find at least 50 high-valued customers in the sample to target with a specially designed campaign. How likely is it that a random sample of 200 customers will contain at least 50 high-valued customers? Explain briefly.

---

[7]Bank of America hit with setback in MBIA Insurance mortgage liability lawsuit, *The Washington Post*, Friday, December 31, 2010, by David S. Hilzenrath.

35. Plunge in obesity. The National Health and Examination Survey found that the percentage of preschoolers who are obese declined over forty percent from 13.9% to 8.4% between 2004 and 2012. Each of these figures is based on a random sample of approximately 800 preschool students. First lady Michelle Obama interpreted these results as a sign that efforts to combat the national obesity epidemic were paying off. But a an epidemiologist looking at the same numbers said "there might have been no change in preschoolers' obesity rate. Even an increase is a statistical possibility." Looking at the numbers, which viewpoint do you think is more reasonable?[8]

---

[8] "A plunge in U.S. preschool obesity? Not so fast, experts say", by Sharon Begley, Reuters, Mon, Mar 17 08:11 AM EDT, http://www.reuters.com/article/2014/03/17/us-usa-health-obesity-insight-idUSBREA2F0CX20140317?irpc=932

Fun Problem: A Test for Nursery School Kids

Here is a nursery school entrance exam. There are a bunch of symbols below. What comes next in the progression?

What symbol comes next here?

Solution. Don't read any further until you think about the puzzle for a while!

In the following diagram, just look at the part highlighted with grey—ignore everything else. Do you see the numbers one through six? The last symbol would be a seven along with its reflection. I'll let you fill it in yourself!

I bet you can guess what goes here now.

Chapter 11

Hypothesis Testing

1. Introduction

On Saturday, December 2, 2006 the pharmaceutical giant Pfizer made an announcement that shook global stock markets and immediately lowered its stock price by nearly 13%, slashing almost $21 billion off the market value of the company.[1] Pfizer announced it was canceling development of one of its most promising and potentially blockbuster cholesterol drugs, torcetrapib, on which the company had already spent nearly $1 billion over a decade to develop. The company also announced that all patients in ongoing clinical trials should stop taking the medication immediately.

Why did Pfizer do this? The reason is that they had seen some deaths of patients in a clinical trial: in a study involving 15,003 people, split about evenly between treatment and control, 82 people taking the new drug died, while 51 people in the control group died. Even though this means there were 60% more deaths in the treatment group, it is only a difference of 31 people out of 15,003 people. Couldn't this small difference be just ordinary chance variation? Was it really big enough to justify immediately abandoning the drug it had invested so heavily in? The answer is that this difference had just crossed the threshold of what statisticians call **statistical significance** and

---

[1] "Pfizer's Bitter Pill: Shares plummeted Monday after the company halted work on its highly touted cholesterol drug, throwing future growth into question," *BusinessWeek Online*, December 5, 2006. http://www.msnbc.msn.com/id/16055205/

it strongly indicated that further studies would give similar results.[2] This chapter discusses the concept of statistical significance and shows how you can use it to rule out chance as an explanation for some trend in your data—so you can start to believe other more important theories for the trend.

2. Null and Alternative Hypotheses

When researchers find a potentially important trend in their data they must decide between two theories, or hypotheses, to explain what they found. The first theory is called the **null hypothesis** (written as H_0 in statistical notation) and it is the theory that what they found was nothing but chance variation. The second theory is called the **alternative hypothesis** (written as H_A) and it says they made a real discovery—they found something beyond just chance variation. Scientists usually have many theories for trends they discover in their data, but before they entertain any of these theories they first need to rule out ordinary chance variation as the explanation for what they have seen. Scientists are often hoping the alternative hypothesis is true, although in Pfizer's case the company was really hoping for the null hypothesis—the theory that the difference in deaths was just chance variation.

> The **null hypothesis** (H_0) says the trend you found in your data was caused by nothing but ordinary chance variation.
>
> The **alternative hypothesis** (H_A) says the trend you found in your data was caused by something beyond chance variation and that you made a real discovery.

The logic good researchers use to fairly decide between the two hypotheses is to assume the null hypothesis is true until they have strong evidence otherwise. In our legal system we have "innocent until proven guilty"—so the null hypothesis is "innocent" and the alternative hypothesis is "guilty."

---

[2] "Relatively Small Number of Deaths Have Big Impact in Pfizer Drug Trial," by Carl Bialik, December 6, 2006, *The Wall Street Journal*, http://online.wsj.com/public/article/SB116535192161641418.html

To then decide which of the two hypotheses is more reasonable, the first step is to see how well the data fits the null hypothesis—the default assumption. If it fits reasonably well—before even considering the alternative hypothesis—then you can't rule out chance variation as a reasonable explanation for the data. If the trend in the data would be very unusual under the null hypothesis, this is evidence of something beyond chance variation and you should start to think about the alternative hypothesis. Notice that you are not really proving the case for any particular alternative hypothesis but you are giving evidence against the null hypothesis by trying to rule out chance as a reasonable explanation. Testing the two hypotheses is usually called a **test of statistical significance** or a **hypothesis test**.

We will first give two simpler examples of hypothesis tests and then we move to the example in the introduction from Pfizer.

3. Conducting a Hypothesis Test

Telephone telepathy.   An experiment was conducted to see if people had "telephone telepathy," the ability to tell who was calling before answering the telephone.[3] Several volunteer subjects who believed they had this ability were recruited through an online advertisement that read "Do you know who is ringing before you pick up the phone? Good pay for fun and simple experiments as part of psychic research project." Each volunteer subject picked four friends to be potential callers and experimenters randomly asked one of the four friends to call the subject. This was repeated 271 times total over all subjects and in 122 cases the subject correctly guessed the caller before picking up the phone. Is this evidence of telephone telepathy?

We can calculate that 122 correct out of 271 is 45% correct. Since each time there were four friends who could have been calling, we would have expected 25% correct by just luck. Could this difference be just lucky guessing? Or is this evidence of something beyond just lucky guessing?

---

[3] "Videotaped experiments on telephone telepathy," by Rupert Sheldrake and Pamela Smart, *Journal of Parapsychology* 67, 187–206, June 2003 and "Phone telepathy: You knew it was true," CNN.com, September 5, 2006. See `http://www.skepticalinvestigations.org/currentresearch/calls_video.html` and See `http://www.cnn.com/2006/TECH/science/09/05/telepathy.reut/index.html`.

Solution.   The null hypothesis is that they had no telepathy and were just lucky in their guessing, and the alternative hypothesis is that something beyond pure luck was happening—perhaps it was telepathy.

We first assume the null hypothesis. How rare would it be to get as high as 45% correct answers without using telepathy? We can imagine that, without telepathy, 271 guesses are like a sample of 271 answers from a population which has 25% correct answers. We can then use the formula for the standard error of the sample percentage to get

$$\text{SE for \%} = \sqrt{\frac{25 \times (100 - 25)}{271}} \approx 2.7\%$$

Since we observed 45% correct answers, this is $45\% - 25\% = 20\%$ more than we would have expected. And since the SE for % is 2.7%, this means we observed something that was $20\%/2.7\% = 7.4$ standard errors above what we would have expected to see if they were just guessing.

Is 7.4 standard errors above what we expected to see an unusually high number of standard errors? If so, it would mean that these test results are really beyond chance—lucky guessing is not a good explanation. If it's not so unusual, then we can't really rule out chance—lucky guessing could be a reasonable explanation.

Some very rough rules of thumb you can apply here are the following. Something that is more than two standard errors away from what you would have expected is often viewed as being beyond the normal range for chance variation and is called **statistically significant**. Something more than three standard errors away is often called **highly statistically significant**. Something well within two standard errors of what was expected is usually viewed as being consistent with chance variation. If something is somewhere around two standard errors away from what was expected it can be viewed as near the border-line for statistical significance. These "cutoffs" are not hard-and-fast rules but just rough rules of thumb you can use.

Something that is...

...more than two standard errors away from what you would have expected is often called **statistically significant** and is usually considered beyond the reasonable range of chance variation.

...more than three standard errors away is often called **highly statistically significant**.

...within two standard errors of what you expected is often viewed as being consistent with chance variation.

Since the 45% was 7.4 standard errors away from the 25% we would have expected them to get by guessing, this can be viewed as highly statistically significant and therefore as strong evidence that something beyond lucky guessing was happening. In other words, if they were just purely guessing it would be extremely unlikely for them to get so many correct. This is the evidence the researchers needed to persuasively reject the null hypothesis in favor of an alternative hypothesis. The researchers were delighted to propose telepathy as the alternative hypothesis and they published their results. There are many other things to scrutinize about their study but they very persuasively ruled out chance variation as the explanation.

Significance levels

Statisticians can make the above rough rules of thumb more precise. It is possible to use the central limit theorem to estimate how rare it would be to get a given number of standard errors away from what you would have expected. For the above example the chance of getting more than 7.4 standard errors above what was expected can be estimated by the area to the right of 7.4 under the normal curve. Looking up 3.72 in the normal table, the closest number listed to 7.4, we see that it corresponds to 99.99%. So the area to the right is less than $100\% - 99.99\% = 0.01\%$. The statistical jargon is to call the 7.4 the z-**value** and the corresponding area to the right of it is called the p-**value**, or the **observed significance level**. In this case, the p-value is less than 0.01%.

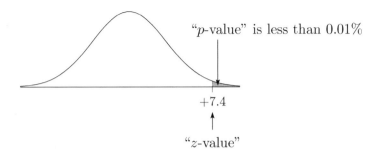

People often use the rule that a p-value below 5% can be viewed as "statistically significant." In this case the technical jargon is to call the 5% cutoff the **level of significance** for your hypothesis test. In different settings there generally are different appropriate levels of significance to use. In medical research, for example, some studies may be so important that people prefer that you have a significance level of 1% or 2% before they will believe your results. In business applications you may get away with using 5% or even 10%. Generally speaking, if the consequences of mistaking chance variation for a real discovery are very bad, people will use a very strict cutoff (a low number like 1%) for their level of significance. If the consequences are not so terrible, as they would be in an exploratory data analysis, you can use a more lenient cutoff (like 5% or 10%). In medical research the consequences of this type of mistake may be very bad and could, for example, mean that an ineffective or dangerous drug gets approved by the government and given to millions of people.

The p-value is often quoted in research journals, where a researcher may write something like "we found that patients got better after taking our new drug $(p < 1\%)$" to mean that benefits of the drug were found to be statistically significant with a p-value of less than 1%.

The advice to the manager is not to worry so much about exactly which level of significance to use and that the rules of thumb above are decent rules to use. If the conclusions of your study change drastically if you adopt a slightly different level of significance, then your results are probably close to the borderline of statistical significance—and you should consider this when basing decisions on the study.

Two-tailed versus one-tailed p-values

In many situations people use double the p-value we described above as their p-value. This is called a "two-tailed" p-value since you are adding up twice the area in the tail of the normal curve. People use this when it would be supportive of the alternative hypothesis if there was significant variation in either the upward or downward direction from what was expected. For example, if you are looking at a machine parts manufacturing process you might be equally concerned if the manufactured parts were either significantly larger or significantly smaller than what you designed them to be. This would be a case where a two-tailed p-value would be more appropriate, since both directions of variation are important to you. In our example of telepathy, we would not get excited if the subjects guessed significantly worse than what we would have expected—we would only get excited if they guessed significantly better than what we expected. A one-tailed p-value is therefore appropriate for the telepathy example (although you could make the argument that guessing significantly worse than expected is also an indication of some sort of telepathy). In business practice this subtle distinction is not so important. If it makes a big difference whether you use a one or a two-tailed p-value, your results were probably of borderline significance anyway. If your results are borderline, you should be aware of this when you draw your conclusions. It is not wise to rely on a single strict p-value "cutoff" for statistical significance when making business decisions—a more flexible approach is usually wiser.

Type I and Type II errors

If a hypothesis test indicates something is statistically significant when in reality it's only chance variation, this is called a **Type I error**. When there is something beyond chance variation happening but your hypothesis test doesn't indicate statistical significance, this is called a **Type II error**. These can be summarized in the following diagram.

Hypothesis test says

		Accept H_o	Reject H_o
	H_o True	◡	Type I error
The reality	H_o False	Type II error	◡

The rule of thumb we give in this chapter is that whenever you see something more than two standard errors from what you would have expected, this can be viewed as statistically significant. This cutoff of two standard errors is a convention that many people use. If Type I errors are very costly you might consider using a larger cutoff, such as three or four standard errors. In the case of testing the effectiveness of a new drug for a common ailment, a Type I error means that an ineffective drug may be prescribed to millions of people. This is very bad and so it could be appropriate to use a cutoff larger than two standard errors. In the case of an experimental treatment for a fatal but otherwise incurable disease, a Type II error is probably worse than a Type I error. Now a Type II error means that someone could miss out on the opportunity to be cured, and a lower cutoff for statistical significance could be better. In fact, you might consider giving this treatment to someone even if it has only been tested on a few people in the past; waiting until you have enough data for the usual level of statistical significance may be too conservative and your patient may die in the meantime.

Another example

New York City Blackout.   At approximately 4:20 pm on August 14, 2003 the largest blackout in US history occurred.[4] The power was completely out for almost 30 hours over an area from New York City to Toronto to Detroit and temperatures reached into the 90s. A few days later, the Mayor of Toronto predicted the "biggest baby boom we've ever seen" would result from the blackout. In fact, *Summary of Vital Statistics* published by the New York City Health Department in 2004 had a special section on blackout-related births. Usually on average 340 babies are born each day with an SD of 18, but the 28 day stretch between 9 and 10 months after the blackout

---

[4]http://www.camworld.com/archives/001236.html   and   http://www.time.com/time/covers/1101030825/

had an average of around 343 births per day: up by around 1% from the usual.[5] Is this evidence of a small baby boom?

Discussion.   Before we discuss any theories for why a baby boom occurred, we must first rule out the boring and simplest explanation that this was just ordinary chance variation in the birth numbers. Here the null hypothesis says that this is just ordinary chance variation and the alternative hypothesis says that there really was a baby boom.

Notice that even though the average for this stretch of 28 days is much less than one standard deviation away from the usual average, it does not tell us anything because the standard deviation is not the right measure of variability to look at. We really need to find out how variable the average of 28 days typically tends to be. The standard deviation of 18 tells us how variable a single day tends to be; the average of 28 days will generally be much less variable than this.

To do this we need to look at the standard error for the average of 28 days. If we took 28 ordinary days chosen at random, the standard error for the average would be

$$\text{SE for average} = \frac{18}{\sqrt{28}} = 3.4 \text{ babies.}$$

This means we would expect variability of about 3.4 babies in the average of 28 days. Since this particular stretch of 28 days averaged out to 343 babies per day, it was only $(343 - 340)/3.4 = .9$ of a standard error away from what we would ordinarily have expected. Since this is less than two standard errors, the rule of thumb above says this is not a statistically significant baby boom. Ordinary chance variation could easily explain what happened. This means we should accept the null hypothesis and we don't need to brainstorm for the interesting alternative hypotheses.

Nine months after a previous serious blackout in 1965, there was a front-page *New York Times* article with the headline "Births up 9 months after the blackout." A careful statistical analysis later showed that this also was just chance variation and not a statistically significant baby boom.

---

[5]Data from `http://www.nyc.gov/html/doh/downloads/pdf/vs/2003sum.pdf`. See also A. Izenman, "Babies and the Blackout: The genesis of a misconception," *Social Science Research* 10, 1981: 282–299, and also *New York Times*, Wednesday, August 10, 1966, "Births up 9 months after the blackout."

Side note. It is interesting to note that the four Sundays during those 28 days averaged out to 250 babies per day. Repeating the calculation above this time for a sample of 4 days we get

$$\text{SE for average} = \frac{18}{\sqrt{4}} = 9 \text{ babies}$$

and so these four days averaged $(340-250)/9 = 10$ standard errors below the usual average. Since this is more than 3 standard errors, the rule of thumb above says this is a highly statistically significant difference! Even though it's only four days, the average is off by way too much to be just a chance variation. What could explain this?

Discussion. A large fraction of births are actually induced births and doctors don't usually schedule these types of procedures on Sundays!

We look at a business problem next.

Process redesign. A manufacturing company is studying the time it takes to complete and deliver an order in one of their product lines. In the past it was taking 140 days on average to fill orders but they recently redesigned the process hoping it might reduce delivery times by about a month. In fact, the first 25 orders processed after the change was made were filled in 100 days on average. At first glance this seems like it was a success, but could chance variation reasonably explain this drop in the average?

Answer. We first need to know something about the variability of the time it takes to fill an order. Company data shows there was a standard deviation of 60 days. This means the variability in the average of 25 random orders would be

$$\text{SE for average} = \frac{60}{\sqrt{25}} = 12 \text{ days.}$$

The decrease in the average of $(140 - 100) = 40$ days corresponds to a decrease of $40/12 = 3.33$ standard errors. The rule of thumb says this is beyond the usual limit of chance variation, so this is reasonable evidence the change was not just chance variation. As long as there were no other major changes at the same time, this process change looks effective.

Exercises

1. **Workplace absenteeism.** Over the last year the absentee rate at a large corporation averaged 8.2 days absent with a standard deviation of 6 days. One department with 40 employees had an absentee rate of 12 days per employee. During an investigation the department head argued as follows: "If you took 40 employees at random from the corporation, there is a pretty good chance the average number of days absent would be 12 or more. That's what happened to us—chance variation." Is this a good defense? Justify your answer.

2. **Cereal packaging.** A machine being used for packaging cereal needs to be set so that 15 ounces of cereal on average will be packaged per box. After the machine has been set, the quality control engineer wishes to test the machine setting and selects a sample of 36 random packages filled during the production process. The standard deviation of the 36 weights is .42 ounces and the average is 14.8 ounces. (a) Is there evidence here that the machine is not set properly? Justify your answer using a test of statistical significance. (b) How large a sample would the engineer need to have a margin of error for the average of roughly .035?

4. Interpretation of the *p*-value

In the telephone telepathy example above, the *p*-value turned out to be less than .01%. Does this mean that there was more than a $100\% - .01\% = 99.99\%$ chance that the people had telepathy? No, the correct interpretation of the *p*-value is the following: if we ran a similar test on a group of people who did not have telepathy, there would be less than a 0.01% chance we would have at least 45% correct guesses. The *p*-value does not tell us the chance of having telepathy because it is computed under the assumption that the people do not have telepathy.

> If the p-value for a study equals p, it means the
> following: if we repeated the study in a situation
> where the null hypothesis was actually true, there
> would be a p chance of getting results at least as
> convincing (against the null hypothesis) as the re-
> sults we saw in the original study.

More useful interpretations of the p-value depend on the specifics of the
particular situation you are looking at. Here we give a few more examples.

New drug development. Suppose a new cold medicine is being tested and
researchers find it statistically significantly reduces the number of colds people get. Researchers compute a p-value of 1% for their study. How do you
interpret this number?

Solution. The correct interpretation is that if we ran a similar test on
medicine that has no effect we would have a 1% chance of seeing a difference between the treatment and control group that is at least as large as
what these researchers got. This does not mean that there is either a 1% or
a 99% chance that this drug works. The p-value doesn't tell you the chance
the drug works—it tells you the chance of similar or better results using a
drug that doesn't work.

Stock picking. Financial analysts would like to see if a specially trained
monkey can pick successful stocks to buy. They put the names of a large
number of stocks randomly on different bananas and observe which bananas
the monkey eats. The analysts find that the average return of the stocks the
monkey picks is pretty high and they compute a p-value of 0.2. Does this
mean there is an 80% chance that the monkey is a good stock picker? Does
this mean there is a 20% chance that the monkey is a good stock picker?[6]

---

[6]For other surprising animal performances, see the story of Clever Hans (`http://www.`
`kbrhorse.net/tra/hans.html`), the 19th century horse who seemed to be able to do math
problems by tapping out the answer with his hoof until researchers realized he was somehow
picking up subtle unintentional cues from the observers as he neared the correct number.
See also the story of Ziggy, the parrot who revealed the affair his owner's wife was having
by repeatedly saying the wife's lover's name: `http://www.cnn.com/2006/WORLD/europe/`
`01/17/uk.parrot/index.html`

Solution.   No, the *p*-value tells us that if we simply picked random stocks we would have a .2 chance of doing at least as well as the monkey did. Since .2 equals 1 out of 5, it also means one out of five monkeys grabbing random bananas would be expected to do at least this well picking stocks. The *p*-value does not tell us the chance that a monkey is or is not a good stock picker. (Incidentally, this *p*-value of 20% is not statistically significant, but if the monkey instead came up with a *p*-value of 1% we might consider offering him a job.)

Telepathy test.   Suppose that the telepathy study above was replicated and the subjects guessed 27.7% correctly. Since the standard error for the sample percentage was 2.7%, this corresponds to a *z*-value of 1 and thus a *p*-value of about 16%.

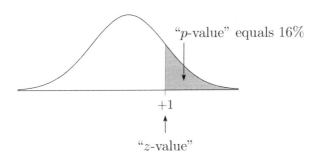

By the usual rule of thumb (a *z*-value of more than 2 is "statistically significant") this is not a statistically significant number of correct guesses so we can't conclude that anything other than chance is operating. But how do we interpret the *p*-value of 16%? Does this mean that there is a 16% chance the people have telepathy? Does this mean there is an 84% chance the people have telepathy?

Solution.   No, and no. The correct interpretation is to say that if we again repeated the experiment on people with no telepathy there would be a 16% chance of getting at least 27.7% correct guesses.

Exercises

3. **Soft drink preferences.** Market researchers would like to know if customers prefer a well-known brand over a generic brand of soft drink. They give a large sample of people the two drinks to taste in a random order and ask them which one tastes better. They find that 70% of people say the branded one and researchers compute a p-value of .02. Interpret this number.

4. **Drug side effects.** A pharmaceutical company testing a new drug is worried about whether or not it may have one particular side effect that would drastically reduce the potential market size for the drug. After conducting a study they find that a statistically insignificant number of people taking the drug have this particular side effect (compared with a control group that takes a placebo) and the researchers compute a p-value of .4. Interpret this number.

5. Two-Sample Tests of Significance

When you're comparing two averages from two independent groups, calculating the standard error for their difference is slightly more complicated than if there was just one random group. In fact, the variability in the difference between two random quantities depends on the correlation between them. For example, the difference in weekly stock returns you see between two companies in the same industry tends to be a lot less variable than it is for two companies in different industries. This is because returns for companies in the same industry often tend to be positively correlated: they tend to be high together and low together so the difference in their returns tends to be small. When the companies are in completely different industries they can easily move in different directions and the difference can be quite large. This of course is not an ironclad rule but is just an example to illustrate that the difference between two random quantities depends on how correlated they are.

In the case where we are comparing two averages (or percentages) A and B that are based on completely independently operating groups, the formula for the standard error of the difference between A and B is straightforward:

> Given two independent random quantities having respective standard errors SE_A and SE_B, the standard error for the difference between them equals
>
> $$\text{SE for difference} = \sqrt{(\text{SE}_A)^2 + (\text{SE}_B)^2}.$$

For example if Stock A has a variation in its weekly return of about 5% and Stock B has about the same but is completely independent of A, the variation we will see in the difference between the two returns will be approximately $\sqrt{5^2 + 5^2} \approx 7$, or 7%.

Now this has major implications in the analysis of data from studies where you are comparing a treatment group and a control group and the responses are essentially independent. In the introduction we discussed Pfizer's announcement to cancel development of one of its most promising cholesterol drugs because of patient deaths in a clinical trial. Here we will discuss how statistical significance can be assessed in that setting using the formula for the standard error for the difference.

Researchers saw that 82 people out of 7501 people (about 1.1%) in the treatment group died and 51 out of 7501 people (about 0.68%) in the control group died. This difference of $1.1\% - 0.68\% = 0.42\%$ may seem like a small difference but is it statistically significant?

We can calculate the standard error for the percentage in the treatment group as

$$\text{SE for \% for treatment group} = \sqrt{\frac{1.1 \times (100 - 1.1)}{7501}} \approx 0.12\%$$

and similarly for the control group we get

$$\text{SE for \% for control group} = \sqrt{\frac{.68 \times (100 - .68)}{7501}} \approx 0.095\%$$

and so

$$\text{SE for difference} = \sqrt{(0.12)^2 + (0.095)^2} \approx 0.15\%.$$

This means the difference we saw of 0.42% represents about $0.42/.15 = 2.8$ standard errors of variation. This is a statistically significant amount of variation, and way beyond what is usually expected from chance variation.

Although the difference in the number of deaths between the treatment and control groups only amounted to 30 cases out of nearly 15,000 people, it turns out to be way beyond what would normally be expected from chance variation. This unfortunately was the strong and convincing evidence Pfizer needed to see in order to pull the plug on the project. Such a difference was so rare that Pfizer concluded they would see similar results if they repeated the study. Although they had already invested a lot of money in the drug, any drug that can be shown to cause death in its patients is a sure loser.

Exercises

5. **Drug testing.** Researchers testing a drug divide a group of 200 subjects randomly into treatment and control groups consisting of 100 subjects each. In the treatment group we see 75% of the people get cured and in the control group it's 40% of the people. Is this different statistically significant?

6. The t-Test

When a study is based on a very small sample, p-values computed using the normal table and the methods above can be inaccurate. If you have a very small sample from a population with normally distributed data, you can use a different table, Table 11.1, called the "t-table." Extra variation introduced when you estimate the standard deviation of the population using the standard deviation of the sample is accounted for in the t-table. When you use this table for a hypothesis test, it's called a "t-test." Using the t-table is not so important when the sample size is more than 20 or so, as it usually will be in practice. If your data do not follow a normal distribution, the t-table does not apply and you should seek the advice of a statistician.

The usual rule of thumb for statistical significance is to look for at least two standard errors of variation from what was expected. This will give a p-value of 2.5% for a 1-tailed test because the area above $+2$ under a normal curve is 2.5%. This rule of thumb applies to large samples. With a sample of size 10 the t-table tells us that we need to see 2.26 standard errors of variation instead of just 2 standard errors. To see where this number 2.26 comes from, just subtract one from the sample size 10 to get the "degrees of freedom" (degrees of freedom is a technical term for the number of dimensions of variation in the data if the sample average or sample percentage were known) and then look in the column labeled "p-value $= 2.5\%$" under $10 - 1 = 9$ degrees of freedom and we see the number 2.26. Then only consider something statistically significant if it is more than 2.26 standard errors.

Now we will work out an example. Suppose last year the average tax paid for people in a large population was \$15,000 and follows a normal distribution. Someone samples 10 random tax returns this year from the population and finds the average tax paid this year is \$20,000 with a standard deviation of \$10,000. Is this average statistically significantly higher than \$15,000?

First of all, since the sample is very small a better estimate of the standard deviation of the population is $\sqrt{10/9} \times 10,000 = \$10,541$ (see the technical note about the *sample standard deviation* in Chapter 2.) We can then calculate

$$\text{SE for average} = \frac{\$10,541}{\sqrt{10}} \approx \$3,333,$$

and so this means we see a difference of $(20,000 - 15,000)/3,333 = 1.5$ standard errors. With a sample size of 10 this means we look up in the t-table in the row where it says the degrees of freedom equals 9 and we see that 1.5 standard errors is not beyond the threshold for statistical significance to give a p-value below 5% (the cut off for this is 1.83 standard errors) but it is beyond the threshold for a p-value below 10% (the cut off for this is 1.38 standard errors). So we can conclude that the average this year is significantly higher at the 10% level but not at the 5% level.

7. Statistical Significance versus Practical Importance

When people try to convince you that the results of some study are significant, they may naturally use the term "statistically significant." But statistical significance doesn't tell you whether or not the results are of practical significance. Something is statistically significant if it is clearly more than just a chance occurrence, whereas something is practically significant if it would have an important impact. If you gather enough data everything looks statistically significant because then there is very little room for chance variation.

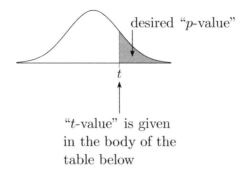

desired "p-value"

t

"t-value" is given
in the body of the
table below

Degrees of freedom	p-value = 10%	p-value = 5%	p-value = 2.5%	p-value = 1%
1	3.08	6.31	12.71	31.82
2	1.89	2.92	4.30	6.96
3	1.64	2.35	3.18	4.54
4	1.53	2.13	2.78	3.75
5	1.48	2.02	2.57	3.36
6	1.44	1.94	2.45	3.14
7	1.41	1.89	2.36	3.00
8	1.40	1.86	2.31	2.90
9	1.38	1.83	2.26	2.82
10	1.37	1.81	2.23	2.76
11	1.36	1.80	2.20	2.72
12	1.36	1.78	2.18	2.68
13	1.35	1.77	2.16	2.65
14	1.35	1.76	2.14	2.62
15	1.34	1.75	2.13	2.60
16	1.34	1.75	2.12	2.58
17	1.33	1.74	2.11	2.57
18	1.33	1.73	2.10	2.55
19	1.33	1.73	2.09	2.54
20	1.33	1.72	2.09	2.53

Table 11.1. The t-table.

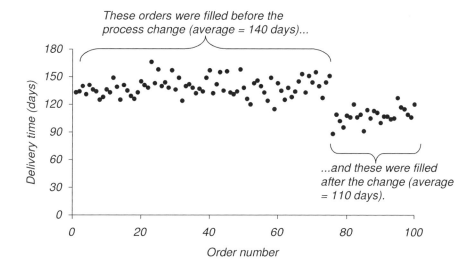

Figure 11.1. A change that is both statistically significant and practically significant.

> Something is **statistically significant** if it is clearly more than just a chance occurrence.
>
> Something is **practically significant** if it would have an important impact.

Question.   A manufacturing company is studying the time it takes to complete and deliver an order in one of their product lines. In the past it was taking a few months to fill orders but they recently made a process change hoping that it might reduce delivery times by about a month. Figure 11.1 is a graph of delivery times for the most recent 100 orders: the first 75 are before the process change was made and the last 25 are using the new process. Does it appear this change was effective? Was this change practically significant? Was it statistically significant?

Solution.   Well, looking at the graph there is a clear-cut difference between before and after the change. The average went down from 140 days to 110 days: a drop of 30 days on average. It seems as though the process change was effective in doing what they had hoped to do. This change was both statistically significant and practically significant.

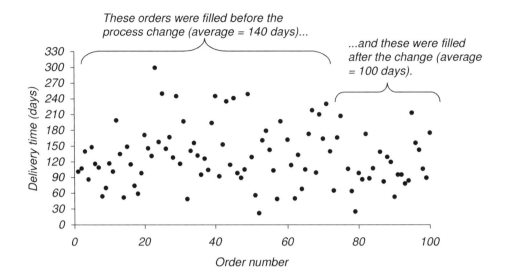

Figure 11.2. A difference that is practically significant but not statistically significant.

Question. The same company decides to investigate another product line where delivery times have also been around 140 days on average. They institute a different process change on a trial basis and the average goes down to 100 days. They are very happy with this reduction but it would be expensive to implement this change on a permanent basis. Was this change effective in reducing average delivery times? Was this reduction practically significant? Was it statistically significant? Figure 11.2 is a graph of delivery times.

Solution. This time there is a lot more variability involved and in the past some delivery times were almost a whole year. Though the average has reduced from 140 days to 100 days, there does not appear to be such a clear-cut difference between before and after. It's difficult to tell if the reduction in delivery times after the change was just due to luck or due to the new process. This drop in the average delivery time is practically significant but it doesn't look statistically significant.

Question. A graph of delivery times for a third product after instituting a change is shown in Figure 11.3. Is the change in average delivery times practically significant? Is it statistically significant?

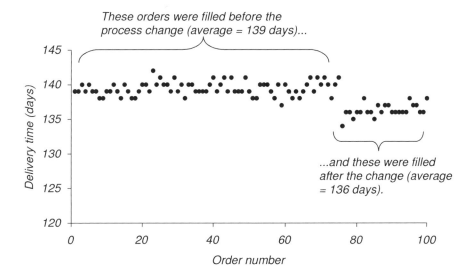

Figure 11.3. A change that is statistically significant but not practically significant.

Solution.   This time the average drops from 139 days to 136 days: only a three day drop. This small drop is of not much practical significance, because customers would barely notice the difference. But in the graph there is clearly a difference between before and after the change was implemented, so the drop is most likely statistically significant. This time we have something that's statistically significant, but not practically significant.

8. Statistical Power

The chance of avoiding a Type II error using a given cutoff for statistical significance and a given alternate hypothesis is called the **statistical power** of the test. We illustrate with an example.

Statistical power.   A hospital sees that 20% of its smoking patients do not receive advice to quit smoking. An awareness campaign for staff is implemented to decrease this percentage, and management will be happy with its performance if this percentage can be decreased to 10%. In a sample of 100 patients and a cutoff of two standard errors for statistical significance,

The null hypothesis is rejected here...

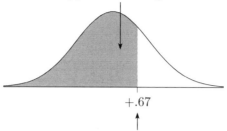

...using this cutoff for significance.

Figure 11.4. Computing the statistical power of a test.

how low does the percentage need to be in order for it to be considered statistically significant? And what is the statistical power here?

Discussion.   The null hypothesis is that the awareness campaign does not reduce the percentage below 20%, and so under this hypothesis the standard error can be computed as

$$\text{SE for \%} = \sqrt{\frac{20 \times (100 - 20)}{100}} \approx 4\%.$$

Since we want the cutoff for statistical significance to be a decrease of two standard errors, anything below $20 - 2 \times 4\% = 12\%$ could be considered statistically significant.

To compute the power, we assume the alternative hypothesis that the percentage is reduced to 10% and we can compute the standard error as

$$\text{SE for \%} = \sqrt{\frac{10 \times (100 - 10)}{100}} \approx 3\%.$$

Using the 12% cutoff we computed above, under the alternate hypothesis this corresponds to $(12 - 10)/3 = .67$ standard errors. As shown in Figure 11.4, looking up .67 in the normal table tells us that the statistical power is 75%. This means there is a 75% chance of correctly rejecting the null hypothesis if in fact the alternate hypothesis is true.

Table 11.2. An excerpt from the file `stocknames.xls`.

Company name	% return 2004	# of c's in name
@Road Inc	-2.24%	1
1-800 Contacts Inc	2.99%	3
1-800-ATTORNEY Inc	0.00%	1
1-800-FLOWERS.COM	-0.62%	1
1mage Software Inc	0.00%	1
1st Centennial Bancorp	-1.66%	2
1st Colonial Bancorp Inc	3.98%	3
1st Independence Finl Group	1.70%	1
1st Pacific Bank of California	-8.57%	3
1st Source Corp.	-0.79%	2

Exercises

6. **Adverse drug reactions.** A medical center is interested in studying the impact an electronic medication information system could have on the percentage of patients who experience an adverse drug reaction (ADE) due to inadvertently prescribed medications that are not supposed to be taken at the same time. Currently 40% of patients experience an ADE, and administrators hope the new system will reduce this to 30%. Administrators plan to implement the new system and analyze a random sample of patient records to determine if the percentage has decreased. As it is expensive and very time-consuming to analyze the patient records, they would like to use as small a sample as possible in order to have a statistical power of at least 80%. Is a sample of 200 patients enough?

9. Data Snooping

Figure 11.5 is a graph showing how the average return for stocks in 2004 relates to the number of times the letter "c" appears in the company's (abbreviated) name. For example, the company listed as "Advanced Environ Recycling Tec" has the letter "c" appearing four times and the return was 6.55% The data for both 2003 and 2004 appear in the file `stocknames.xls` (an excerpt is in Table 11.2). What story does this tell? Well, it certainly tells the story that the more times the letter "c" appears the higher the average return; the pattern is pretty clear. Does this mean this could give

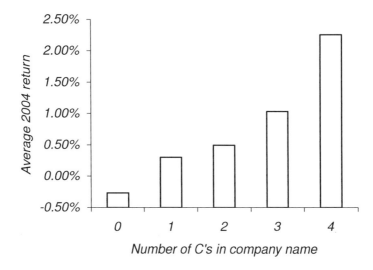

Figure 11.5. Average return as a function of the number of times the letter "c" appears in the company name.

us a good investment strategy?

The graph looks pretty impressive but you didn't get to see the whole story. I also created separate graphs for each of the 26 letters and this is the only one that had such a striking pattern—see Figure 11.6. And actually if you look at the graph for the letter "c" in a different year, you don't see any pattern like this—see Figure 11.7. This pattern was probably just a fluke but it seems much more impressive when you don't get to see all the other graphs. It's often called "data snooping" when you look at many different things and only report the most favorable one.

The problem of **data snooping** in statistical studies can be as misleading as the previous example. It says that if you conduct many tests of statistical significance you're bound to see something that appears to be significant just by pure luck. In other words, rare events are very likely to happen if you wait long enough.

To correct for this, researchers should use a higher standard for what constitutes statistical significance. For example, the usual rule of thumb for statistical significance is finding something two or three standard errors away from what was expected. If you are conducting a large number of statistical tests you might consider using the higher cutoff of three or four standard

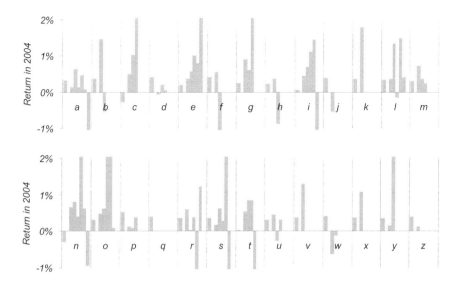

Figure 11.6. Average stock return in 2004 as a function of the number of times each letter appears in the company name.

errors instead of two or three standard errors.

There are many mathematical approaches for handling this. One is to say that if we have decided to view any p-value below 5% as statistically significant, when we conduct n different statistical tests we should look for a p-value below $5\%/n$. In other words, if we conduct 10 different statistical tests we should look for a p-value below $5\%/10 = 0.5\%$.

> The problem of **data snooping:** something that is statistically significant in a single experiment may easily occur by chance if the experiment is repeated many times.

Technical Notation

The common statistical notation is to write the null and alternative hypotheses as statements about an unknown parameter. For the telepathy example

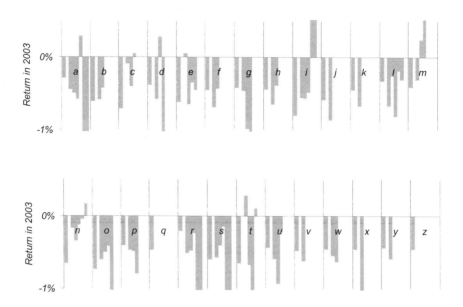

Figure 11.7. Average stock return in 2003 as a function of the number of times each letter appears in the company name.

above we suppose that the probability each question is guessed correctly is an unknown parameter p. The null hypothesis says that $p = 1/4$ and the alternative hypothesis says that $p > 1/4$. This is usually written as

$$H_0:\ p\ =\ 1/4$$
$$H_a:\ p\ >\ 1/4.$$

A two-tailed alternative hypothesis would be written as

$$H_a:\ p \neq 1/4$$

and would only be appropriate if getting significantly below $1/4$ was just as interesting as getting significantly above $1/4$.

For a significance level of 5%, people write $\alpha = .05$. The notation z_α and $z_{\alpha/2}$ are used to represent the cutoff, in terms of standard errors, for the one and two-tailed tests respectively. For example, $z_{.025} = 2$ since looking up 97.5% on the normal table gives $z = 2$.

10. Exercises

7. **Nurse hotline effectiveness.** A health insurer wants to know if having a nurse available to answer simple questions over a telephone hotline could cut the costs of unnecessary doctor visits. Records show the average cost per family is typically around $900 with an SD of $1,000 and the cost distribution is skewed with a tail to the right. They decide to conduct an experiment in which they randomly select 2,500 families, give them access to the hotline and record their health-care costs. After doing this, in the sample of families with hotline access they observe that the average cost per family is $850. Is there evidence here that the hotline actually reduces costs?

    (a) Justify your answer with a test of statistical significance.

    (b) Compute and give a brief interpretation of the corresponding p-value in this setting.

    (c) Should the fact that the cost distributions were skewed affect the calculations you made in Part (b)? Explain briefly.

8. **Summer employment.** A university financial aid office polls a simple random sample of 1,600 undergraduate students at Boston University to study their summer employment. There are 800 women and 800 men in the sample. Only 80 of the men say they were not employed over the summer, while 160 of the women say they were not employed over the summer. Could the fact that a smaller percent of men say they were not employed be easily explained by chance variation? Answer yes or no and give a justification using a statistical hypothesis test.

9. **Type 1 and Type 2 errors.** One commonly used cutoff for a p-value to be statistically significant is 5%. If a Type 1 error is much more serious than Type 2 error, should you use a higher or lower cutoff?

10. **Taste test.** Market researchers would like to know if consumers can tell the difference between a product made with low calorie oil and the same product made with regular oil. A large representative sample of 400 people are blindfolded and given both products to taste. Overall 230 people correctly identify which of the two products is low-calorie and which is regular.

    (a) Is this evidence of a difference in taste? Justify your answer with a test of statistical significance.

(b) Compute and give an interpretation of the corresponding p-value in this specific setting.

(c) Is this evidence of a big difference in taste? Briefly explain the role of statistical tests in answering this type of question.

11. **Cereal packaging.** A manufacturer of cereal has a machine that, when working properly, puts 20 ounces of cereal on average into a box with a standard deviation of 1 ounce. Every morning workers weigh 25 filled boxes. If the average weight is off by more than 1 percent from the desired 20 ounces per box, company policy requires them to re-calibrate the machine. In a sample of 100 days where the machine is working properly all day, on how many of the days is it expected the machine will be re-calibrated?

12. **Identifying costly physicians.** A health insurer's records show that average yearly patient medical expenses are $800 per year with an SD of $1,000. Under their coverage plan each patient is assigned to a primary care physician. The insurer tries to control costs by tracking the average yearly per-patient medical expenses for each of the eighteen hundred affiliated primary care physicians. They see that the physician with the highest average per-patient expenses had one hundred patients assigned to her and a per-patient average expense of $1,100. Could this high average be reasonably explained by chance? Or is this some evidence of a problem? Explain briefly.

13. **Election irregularities.** In an election between two candidates the regular vote count came out in favor of Candidate A. When the absentee ballots were counted, the percent in favor of Candidate B was so high that B came out ahead overall. Someone conducts a statistical test to see how rare such a difference between the two vote counts would be just by chance and calculates the p-value to be .08. A newspaper writes "this shows that there is roughly a 92 percent chance that the difference between the two counts is due to some irregularity other than simply chance alone." Do you agree with the newspaper's interpretation of the p-value? Explain briefly.

14. **Powerball lottery.** In the Powerball lottery between 1997 and 2002, five different numbers between 1 and 49 were chosen in each drawing.

(a) What is the chance all the five numbers in a drawing are odd?

(b) Over 515 games all five numbers came out even on 15 occasions.[7] Is this statistically significantly more than what was expected? Justify your answer with a test of statistical significance.

15. **Game to play.** Player I should write down a number from 1 to 4, and Player II should try to guess it. Make sure Player I tells Player II the number after each guess. Repeat this forty times. Can Player II guess statistically significantly better than chance?

16. **Perceived waiting time.** A large bank would like to know if installing televisions where customers are waiting would reduce the perceived waiting time.[8] They pick one bank branch to be the "treatment" branch and receive the televisions and use the other as the "control"—it receives no televisions. They then interview all people who had to wait more than 5 minutes in both branches both before and after the televisions were installed. In the "treatment" branch the percentage of these people who overestimated their waiting time drops from 32% before the televisions to 15% after they install the televisions—a 17 percentage point drop. The article then says that "the control branch actually saw an increase in overestimated wait times, from 15% to 26%."

(a) What might explain the 11 point rise in the control group?

(b) Does this rise in the control group make the 17 point drop in the treatment group seem more impressive or less impressive?

17. **Order fulfillment.** At a company it historically took 147 hours on average from when an order was received until the order was shipped. The standard deviation was 50 hours and order times followed a normal distribution. To reduce this processing time the processing system was redesigned. Based on a sample of 10,000 orders since the redesign, the average time is now 145 hours (still with a standard deviation of about 50 hours and following a normal distribution). To see if this was a significant improvement, a consultant conducts a test of statistical significance and finds a z-value of 4 and a corresponding p-value of less than .0001%. In a report to management the consultant writes "this is evidence of a highly significant reduction in the order processing times—this redesign was highly effective." Do you agree? Explain briefly.

---

[7]http://www.powerball.com/powerball/winnums-text.txt

[8]"R&D Comes to Services—Bank of America's Pathbreaking Experiments," by Stefan Thomke, *Harvard Business Review*, April 2003, p. 71–83.

18. **Drug comparisons.** A pharmaceutical company would like to advertise that its product (Drug A) works faster than the competition's product (Drug B) but government regulations require that such claims be backed by convincing statistical evidence. Many large studies show that Drug B works in 8 hours on average with a standard deviation of 5 hours, and the histogram for this data has a large tail to the right. In a study on Drug A using a random sample of 400 patients, it was seen that it worked in 6 hours on average with a standard deviation of 5 hours.

    (a) Is this statistically significant evidence that Drug A works faster than Drug B? Or would a larger sample be needed in order to conclude this? Explain.

    (b) Compute and interpret the p-value in this setting.

19. **Petition signatures.** A candidate must gather at least 8,000 valid signatures on a petition before the deadline in order to run in an election. One candidate turns in 10,000 signatures right before the deadline, but it's always expected that some percentage of them are invalid. Election officials take a random sample of 100 signatures and thoroughly investigate them to find that 84% are valid. Is this statistically significant evidence that the candidate has enough valid signatures overall? Explain.

20. **Tripled cancer risk.** The stock price of Teva Pharmaceutical Industries plummeted after results of a study of 892 women came out showing one of the company's drugs tripled the risk of breast cancer in women.[9] But an analyst says that this result is "not statistically significant." Assuming the study was a properly randomized experiment, does this mean the analyst thinks that tripling the risk of cancer is not significant? If not, what does the analyst mean by "not statistically significant?"

21. **DNA evidence.** Investigators gather DNA evidence from the scene of a crime and use a large database of ten million DNA samples and find a suspect who matches as closely as they can measure. A statistician computes that the chances a random person matches this closely as being around one in ten million. Does this small chance make it statistically significant?

---

[9] "Teva Shares Fall on MS Drug and Cancer Paper," Reuters (Chicago), Tue Jan 18, 2005 06:05 PM. See http://www.reuters.com/newsArticle.jhtml?type=topNews& storyID=7361089

22. **Vioxx lawsuits.** In September 2004 Merck & Co. removed its best-selling painkiller Vioxx from the market citing evidence it increased the risk of heart attacks or strokes.[10] Hundreds of lawsuits with potential liability costs of up to $20 billion were then filed claiming the company knew about the dangers of the drug for some time before they notified the public. While the news of each heart attack and stroke case arrived from an experimental trial of the drug, Merck executives said they waited just until they had results that "crossed the threshold of statistical significance" before notifying the public. A key issue in the lawsuits is whether this was in fact true or whether the company waited until they had results far beyond this threshold. When the company notified the public they knew that 45 of the 1,287 patients who took Vioxx experienced heart attacks or stroke, as did 25 out of the 1,299 patients in the control group. Do these results just barely cross the threshold of "statistical significance" or do these results go far beyond the threshold? Justify your answer.

23. **Call centers.** A customer service call center employs 200 phone operators and the average call handled lasts 6 minutes with a standard deviation of 4 minutes. To monitor operator performance and improve efficiency, a system randomly samples 36 calls from each of the operators and calculates the average time it takes them to complete the calls. Management would like to identify operators who are unusually slow (in order to either retrain or fire them) and those who are unusually fast (in order to either learn from them or check to make sure the quality of their work is good).

    (a) Using reasonable assumptions (specify what these are), give a cutoff for the average time you would use to define "unusually slow" and "unusually fast."

    (b) After the study, a manager was upset to discover that 6 employees had average times that were higher than the cut-off used to define "unusually slow" and suggested firing these employees. Is this a reasonable suggestion? Explain briefly.

24. **Terrorism loan fraud.** In 2005 the Small Business Administration became concerned that a high percentage of the billions of dollars in spe-

---

[10] "Merck Documents Shed Light on Vioxx Battles: Records Show Safety Panel Had Early Data Indicating Higher Heart-Problems Risk," by Barbara Martinez, *The Wall Street Journal*, Monday, February 7, 2005, page 1.

cial loans earmarked for 9/11 terrorism recovery may have been given to unqualified businesses.[11] Out of 59 randomly sampled loans, only nine appeared to be to qualified businesses. One analyst says that this small sample size is not enough to conclude anything, since there were an enormous number of loans given. Do you agree with the analyst? Justify your answer with relevant statistical calculations.

25. **Employee motivation.** A survey of employees at a company measures motivation on a scale from one to five, where five represents the highest level of motivation. The company-wide average turns out to be 3.2. As part of a study a group of 200 randomly selected employees are given a special motivational program, and the average level of motivation afterwards for them turns out to be 3.3. A statistical test is made and the p-value associated with this difference is computed to be .0013, or around 1/10 of 1%.

    (a) Explain what the p-value means in the context of this problem by completing the following sentence without using any technical jargon: There is approximately a .0013 chance that _____

    (b) A human resources manager comments that the difference between 3.2 and 3.3 seems much too small to be significant. How would you respond to this comment? Explain briefly.

    (c) Give an estimate of the SD of the motivation scores.

26. **Emotional intelligence.** A newspaper article reports on the booming business of education on emotional intelligence for business people: training in the ability to monitor one's own and others' emotions in business situations and act sensibly from that knowledge.[12] A study by the Air Force found that using an emotional intelligence assessment to select recruiters increased threefold its ability to identify successful recruiters. Another study found only a very small performance benefit from emotional intelligence education, but the p-value for the study is reported as .001, or 1/10 of 1%. Despite the small performance benefit an an-

---

[11] "SBA Finds 9/11 Loan Recipients Ineligible," by Larry Margasak, *Associated Press*, Thu Dec 29, 7:14 AM ET. `http://news.yahoo.com/s/ap/20051229/ap_on_go_ot/sept_11_lax_loans`

[12] "How best to deal with difficult people. Experts say a firm grasp of 'emotional intelligence' helps," by Judy Foreman, *Boston Globe*, November 13, 2005, `http://bostonworks.boston.com/college/articles/2005/11/13/how_best_to_deal_with_difficult_people_experts_say_a_firm_grasp_of_emotional_intelligence_helps/`

alyst comments that, based on the low p-value, the benefit is highly statistically significant and is thus consistent with the Air Force study.

(a) Does the analysts comment seem sensible?

(b) The authors of the study interpret the p-value saying that "there is less than one chance in a thousand that emotional intelligence was not the cause of the jump in the score." Is this a correct interpretation of the p-value? If it is not, give a correct interpretation in the context of this problem.

27. **Rising unhappiness.** A study from the University of Chicago's National Opinion Research Center finds that unhappiness has risen over the past decade.[13] They surveyed random samples of 1,340 people in 1994 and 2004 and found that the percentage of people who reported at least one significant negative life event rose from 88% to 92%. Also the percentage who reported breaking up with a steady partner doubled from 4% to 8%. The study's authors say this is surprising because the good economic years of the 90s were expected to have brought more happiness—not less happiness. Are these increases likely to be just a chance fluctuation or are they statistically significant? Explain briefly, showing any necessary calculations.

28. **Cancer sniffing dogs.** Can a dog detect cancer by sniffing the breath of a patient? Dogs who had been trained to "sit" when they smelled cancer did so 565 times out of 575 breath samples from patients who had been already diagnosed with cancer, but also 4 times out of 712 breath samples from people who did not have cancer.[14] Is this statistically significant?

29. **Magical duct tape cure.** A CNN.com news article titled "Duct tape no magical cure for warts, study finds" summarizes a study of nearly 100 children (assume half in a treatment group that used duct tape and half in a control group that used no duct tape) that finds that after 6 weeks the warts of 8 children (about 16 percent) in the treatment group and the warts of 3 children (about 6 percent) in the control group

---

[13] "Unhappiness has risen in the past decade," by Sharon Jayson, *USA Today*, Mon Jan 9, 7:23 AM ET, see `http://news.yahoo.com/s/usatoday/20060109/ts_usatoday/unhappinesshasriseninthepastdecade`

[14] "Dogs Excel on Smell Test to Find Cancer," by Donald McNeil Jr., *The New York Times online*, January 17, 2006, `http://www.nytimes.com/2006/01/17/health/17dog.html?pagewanted=print`

had disappeared.[15] Is this difference statistically significant? Justify your answer with relevant calculations and discuss the conclusions you can draw from this study.

30. **Rising crime rate.** In 2006 CNN reported "violent crime was up 2.5 percent for the year, marking the largest annual increase in crime in the United States since 1991, according to figures released Monday by the FBI."[16] The article gave several possible reasons for the increase but also quoted a Justice Department statistician as saying "It's certainly a matter of concern. But the question is this—'Is this a real increase or is it...statistical noise, which you see with year-to-year changes?'" Based on the numbers given, does the Justice Department statistician seem to have a valid point?

31. **Antigenics drug trial failure.** A recent front-page *Wall Street Journal* article tells the following story: "Garo Armen froze as colleagues broke the bad news last year about a kidney-cancer vaccine he had championed for over 12 years at a cost of $300 million: The clinical trial had failed. Shares of his tiny biotechnology firm, Antigenics Inc., plunged. His ambition and patients' hopes were dashed. Now Mr. Armen, 54 years old, is on the move – traveling the world on a quest to persuade regulators to approve the vaccine despite the initial results. His argument: The vaccine worked well in a subgroup of less-sick patients. Yet the U.S. Food and Drug Administration has never approved a drug based on efficacy in a subgroup of patients, defined after a trial."[17] Then the article then quotes a statistician from the Food and Drug Administration who explains why the agency has this policy. What do you think he says?

32. **Lipitor versus Pravachol.** An article writes "In the battle for the $22 billion global market for cholesterol-lowering drugs, Bristol-Myers Squibb just shot itself in the foot. A study funded by Bristol-Myers shows clearly that the highest dose of Pfizer's Lipitor prevented heart problems better

---

[15] "Duct tape no magical cure for warts, study finds," *Reuters*, Posted: 5:47 p.m. EST, November 6, 2006, http://www.cnn.com/2006/HEALTH/11/06/warts.duct.tape.reut/index.html

[16] "Violent crime takes first big jump since '91:Murder numbers climb in smaller cities," Terry Frieden, CNN, Monday, June 12, 2006. See http://www.cnn.com/2006/LAW/06/12/crime.rate/index.html

[17] "Burden of Proof: Cancer Drug Fails, So Maker Tries New Pitch," by Geeta Anand, *The Wall Street Journal*, New York, N.Y.: Aug 2, 2007. pg. A.1.

than a high dose of Bristol's own Pravachol....A study carefully designed by Bristol-Myers Squibb to be too small and too short backfired badly and wound up demonstrating their competitor's superiority."[18]

(a) After two years, the people on Pravachol had a combined rate of heart attack, bypass surgery, angioplasty, stroke and death of 26.3% compared with 22.4% for people on Lipitor. The death rate from heart disease was 1.1% for the Lipitor group compared to 1.4% for the Pravachol group. The rate of death from any cause was 2.2% for people on Lipitor and 3.2% for people on Pravachol.[19] If we assume the 4,162 people in the study were split equally between a group receiving Lipitor and a group receiving Pravachol, do the differences here appear to be statistically significant? Explain briefly.

(b) The article goes on to quote one of the researchers as saying "This is one of the most extraordinary and improbable outcomes anyone could have anticipated." Do you agree?

33. **Live in the moment.** Do people "live in the moment" or are they usually distracted by other thoughts? A psychology experiment conducted on the Cornell University campus has the researcher approaching and asking a subject for directions to a campus building. While the subject is giving directions, two people carrying a door pass between them and the researcher secretly switches places with one of the people carrying the door. After the door goes by, a completely different looking person is now standing there listening to the directions.[20] Figure 11.8 illustrates what happens. Only seven out of 15 people said they noticed they were speaking with a different person. In a later study, researchers asked a large number of subjects if they thought they would notice – and 83% said they thought they would.[21] Is the difference between the two studies statistically significant?

[18] "Bristol Study Proves Rival Drug Better," by Matthew Herper, 03/08/04, *Forbes.com*, http://www.forbes.com/sciencesandmedicine/2004/03/08/cx_mh_0308bmy.html.

[19] http://www.medicalconsumers.org/pages/CardiologistspoisedtogiveeveryoneLipitor.html

[20] "Failure to detect changes to people in a real-world interaction.," by Simons, D.J., & Levin, D.T., 1998, *Psychonomic Bulletin & Review,* 5, 644-649. See also http://scienceblogs.com/mixingmemory/2006/12/coolest_experiment_ever.php.

[21] Levin, D.T., Momen, N., Drivdahl, S.B., & Simons, D.J. (2000). Change blindness blindness: The metacognitive error of overestimating change-detection ability. *Visual Cognition,* 7, 397-412.

Figure 11.8. A psychology experiment: the experimenter first asks someone for directions, then a door passes and the experimenter is secretly swapped with another person, and finally the new person asks the subject if he notices he's talking to a different person.

34. **Slow customer payments.** A large retail chain has changed some policies and the controller is concerned about the policy change's effect on how long it takes customers to make payments. She measures payment time in terms of the average number of days receivables are outstanding. Historically, payment time has averaged 50 days with a standard deviation of 10 days. Since it is too expensive to analyze all receivables, she draws a random sample of 75 accounts. The average payment time in the sample is 53.

(a) Is the difference in payment time statistically significant? Briefly explain.

(b) Calculate the p-value for the statistical test in part (a) and explain what it means in the context of this problem.

(c) The controller's boss points out that there are major financial implications from losing an average of three days interest on the usually tens of millions of dollars of receivables outstanding. The boss says there is no need to conduct any statistical test here because, regardless of the results of the statistical test, we already know three days has major financial significance. How would you respond to this comment?

35. **Changing drug study goals.** The cholesterol-lowering drugs Zetia and Vytorin, made by Merck and Schering-Plough, are used by almost a million Americans every week at a cost of around $4 billion per year. Despite tremendous growth in their use, there were no published clinical trials that compared these two drugs to other competing medicines. After completing an important study of the effectiveness of these drugs, but before publishing the data, the companies announced they had decided to change the goal of the study from measuring the drugs' effectiveness in reducing cholesterol to measuring the drugs' effectiveness in reducing harmful arterial plaque formation.[22]

   (a) A spokesperson for the companies say that this goal change should not impact the credibility of the study as long as the p-value for the new goal is less than 5% – the traditional cut-off for statistical significance. Another researcher responds, says that goals for a study should be decided in advance and never changed. Assuming both goals are of similar medical importance, do you agree with the spokesperson? Do you agree with the researcher? Explain briefly.

   (b) The article goes on to say that even though the study had already been completed, "Merck and Schering said they did not yet know the results of the trial. They said they were changing the [goal] only because they want to be able to analyze the data more quickly." If this were really true, how would this affect the credibility concern in Part (a) above? Comment briefly.

36. **Healthcare quality.** Analysts profiling the quality of health care facilities randomly select 225 different health care facilities nationwide. In each of these facilities, they then examine the records of 100 randomly chosen patients treated there. They use these 100 records to estimate the percentage of patients who successfully receive all of the treatments recommended by their doctors at the corresponding facility.

   (a) Overall, approximately 80% of patients nationwide successfully receive all of the treatments recommended by their doctors. If we as-

---

[22] "Cardiologists Question Delay of Data on 2 Drugs," by Alex Berenson, *The New York Times*, page C1, November 21, 2007. http://www.nytimes.com/2007/11/21/business/21drug.html?_r=1&ref=business&oref=slogin

sume that there are no important differences between the facilities, how much variation would you expect to see in the percentages estimated above across the 225 different health care facilities? Explain briefly.

(b) Three of the facilities have a special bonus system that rewards its staff according to the quality of care delivered. In the best-performing one of these, the sample percentage is 90%. Is this statistically significantly higher than the national percentage? Explain briefly, conducting a test of statistical significance and interpreting the results.

37. **Catalog redesign.** Market researchers at a nationwide retailer would like to know if a redesigned mail order catalog would increase sales. A random sample of 900 households will receive the redesigned catalog, and the remaining (millions of households) will receive the original catalog. The average sales amount per customer is $29 with the redesigned catalog, compared with an average of $25 per customer for the original catalog. The standard deviation is approximately $30 in each group.

(a) Is this evidence that the redesigned catalog significantly improves sales? Explain briefly, using a test of statistical significance.

(b) Compute a p-value for the statistical test in Part (a) and give a one sentence managerial interpretation for it.

(c) Someone comments that a four dollar increase in sales is not significant enough to justify the additional expense of the redesigned catalog. In light of your answer to Part (a), how would you respond to this comment?

(d) Is it reasonable to say that approximately 95% of the households with the redesigned catalog had sales in the range from $27-$31? Explain briefly.

(e) In reality, the distribution of sales does not follow a normal curve because the majority of customers buy nothing. How should this fact affect your answer to Part (b)? Explain briefly.

38. **Unaided product recall.** To create awareness of a redesigned brand, a firm develops a new TV ad. Before launching the ad, a study of "unaided recall" is conducted where subjects watch a TV show containing the ad and then are asked open-ended questions such as "what ads do you remember seeing during the show?" to test if they remember seeing the

ad without any prompting. In a random sample of 200 people, 23% of them were able to recall the ad without prompting.

(a) Calculate a 95% confidence interval for the proportion of people who can recall the ad without prompting.

(b) Explain what the confidence interval tells us, in the context of this problem.

(c) The norm for TV ads for this product class is that 17.5% of people who see an ad are able to recall it without prompting. Calculate a p-value for the difference between the 17.5% and the 23%.

(d) Explain what the p-value tells us in the context of this problem.

39. **Customer service system redesign.** A firm would like to test the effectiveness of a redesigned customer service system. After the redesign is implemented, a sample of customer wait times is measured and the average wait time in a sample was reduced by 16% compared with the historical average wait time before the redesign. Management views a 16% reduction as significant, but the analyst who took the sample and did the analysis reported that the p-value associated with the reduction was 0.24, or 24%.

(a) What does the p-value mean in the context of this problem? Explain briefly.

(b) Would you conclude from this that the redesign did not work? Explain briefly, using the concepts of Type I and Type II errors, if appropriate.

40. **Drug effectiveness.** Three different studies are conducted to measure the effectiveness of the same drug and each study is designed so that there is a 5% chance of making a Type I error.

(a) If we assume the drug is not effective and the three studies are independent, what is the chance at least one of the studies makes a Type I error? Explain briefly.

(b) Suppose that, instead of the studies being completely independent, we learn that there are a number of patients who participated in all three studies. Would this knowledge increase or decrease your answer to Part (a) above? Explain briefly.

41. **Crime and astrology.** A newspaper article titled "The evil of Aries: Could astrology help point to future jail time?" reports that of the 1,986

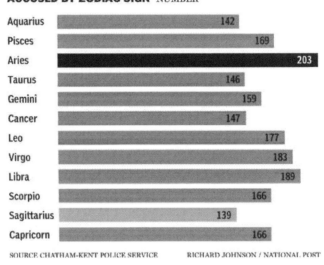

ZODIACAL CRIMINALS

The Chatham-Kent Police Service has provided a zodiacal breakdown of the 1,986 people arrested in the area in 2011.

ACCUSED BY ZODIAC SIGN NUMBER

Sign	Number
Aquarius	142
Pisces	169
Aries	203
Taurus	146
Gemini	159
Cancer	147
Leo	177
Virgo	183
Libra	189
Scorpio	166
Sagittarius	139
Capricorn	166

SOURCE CHATHAM-KENT POLICE SERVICE   RICHARD JOHNSON / NATIONAL POST

Figure 11.9. Are Aries more prone to being criminals?

people arrested in 2011 in the Canadian city of Chatham-Kent, 203 were Aries whereas just 139 were Sagittarius. The article goes on to quote Georgia Nicols, who writes the National Posts horoscope, that the results in Chatham-Kent make some sense because "Aries is the sign of the warrior, Aries rules the military. Aries jump in head first, and love adventure. Sagittarius stays out of the crime stats because they don't get caught. They are smooth. They can talk anybody into anything." Is the difference between the two science as statistically significant? What do you conclude? Comment briefly. The rest of the data can be found in Figure 11.9.[23]

42. **Zicam investor lawsuit.** A newspaper article writes about a recent unanimous Supreme Court decision involving a case where investors were su-

---

[23] "The evil of Aries: Could astrology help point to future jail time?" by Peter Kuitenbrouwer, Dec 28, *National Post*, http://news.nationalpost.com/2011/12/28/astrology-as-a-sign-of-future-jail-time-experts-remain-skeptical/

ing the drug company Matrixx Initiatives company for securities fraud since it did not disclose scattered reports of patients permanently losing their sense of smell (a condition known as anosmia) from Zicam, a homeopathic nasal spray and cold remedy made by the company; the company's stock plummeted when this information was eventually disclosed and the Food and Drug Administration warned consumers not to use the product. Lawyers for Matrixx argued that the reports had been collectively statistically insignificant and therefore it was not necessary to have disclosed them. Do you think the Supreme Court ruled in favor of the company or the investors? And what reasoning do you think the Court gave? As part of your answer, discuss the relative importance of Type I and Type II errors in this setting.[24]

43. **Cheating at cards.** A person claims she can guess the color of playing cards before they are turned over. An experiment is conducted where she makes a number of guesses and comes out to be 90% correct. A p-value equal to 4% is computed to test whether or not she is cheating or just making lucky guesses. Which two of the following choices give the correct interpretation for this p-value?

   (a) It's the chance she is not cheating.

   (b) It's the chance she's not cheating given she gets 90% correct.

   (c) It's the chance she gets at least 90% correct given she's not cheating.

   (d) It's the chance she's not cheating if she repeats the experiment.

   (e) It's the chance she gets at least 90% correct if she repeats the experiment.

   (f) It's the chance she gets at least 90% correct if she repeats the experiment without cheating.

   (g) It's the chance she gets at least 90% correct if she repeats the experiment.

---

[24] "Supreme Court Rules Against Zicam Maker," by Adam Liptak, *The New York Times*, March 22, 2011.http://www.nytimes.com/2011/03/23/health/23bizcourt.html

Fun Problem: Rent-Free Space

How can this be true?

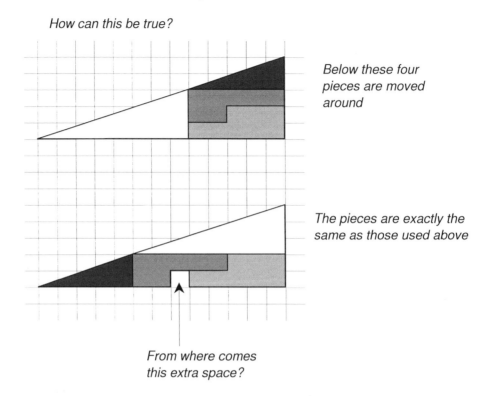

Below these four pieces are moved around

The pieces are exactly the same as those used above

From where comes this extra space?

Discussion.   This has important business implications. Suppose you have a warehouse and need some extra space but can't afford the extra rent. Just divide up the warehouse space as above and re-arrange to get the extra space. Or you could rent out this extra space and keep the money as profit. If you need a second free space, just divide the bottom shape up again as in the top figure and re-arrange to get two free spaces. Very nice, isn't it?

Solution.   No solution this time—you've got to figure this one out on your own.

Chapter 12

Building Multiple Regression Models

1. Introduction

Suppose you are trying to forecast some variable Y by regression using either Variable A or Variable B. Figure 12.1 shows graphs of some sample data collected for each variable and the regression line; the equation and correlation is also computed for each. It turns out that there is a much stronger correlation between Y and A, and so using A to forecast Y works better in this data set. But which of these two regression lines do you think would give more accurate forecasts for Y in the future?

If you said A, I would say you like to gamble. If you said B, I would probably be more comfortable. Variable A has a higher correlation with Y so it may seem as though its line should give more accurate forecasts. But since there are so few data points, it makes me worried that the line might come out quite differently if we had more data—and also that the future may look very different from this sample. I am worried this very nice correlation might be just a coincidence in this particular small sample. Variable B does not have such a strong correlation, but at least we can confidently estimate the slope and intercept of the line. I am not so worried that the future will look a lot different from this sample, since we have so much data here. Using the line for A is more of a gamble, while the line for B is more of a safe bet.

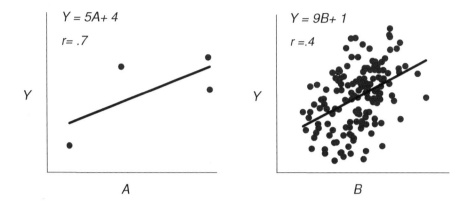

Figure 12.1. Two regression lines for predicting a variable. Which one do you think would be better to use in the future?

The point here is that to decide which variables to use for forecasting another variable, you must consider not only how strong you think the relationship is with what you are trying to forecast but also how much confidence you have in your estimate of the relationship. In the case of Variable A above, we saw a stronger relationship but we did not have much confidence in estimating exactly what the relationship was. We didn't see as strong a relationship with Variable B but we were very confident in our estimation of the relationship.

2. The Standard Error for a Regression Coefficient

One way people measure the confidence they have in their estimate of a regression line is by attaching standard errors to the regression coefficients. The formulas for computing these standard errors are not as simple as the ones for sample averages and percentages, so we will leave the computational details to the computer. We will first focus on interpreting these standard errors and then show you how to get them from the computer.

The standard error for a regression coefficient tells you about how much error you believe you may have in estimating that coefficient. The margin of error equals twice the standard error, and is roughly the outer limit on the amount of error you would expect to see in your estimate of this coefficient.

> The **standard error for a regression coefficient** tells you about how much error you believe you may have in your estimate of that coefficient.
>
> The **margin of error** for a regression coefficient equals twice the standard error for that coefficient and is roughly the outer limit on the amount of error you would expect to see in your estimate of this coefficient.

We will next give an example of the interpretation of the standard error for a regression coefficient. For the regression lines in the example above, suppose the standard error for the coefficient in the first equation is 4, and the standard error for the coefficient in the second equation is 2. How do we interpret these numbers?

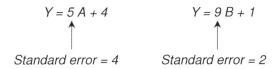

In the first regression line, the margin of error for the coefficient is twice 4, or 8. Although 5 is our best estimate of the slope in the first line, we believe there could be an error of up to around 8 units up or down from this. Are 8 units of error such a big deal to worry about? Here is a deal for you: suppose I promise to transfer $5 million into your bank account with a possible error of up to $8 million up or down from that number. Would you be interested in this deal?

That would be a terrible deal for you, since I could give you any amount up to $13 million or down to negative $3 million—or in other words I could take $3 million from you. You can't even have any real confidence here that you'll get anything at all out of this deal and it may even hurt you.

A similar thing is true about the regression coefficient. Since the margin of error is larger than the coefficient in the first equation, you can't even have any real confidence that the slope coefficient is positive, negative or even zero—and zero corresponds to no linear relationship at all! This says it is quite possible there could be no relation between A and Y and the nice correlation we saw was just due to chance. Even though the points fit so nicely around a line, it could easily have been good luck. One reason for this unfortunate fact is that we have so little data and this is why I would be worried to use this line to make forecasts.

In the second regression line, twice the standard error equals twice 2, or 4. This means our estimate for the slope of the line equals 9, but it could be reasonably in the range from $9 - 4 = 5$ to $9 + 4 = 13$. This range doesn't include a slope of zero so we are at least confident that there is some sort of positive linear relationship. There is still error in our estimate of the slope but at least we are confident that B is really linked to Y with a positive correlation. I would feel more comfortable using this line to forecast Y. And I would, incidentally, also feel very comfortable receiving a bank transfer of \$9 million with a possible error of up to \$4 million up or down from this. Either error up or down would be just fine with me, thank you.

A variable is called **insignificant** if it has a regression coefficient smaller in absolute value than the corresponding margin of error. It is unwise to use such variables for forecasting because it means that we can't confidently say if the coefficient is positive, negative, or zero. It would be wise to remove that variable from your regression equation and look for another **significant** variable, one with a coefficient larger in absolute value than the margin of error.

This advice applies to multiple regression equations as well, where there is a separate standard error for each coefficient. In a multiple regression equation, however, a variable can change from being insignificant to significant—or vice versa—as you add or remove other variables from the equation. So it's best to add and remove variables one at a time to see what happens.

If a regression coefficient is smaller in absolute value than the corresponding margin of error, there is not enough accuracy to sensibly estimate the coefficient and the variable is called **insignificant**. You can't even confidently determine if the coefficient is positive, negative, or zero. It is wise to remove the variable from your equation.

Computers that do a regression analysis usually print out standard errors for each of the regression coefficients as well as a *t*-**value**, which equals the coefficient divided by the standard error. The significant variables will thus have *t*-value above $+2$ or below -2. A *t*-value between -2 and $+2$ means the coefficient is smaller in absolute value than the margin of error.

> The *t*-**value** for a regression coefficient equals the ratio of the coefficient to the standard error.

This brings us to the fundamental goal of regression model building. You should try to find an equation that fits your data as closely as possible using the fewest number of variables—and which makes the best physical sense.

> When building a regression equation, you should try to find an equation that fits your data as closely as possible using the fewest number of variables— and which makes the best physical sense.

Question.   Here is a printout of a regression equation, the associated standard errors, and the *t*-values. Which variable would you recommend removing from the regression equation to improve it?

	Coefficient	Std Err	*t*-value
Constant	589.3	135.3	4.4
X_1	1.6	2.3	0.7
X_2	-566.4	81.3	-7.0
X_3	0.022	0.001	18.3
X_4	-297.2	34.0	-8.7

Solution.   We see that the regression coefficient for X_1 is 1.6 with a standard error of 2.3. The *t*-value of .7 tells us that this coefficient is less than one standard error away from zero. This means we can't really even be confident whether this coefficient is positive, negative, or zero. It would be wise to drop this variable from the regression equation and create a new equation. All the other variables have coefficients more than 2 standard errors from zero, as indicated by *t*-values of greater than $+2$ or less than -2. Since it is possible for the standard errors, coefficients and *t*-values for the other variables to change when you remove a variable from the regression equation, it is wise to add or remove one variable at a time.

Interpretation of the p-value in this setting.   The p-value for a regression coefficient is twice the area beyond the corresponding t-value under the tail of a curve very similar to the normal curve.[1] Suppose the multiple regression coefficient for X_1 equals 10 and the corresponding p-value equals .02. This means that if there were no link between X_1 and Y, the chances of getting a coefficient in absolute value of at least 10 is .02. It tells us how rare it would be to see a coefficient this large (in absolute value) if there really was no link between that variable and what we're trying to forecast.

Often instead of looking for the "insignificant" variables that have a t-value between -2 and $+2$, people often equivalently look for variables that have a p-value above .05.

Exercises

1. An analyst creates a multiple regression equation that uses four different variables (X_1, X_2, X_3 and X_4) to forecast sales with the same amount of data gathered for each. Below is a printout of the regression coefficients and the corresponding standard errors for each. Give some advice on how this forecasting equation could be improved.

	Coefficients	Standard Error
Intercept	87	66
X_1	-6	22
X_2	28	9
X_3	10	16
X_4	-74	34

2. An analyst would like to build a multiple regression equation to forecast sales and has four different variables available to use for this: X_1, X_2, X_3 and X_4. Below is a printout of the correlation between each of these variables and sales:

	Sales	X_1	X_2	X_3	X_4
Sales	1.00				
X_1	0.66	1.00			
X_2	0.79	0.80	1.00		
X_3	0.41	0.44	0.50	1.00	
X_4	-0.61	-0.38	-0.56	-0.16	1.00

---

[1]The actual curve commonly used by software packages is called the "t-distribution" and for large samples it looks very similar to a normal curve.

(a) Which variable by itself looks like it would be a better predictor of sales in a simple regression equation: X_1 or X_4?

(b) The analyst decides to create a multiple regression equation using the two variables having the strongest correlation with sales: X_1 and X_2. The coefficients and corresponding standard errors are listed below.

	Coefficient	Std Err
Constant	31	47
X_1	−19	22
X_2	40	8

Comment on how this equation could be improved.

(c) The analyst decides to try another equation using just variables X_2 and X_4 and gets the following:

	Coefficient	Std Err
Constant	102	55
X_2	28	4
X_4	−73	31

(d) Do you think this equation would give better forecasts than the one in (b)? Explain.

(e) Compare your answers for questions (a) and (c). Do they contradict each other? Explain.

3. Building a Model

A retailer has 45 stores spread across the country. They're interested in forecasting sales for a potential new store location under consideration. Using data from the current stores, how can they forecast sales? In this section we will illustrate the process of building a good multiple regression model for forecasting. Then we interpret and summarize the results we obtain for this retailer.

The retailer has the following data on current stores: the amount spent on advertising (in $1,000s), the number of miles to the nearest freeway, the square footage of the store, the percentage of people in the area who are homeowners, the median income in the area, and the primary method of promotion for the store (1 = newspaper, 2 = radio, 3 = direct mailing), and the total annual sales (in $1,000s). In reality, many more variables would probably be available but to simplify things for the purpose of illustrating the approach, we just use this small number of variables.

Below are the first four rows of the data (the full data set is in the file `retailchain.xls`):

store	advertising	miles	sqfeet	% owners	income	promotion	sales
1	98.7	1.8	7050	75.1	47600	2	1976
2	122.2	2.2	9744	94.7	49400	2	2673.6
3	98	1.6	5098	74	42000	3	1303.5
4	129.36	2.1	7564	98	64400	3	2127.7

A potential new store site under consideration has 5,000 square feet, is 2 miles from the nearest freeway and has a median income of $50,000. What would be a reasonable sales forecast for this site?

Step 1: *Draw scatter plots to look for non-linearities or outliers.* The first step in good model building is to draw scatter plots to see how each variable relates to what you're trying to forecast—in this case its sales. For multiple regression to be a reasonable tool for forecasting, each variable ideally should have a roughly linear relationship either gently sloping up or gently sloping down. Figure 12.2 shows the graphs in this case.

Everything looks roughly linear except for the last graph on the right, where there is clearly no linear relation between "promotion" and sales. In this case "promotion" is a categorical variable and this graph reminds us we should properly convert it into dummy variables. This is our next step.

Step 2: *Convert categorical variables into dummy variables properly.* This means for each category we must create a variable which equals one for that category and zero otherwise. And we always must omit one of the categories from a regression equation to avoid a multicollinearity problem. In this case we will arbitrarily choose to omit the third category so it will be viewed as the baseline category (and coefficients for the other two categories will be interpreted as relative to this third category). To do this we create two new columns:

store	advert	miles	sqft	own	income	promotion	sales	prom1	prom2
1	98.7	1.8	7050	75.1	47600	2	1976	0	1
2	122.2	2.2	9744	94.7	49400	2	2673.6	0	1
3	98	1.6	5098	74	42000	3	1303.5	0	0
4	129.36	2.1	7564	98	64400	3	2127.7	0	0

The numbers in the column labeled "prom1" equal one if we are looking at promotion category one and zero otherwise. The numbers if the column

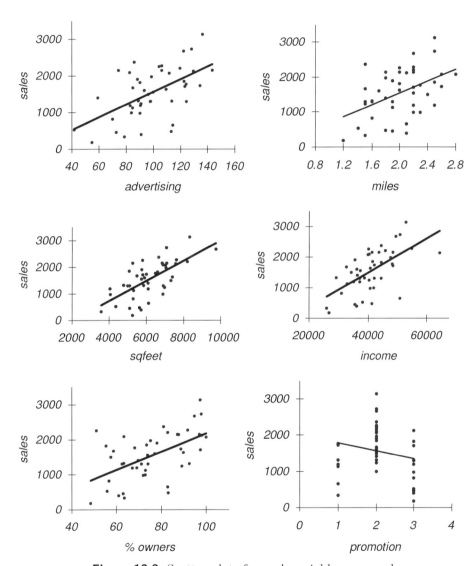

Figure 12.2. Scatter plots for each variable versus sales.

labeled "prom2" equal one if we are looking at promotion category two and equal zero otherwise.

Step 3: *Run a stepwise or backwards procedure.* There are a number of commonly used procedures for selecting the best variables to use in a multiple regression. We will illustrate two of them: backwards regression and stepwise regression. We first illustrate the backwards regression procedure.

In the backwards approach, we start off with all variables included in the regression equation and look for any insignificant variables—those which have coefficients smaller in absolute value than the corresponding margin of error for that coefficient. When this is the case, it means we don't have enough accuracy to tell if the coefficient is really positive, negative, or zero. Since the "t-value" is the ratio of the coefficient to the standard error, we can look for variables which have a "t-value" between $+2$ and -2. If there is more than one variable like this, we pick the one which has a "t-value" closest to 0. Then we eliminate this variable, compute a new regression equation and keep repeating this procedure. It's important to eliminate the variables one at a time because all the coefficients and standard errors can change every time you remove a variable.

We will walk you through step-by-step what happens with the "backwards" procedure and this is summarized in the diagram in Figure 12.3. First we see that variable "prom1" has a t-value closest to zero: it equals 0.2. We remove this variable and afterwards we see that "sqfeet" has a t-value of 0.4. Since this is closest to zero, we remove this variable as well. Finally, after also removing "% owners" and "miles," all of remaining variables have coefficients with t-values either greater than $+2$ or less than -2.

Next, we illustrate the stepwise regression procedure. The Regression Worksheet (`regression.xlsx`) included with this book (and discussed in Section 7.4) has a feature for stepwise regression. To use it, first follow the instructions for pasting in the data and indicating which variable is to be the "y" variable. Then add variables (by entering an "x" next to them) or remove variables one at a time that are suggested to be added or removed in the worksheet; these will be highlighted in red at the left. When the last variable suggested to be added is immediately then suggested to be removed, remove it and you are finished with the stepwise procedure.

In the stepwise approach, we first start off with the variable that is most correlated with the dependent variable. Then at each step we check all re-

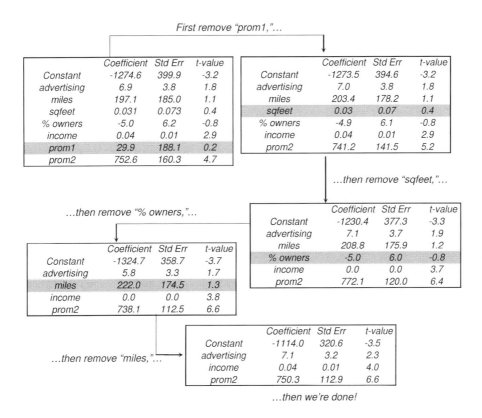

Figure 12.3. The backwards regression procedure.

maining variables, one at a time, to find the one that has the largest (in absolute value) t−value. Then we add this variable to the regression equation. Then we keep repeating this process until no remaining variable would give a t−value larger in absolute value than 2. Also, after each step we check to see if the t−value for any of the previously added variables dips below 2 (in absolute value) And in that case the variable is removed. In other words, we start with the most significant variable and at each step we add the most significant of the remaining variables and remove any variable that has become insignificant.

We will walk you through step-by-step what happens with the "stepwise" procedure and this is summarized in the diagram in Figure 12.4. We first find that the variable "sqfeet" has the highest correlation with "sales." Then we see that the variable "prom2" has a t-value farthest from zero: it equals

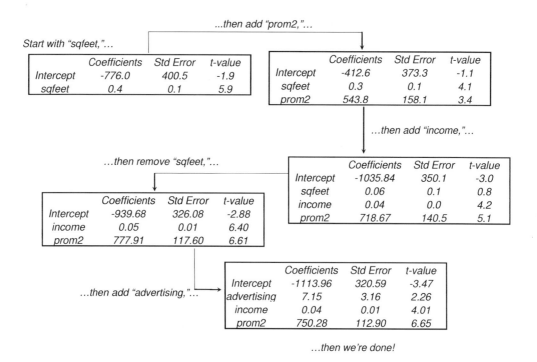

Figure 12.4. The stepwise algorithm in action.

4.1. Then we see that the variable " income" has a t-value farthest from zero: it equals 4.2. We add this variable and afterwards we see that the t-value for "sqfeet" has dropped to 0.8. Since this is smaller than 2 in absolute value, we remove this variable. Finally, we add "advertising" and see that all of the remaining variables would have coefficients with t-values between $+2$ and -2. An illustration of this process can be found in the file `stepwise regression demo.xlsx`.

Step 4: *Check the technical assumptions.* In order to rely on a high t-value as evidence of a link between some variable and what you're trying to predict, there are two main technical assumptions about your data that must hold:

(1) Forecasting errors should approximately follow a normal distribution with roughly the same standard deviation over the whole range of forecasts.

(2) There is no obvious trend or pattern in the forecasting errors.

To check these assumptions, we create a graph of "residuals" versus "fitted"

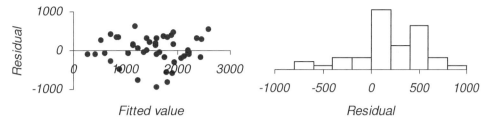

Figure 12.5. Analyzing the residuals.

and we get the graph on the left in Figure 12.5. If we draw a histogram for the "residuals," we get the graph on the right in Figure 12.5.

The "fitted" is the forecast of sales for each of the current stores using the regression equation and the "residual" is the difference between the actual sales figure and the forecasted sales figure. On the histogram above, we see that the residuals roughly follow a normal distribution. It's not exactly a normal distribution, but it's not drastically different from a normal distribution. On the scatter plot above, going from left to right we see no real pattern in the residuals and the vertical spread seems approximately the same all the way along. Of course, it's somewhat more spread out in the middle than on the ends but this is due to just a few points. The points more or less seem to be scattered randomly. If we saw a line or curve or some other pattern, it might indicate that the data don't follow the regression line's general trend. The regression line may still give reasonable forecasts if the technical assumptions are violated but relying on t-values to see if a variable is significant or not may not be so reliable.

There is no law against using a regression equation for forecasting if these assumptions are not satisfied, but if you are using the "backwards" or "stepwise algorithm" (and hence relying on the p-values as evidence of statistical significance) then these p-values should be accurate. If the technical assumptions don't hold, they may be inaccurate. Essentially this means you may wind up choosing the wrong variables for your multiple regression equation.

Step 5: *Interpret your equation and then summarize the managerial implications.* The final equation we obtained was the following, which has an R^2 of 0.73 and a standard error of 372:

Forecasted sales $= -1114 + 7.1 \times \text{advertising} + 0.04 \times \text{income} + 750.3 \times \text{prom2}$

We can interpret the 750.3 as follows: for stores spending the same amount on advertising and income in the area, stores primarily using radio advertising average about $750,000 more in sales than stores primarily using other promotion methods.

We can interpret the 7.1 as follows: for stores in areas with the same median income which use the same promotion methods, each extra thousand dollars in advertising is associated with approximately an extra $7,000 in sales.

We can interpret the 0.04 as follows: for stores spending the same on advertising and using the same method of promotion, each extra thousand dollars in median income for the area is associated with an extra $40,000 in sales.

We interpret the .73 as follows: variation in advertising, median income, and the primary method of promotion can explain about 73% of the variation in sales.

We interpret the 372 as follows: forecasts using this regression equation generally are around about $372,000 from the actual sales figures, and almost all are within the margin of error of $2 \times 372,000 = \$744,000$. This means forecasts are expected to be accurate to within about three quarters of a million dollars.

The overall picture is that stores that spend more on advertising, are in areas with higher median income, and use radio advertising do the best.

Surprisingly, the size of the store is not a big predictor of sales. Although square footage and sales have the highest correlation of any pair of variables (a correlation of 0.7), this can be explained by differences the stores have in advertising, median income, and method of promotion. In other words, when you look at stores in areas with similar median incomes and using similar promotion methods, you don't see much of a correlation between the size of the store and sales. To see why this is so surprising, here is a regression equation to predict sales using only the square footage of the store:

	Coefficient	Std error	t-value
Intercept	−776.0	400.5	−1.9
sqfeet	0.4	0.1	5.9

We see from the t-value of 5.9 that square footage is significant by itself, but we saw above that in the presence of the other variables it becomes insignificant. This is something a simple regression would never show you but the multiple regression reveals.

This, of course does not mean the retailer should open up small kiosks with just 2 square feet space, since this would fall drastically outside the range of the data. I would, however, recommend they look at some of the larger stores to see how the square footage may not be well utilized and look at some of the smaller stores to see how they do such a good job with such a small space. I suspect they would learn something from doing this.

The potential new store site under consideration has 5,000 square feet, is 2 miles from the nearest freeway, and has a median income of $50,000. What would be a reasonable sales forecast for this site? We have learned that the size of the store and the distance to the freeway are not so important, so our forecast is

$$
\begin{aligned}
\text{Forecasted sales} \quad &= \quad -1114 + 7.1 \times \text{advert} + 0.04 \times 50{,}000 + 750.3 \times \text{prom2} \\
&= \quad 886 + 7.1 \times \text{advert} + 750.3 \times \text{prom2}
\end{aligned}
$$

This means we forecast sales of around $886,000 plus $7,100 for each $1,000 spent, plus an extra $750,300 if the primary method of promotion is radio. Since the forecasting error (indicated by twice the standard error of the regression line) is several hundred thousand dollars, it's a little more realistic if we round things off and say $900,000 plus $7,000 for each $1,000 spent on advertising, plus an extra $750,000 if the primary method of promotion is radio.

Since median income is the only variable in the forecasting equation that is not under direct managerial control, our overall advice for the retailer should be to open stores in areas with high median income and use radio advertising as the primary method of promotion—and not worry about the size of the store.

Instructions using R

Start R-Studio, import the data file "**retailchain.xls**" and create a copy titled "d":

```
d=data.frame(retailchain)
```

To create two new columns to respectively represent promotion type 1 and type 2, type

```
d$prom1=ifelse(d$promotion==1,1,0)
d$prom2=ifelse(d$promotion==2,1,0)
```

To run stepwise regression, type the following (pressing control-enter, or command-enter on the Mac, after each line)

```
install.packages("MASS")
library(MASS)
r=lm(sales ~ . -promotion, data=d)
r2=stepAIC(r,direction="both",k=3.84)
summary(r2)
```

You only need use the first "install" line once to install the stepwise package, and you need the "library" command on the second line every time you want to use stepwise. The third line creates the initial regression using all variables except "promotion," which has been turned into dummy variables. The fourth line runs stepwise. The last line prints out the model below:

```
Call:
lm(formula = d$sales ~ d$advertising + d$income + d$prom2)

Residuals:
Min      1Q  Median     3Q     Max
-956.58 -150.81  -12.67  308.44  624.42

Coefficients:
Estimate Std. Error t value Pr(>|t|)
(Intercept)   -1.114e+03  3.206e+02  -3.475 0.001222 **
d$advertising  7.149e+00  3.161e+00   2.262 0.029082 *
d$income       3.750e-02  9.343e-03   4.013 0.000248 ***
d$prom2        7.503e+02  1.129e+02   6.646 5.23e-08 ***
---
Signif. codes:  0 *** 0.001 ** 0.01 * 0.05 . 0.1   1

Residual standard error: 371.9 on 41 degrees of freedom
(41 observations deleted due to missingness)
Multiple R-squared:  0.7272,Adjusted R-squared:  0.7072
F-statistic: 36.43 on 3 and 41 DF,  p-value: 1.217e-11
```

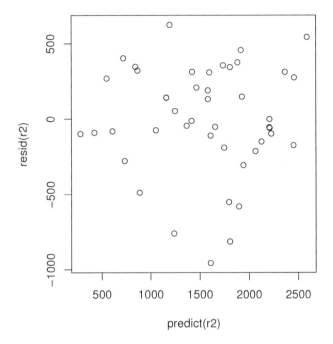

Figure 12.6. Predicted values versus residuals for the final model using R.

You can get the scatterplot in Figure 12.6 of predicted values versus residuals for the final model using

```
plot(predict(r2),resid(r2))
```

and the histogram in Figure 12.7 of residuals for the final model using

```
hist(resid(r2))
```

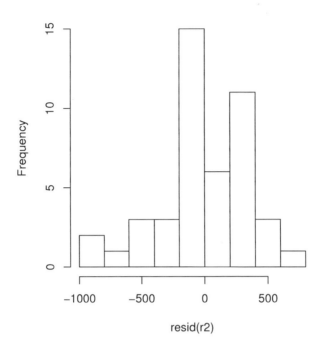

Figure 12.7. Histogram for residuals using R.

Some frequently asked questions about building a regression model

When building a multiple regression equation, is it a good idea to eliminate variables that have a very low correlation with what you're trying to forecast? Not always. It's possible for a variable to have a very low correlation with what you're trying to forecast, yet in the presence of other variables it ends up being statistically significant. For example, Figure 12.8 shows the monthly amount purchased by customers of a hypothetical store as a function of the customer's distance to the store and the customer's gender. You can see from the figure that if you don't know the gender, the distance to the store by itself doesn't tell you much about the amount purchased. But once you know the gender, additionally knowing the distance to the store now tells you a lot

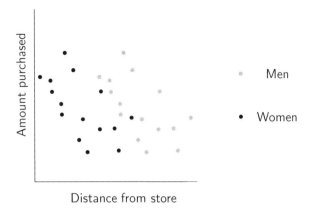

Figure 12.8. A variable with almost no correlation can become a valuable predictor in the presence of a second variable.

about the amount purchased. The stepwise algorithm attempts to consider this by assessing the benefit of a potential new predictor in the presence of other predictors.

What are the units for the standard error for the regression line? For example if the standard error equals 75, does this mean $75, or $75,000, or 75%? The units for the standard error are the same as the units for the dependent variable. The same thing is true about the residuals. If the dependent variable is in thousands of dollars, then so is the standard error as well as the residuals. The same thing is true for the standard error associated with each of the coefficients: each standard error is in the units of that particular variable.

How do you interpret coefficients for dummy variables computed from a categorical variable when not all categories are statistically significant? And do you need to include all the categories in your regression equation even if they're not statistically significant? You probably should not include the statistically insignificant dummy variables in a regression equation. To interpret the coefficients, you can imagine that all the statistically insignificant categories are grouped into a single category viewed as the baseline category. For example, if there are ten possible categories and only dummy variables for the first two categories are statistically significant, then the corresponding coefficients tell you the average difference in the dependent variable between that category and categories 3 through 10 combined – when the other sig-

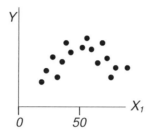

Figure 12.9. A variable with a nonlinear relation to what you're trying to forecast can be a valuable predictor if properly handled.

nificant variables remain unchanged.

If you have a categorical variable in the data set, is it reasonable to create a separate regression equation for each category? No, this is not usually the best approach. It's usually better to create dummy variables for the categories and create one big regression equation. Making separate regression equations can cause problems for several reasons. First, if you have more than one categorical variable then you would need to make separate regression equations for all possible combinations of the two categories; you can very quickly end up with dozens of regression equations. And with separate equations you typically end up with different coefficients for all the variables, and they are each based on a small sample size. The smaller sample sizes can result in inaccurate coefficients with larger standard errors, and it also makes it difficult to make comparisons across the categories. It can be helpful to use an interaction variable to compare how the coefficient for one variable changes from one category to another.

If you happen to have a variable that's not clearly linearly related with what you're trying to forecast, should you eliminate it from the regression model? No, it's possible that having it in the equation could improve forecasts. You may also be able to further improve forecasts by using interaction variables, extra dummy variables or transformations. For example, suppose you see that one of your independent variables has a graph similar to Figure 12.9. You can handle this situation by fitting a curve or using interaction terms, both covered in Chapter 7. Throwing out the variable would not be wise here, since it seems to have a strong relationship.

Is it better to use the stepwise algorithm or the backward algorithm? What about simply picking a set of significant variables using common sense? All three approaches are sensible and, unfortunately, there is no single approach guaranteed to be the best. Since the goal is to forecast the unknown, building a regression model is part science, part art and part luck. It may be reasonable to try several approaches and look to see if the predictive power or if the overall stories they tell are strikingly different. It's good advice in any statistical analysis to see if different approaches all tell the same basic story from the data. If so, that's a good sign and makes the final conclusions more compelling. If not, it's helpful to more carefully look at your data from different angles using charts and graphs to see if you can understand why there are differences between the approaches.

By searching through many different variables, is the stepwise algorithm doing data snooping? If so, doesn't this make the p-values unreliable indicators of statistical significance? Yes, the stepwise algorithm is doing data snooping and this can impact the validity of the p-values. But if there are not too many variables, the effect of this data snooping is usually not too severe. Though in this situation the p-values are not true indicators of statistical significance, they can usually still be used as a guide to help choose good predictor variables in an exploratory setting. It would not satisfy government regulators, for example, if the effectiveness of a new medication was justified using stepwise regression. When there are too many variables, stepwise regression can lead to big problems. In some genetic research, for example, researchers look for genetic indicators that could forecast the development of a particular disease. There may be millions of possible genetic indicators and therefore the data snooping effect could be severe: it's almost certain that by pure luck some genetic indicator will appear to almost perfectly predict the occurrence of the disease when tested on past data – but it will unfortunately have no predictive power on future data. With a moderate number of variables, as is usually the case in many business settings, the data snooping usually does not cause a significant problem.

How many decimal places should be quoted when referring to the results of a multiple regression? Since Excel gives so many decimal places, is it considered more precise to use as many decimal places as possible? No, quoting too many decimal places gives a false illusion of precision when in fact there is a margin of error involved. It is not fair to make estimates appear more precise than the margin of error. Quoting a forecast of \$158,375.31576 with a margin of error of \$9,405.2953 makes little sense from a managerial point of view; it would be more realistic to quote a forecast of around \$160,000 with a margin of error of around \$10,000. Excel is not thinking about being sensible when it computes the output and so it's up to you to make such judgments.

4. Data mining

The use of huge customer databases to target customers for products or services is rapidly growing and becoming an important part of doing business. Data is usually plentiful but the challenge lies in figuring out how to extract value from the data. Netflix, for example, held a contest in 2006 with a \$1 million prize to improve its movie recommendation system. It released its database of hundreds of millions of customer movie ratings and awarded the prize to the first team that could improve the accuracy of its current recommendation algorithm by more than 10%. Over 20,000 teams participated in the contest which was ultimately won in 2009 by a team from AT&T.

Another similar contest was focused on predicting which of an insurance company's customers would be most likely to want to purchase mobile home insurance for a recreational vehicle.[2] Here we will show how the data from that contest can be analyzed using multiple regression to produce almost the same level of accuracy that the winning team obtained using more sophisticated methods.

A portion of the data from 5000 customers of an insurance company is contained in the file `mobile.xls`; each row corresponds to a different customer. There is information about the demographics of the customer, the different insurance policies they currently hold, and information about the demographics in the area where the customer lives. For example, the column "household size" contains the number of people in that particular household and "car" contains the percentage of people in the area where the customer lives who own a car. The last column is the variable we want to predict – whether or not someone holds mobile home insurance. Here the number "1" represents "yes" and "0" represents "no." We will build a regression model

---

[2]Van Der Putten, P., Van Someren, M., A Bias-Variance Analysis of a Real World Learning Problem: The CoIL Challenge 2000, *Machine Learning*, 57, 177–195, 2004. `http://www.liacs.nl/~putten/library/2004vdPvSBiasVariance.pdf`

using this column as the dependent variable.

The contest rules stipulated that contestants must build a prediction model from the given data and then use it to decide which customers are the 800 most likely customers to have mobile home insurance out of data set of 4000 additional customers (that had the information about mobile home insurance omitted). Overall about 6% of the customers purchased mobile home insurance and so if we sample all the customers in this new data set we would expect to find 6% of them with mobile home insurance. If we sample half the cases we would expect to find 3% (half of 6%) and if we sampled 800 cases (1/5 of the cases) we would expect to find $6/5 = 1.2\%$. The contest-winning team was able to increase this to 3.1% by carefully selecting the 800 customers most likely to have insurance rather than just choosing them at random. Here we will show that by using multiple regression to create a forecast and then choosing the 800 households with the highest forecast we can achieve about 2.9%– quite close to the winning percentage of 3.1%.

One concern here is that the dependent variable, whether or not someone holds mobile home insurance, is represented a binary variable which can take only two possible values. Much of what we discussed above about multiple regression is based on the technical assumptions that the dependent variable is continuous and prediction errors from the regression line follow roughly a normal distribution. In the case of a binary dependent variable the prediction errors can only take on two possible values so the assumption of a normal distribution is not a good one. The reliability of p-values and t-values as well as the interpretations of the coefficients and the R^2 all are affected by this. Furthermore, the forecasts themselves may not have any real physical meaning since it may be possible to have negative forecasted values even though the dependent variable can never be negative.

Another concern is that because there are so many rows of data, uncertainty is very low, standard errors become very small and a very large number of variables can appear statistically significant even though they are not practically significant. Yet another concern is that data snooping may be a problem because there are so many independent variables to choose from. Variables that do not really have an impact may wind up appearing statistically significant just by luck and so forecasts on future data may not be as good as forecasts from the current data.

> With large data sets and a binary dependent variable the usual interpretations of a multiple regression model usually are not valid, though forecasts may still be reasonably good.

Even with all of these concerns it often turns out that ignoring the concerns and building a multiple regression equation using the stepwise algorithm discussed above can be quite successful at deciding which customers are the most likely prospects. We can't honestly use the p-values as evidence of statistical significance and the model may fit current data better then it will fit future data, but from a practical point of view it still may be useful for forecasting. Therefore we will use the above procedures only as a guide and without appealing to any theoretical justification to help us select reasonable variables. This also means we should not start telling a story about what the statistical significance of particular variables means because the indicators for statistical significance are not reliable.

With a usual multiple regression we often rely on the R^2 value or the standard error as measures of the quality of the forecasts. Since the dependent variable is binary, this does not have the same physical meaning. We will use a different approach for assessing the quality of the forecasts (lift charts) which we describe next.

A **lift chart** tells you the type of accuracy you would expect when targeting different fractions of the population. In the example above we can use the regression workbook included with this book to follow the stepwise procedure using the number of mobile home policies as the dependent variable, and we end up with about half a dozen significant independent variables out of the 30 or so in the data set. If we then go to the worksheet titled "Lift Chart" we will see Figure 12.10. To create the lift chart, Excel first creates a forecast for the dependent variable for each case in the data set based on the variables selected by the stepwise algorithm. Then cases are selected for targeting in descending order of this forecast. In other words, the cases to be targeted are the ones with the highest forecasts.

The horizontal axis on the lift chart represents the percentage of the data set that is used for targeting. The vertical axis represents the corresponding number of successes as a percentage of the total data set size. There are two lines in the chart: the straight line represents what happens if random

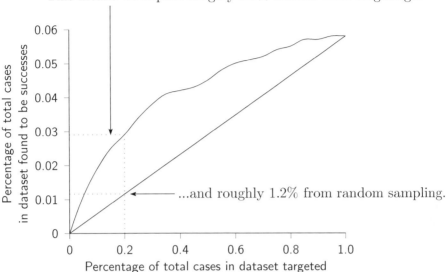

Figure 12.10. Lift chart for evaluating the performance of targeting people who are likely to purchase mobile home insurance.

sampling is used without any targeting based on forecasts and the curved line represents what happens if we use the forecasts from the multiple regression equation to target the most likely cases. For example if we look at the horizontal axis at its rightmost, corresponding to 1.0, this corresponds to 100% of the cases. We see that both of the lines in the chart intersect at 6%, meaning that if in either case (targeting or random sampling) we end up finding all 6% of the cases purchasing mobile home insurance. This doesn't tell us anything very interesting because targeting doesn't make a difference if you target the entire data set.

For the contest above we are interested in seeing what happens if we target just the top 20% most likely cases. This means we should look at the chart where the horizontal axis equals 0.2, or 20%. We see from the straight line that with random sampling we would expect to find 1.2%, which we already knew. If we look at the curved line we see that with targeting using the regression forecasts we can increase this percentage to 2.9%. This is quite close to the contest winning 3.1%.

To implement targeting in practice, one can first create a multiple regression equation using the stepwise procedure. It's a good idea to keep an eye on the lift chart as you are adding variables and it's a good idea to stop adding variables early if the lift chart looks reasonably good for the application you have in mind. Adding more variables than you really need tends to make forecasts on future data worse. Then apply this equation to new data to get a forecast for each case. Finally, decide what percentage of the cases you would like to target and choose the cases in descending order of their forecasts.

There are many other more technical approaches to make these kind of forecasts: logistic regression, discriminant analysis, regression trees, and neural networks are the names of some of these approaches. But multiple regression is still quite a powerful tool in many situations and often gives results in the same ballpark as the more technical approaches.

How large a sample do we need?   In the example above suppose that a database of 1,000 new customers is available. How many of these should be targeted in order to find a total of 40 customers who would be expected to be interested in mobile home insurance?

Discussion.   We would like to find 40 out of 1000, or 4%. Looking at the lift chart if we target 20% of the database we would expect to find 3%. And if we target 40% we would expect to find 4%. This means we should target 40% of 1000, or 400 customers to expect to find 4% of 1,000, or 40 customers.

Exercises

3. **Life insurance.** The file `insurance.xls` contains information on a large number of customers at an insurance company. Build a multiple regression equation to identify the households most likely to own life insurance.

   (a) How would you characterize the type of households most likely to own life insurance?

   (b) In a batch of 3,000 new customers, how many customers are expected to be interested in life insurance if we use random sampling to target one third of these customers?

   (c) In a batch of 3,000 new customers, how many customers are expected to be interested in life insurance if we select one third of these customers using selective targeting based on the regression equation?

4. **Charitable giving.** The file `donation.xls` contains information about households and their charitable giving. Each row represents a household. Some columns are specific to the household and some, expressed as percentages, are from census data for the people in the neighborhood surrounding the household. For example, the column "Percent employed in health services" represents the percentage of workers in the neighborhood who are employed in the health services field. The last column gives data on whether or not the household responded to a particular mailing requesting a donation.

   (a) How would you characterize the type of households that are likely to respond to the mailing?

   (b) In new batch of 10,000 households about how many are expected to respond if the mailing is sent to 2,000 randomly selected households?

   (c) If a regression equation is used to target the top 2,000 households out of 10,000 households, about how many are expected to respond?

5. **Loan applications.** The file `creditrating.xls` contains information about customers at a bank who have applied for a loan. The last column gives data on whether or not the person turned out to have a good credit rating. The bank would like to target new customers for special loan terms who are likely to have good credit, but it is not economically feasible to check the credit of all of them before sending out a mailing. A regression equation to help select people most likely to have good credit would be preferable.

   (a) How would you characterize the type of person likely to have good credit?

   (b) In a batch of 3,000 new customers, how many are expected to have good credit?

   (c) In a batch of 10,000 new customers if 3,000 are selectively targeted using the regression equation, how many are expected to have good credit?

6. **Book club.** The file `bookclub.xls` contains information about some books customers have ordered from a book club. The last column contains information about whether or not a particular customer has purchased a particular book, *The art history of Florence.*

(a) How would you characterize the type of person likely to purchase this book?

(b) In a batch of 300 new customers, how many are expected to be interested in purchasing this book?

(c) In a batch of 1,000 new customers if 300 are selectively targeted using the regression equation, how many are expected to be interested in purchasing this book?

7. **Cell phone cancellations.** A cell phone provider is concerned about losing customers to its competitors. Currently when a customer calls the retention center to cancel their account they are given special deals to entice them to stay. The provider is worried that this may not inspire long-term loyalty and would like to develop a more proactive approach for contacting customers in advance of their call. The file `cellphone.xls` contains account data from a large number of customers for the most recent month; the first column records whether or not the customer subsequently cancels their account. Develop a system that is easy to implement for identifying customers most likely to cancel their accounts. How good is the system and what managerial implications do you see?

5. Exercises

8. **Hotel profits.** The file `hoteldata.xls` (an excerpt is in Table 12.1) contains data from each hotel which is part of a franchise. The hotel's current profit margin (in percent) as well as other data about the hotel and the surrounding area is included. On the second page of the file is a description of the data variables and data on two potential new hotel sites under consideration: Site A and Site B. Which one of the two sites do you forecast would have a higher profit margin? Justify your answer.

9. **Retail sales.** The file `store.xls` (an excerpt is in Table 12.2) contains sales data for a given product from many months in a store, as well as the number of feet of shelf space devoted to the product, the retail price, advertising expenditures and whether or not a coupon was available that month. What impact on sales does the coupon seem to have?

10. **What makes a car beautiful?** An automobile manufacturer would like to know which features of an SUV make it most aesthetically appealing

Table 12.1. An excerpt from the file `hoteldata.xls`.

Hotel #	Profit Margin	Access	Age	College
1	0.58%	2	7	386
2	13.63 %	3	13	601
3	10.53 %	1	7	331
4	6.92 %	1	12	1813
5	0%	4	3	1854
6	9.49%	3	15	3896
7	19.09%	0	11	1971
8	19.22 %	2	13	95
9	11.33 %	2	13	130
10	18.29 %	4	12	108

Table 12.2. An excerpt from the file `store.xls`.

Month	Sales	Advertising	Shelf Space	Unit Price
1	58	$1,770	5	$4.90
2	409	$3,090	7	$4.73
3	258	$2,850	5	$5.16
4	384	$4,470	10	$4.77
5	225	$3,980	7	$5.37
6	401	$4,720	12	$5.37
7	593	$5,070	16	$5.78
8	547	$4,040	11	$5.09
9	241	$5,060	9	$4.75
10	71	$2,970	6	$5.33

to consumers. They use detailed photos of 39 SUV models currently on the market and ask consumers to rate the overall aesthetic appeal of each SUV on a scale from one to ten. They also ask each consumer to separately rate the aesthetic appeal of several specific features of each of the SUVs: the center console, the seats, the front view, the rear view, the color and the size. The file `suv.xls` (an excerpt is in Table 12.3) contains the average consumer ratings for each SUV model. Give some advice to the product designers on which specific features are most important in the overall aesthetic appeal of an SUV to consumers.

11. **Purchasing behavior.** The file `customerdata.xls` (an excerpt is in Table 12.4) contains information on a firm's current customers along with the

Table 12.3. An excerpt from the file `suv.xls`.

Suv Model	Front	Console	Seats	Color	Rear	Size	Overall
1	5	4.5	4.8	3.6	3	5.1	5
2	4.9	2.9	3	4.2	5.2	3.9	3.8
3	5.1	6.8	4.5	4.9	4.5	6	5.3
4	4.8	5.2	3.5	4.3	4.1	4.5	4.2
5	5.1	3.1	6.3	6	6.6	4	5.6
6	4.1	3.9	6.1	7.4	5.7	6.6	4.5
7	3.9	3.3	5.7	6.4	5.2	6	6.2
8	2	3.5	4.9	3.6	3	2.9	2.6
9	3.1	7	5.8	4.8	3.4	3	3.9
10	5.9	3	4.5	5.1	4	4	3.7

Table 12.4. An excerpt from the file `customerdata.xls`.

Customer #	Purchases	Distance (miles)	Gender	Income
1	$3,681	7.2	F	$83,931
2	$9,950	15.4	F	$127,998
3	$6,091	2.6	F	$72,127
4	$8,834	16.6	M	$81,437
5	$404	2	F	$88,913
6	$10,090	16	F	$135,822
7	$9,260	20.2	M	$126,091
8	$7,842	9.4	M	$155,241
9	$2,372	9	M	$91,819
10	$3,357	13.2	F	$48,984

total amount of purchases each customer has made. Describe the type of customer that spends a lot.

12. **Employee performance.** A large bank would like to know which factors affect employee performance. The file `bank.xls` (an excerpt is in Table 12.5) shows data on employees and their performance evaluation ratings. Which factors are linked to performance? What other conclusions can you draw from this data?

13. **Client satisfaction.** A manufacturer conducts a survey of the many client firms that purchase its products. The client firms are asked to rate the quality of various dimensions of the manufacturer (such as de-

Table 12.5. An excerpt from the file `bank.xls`.

Employee #	Performance eval (100=best, 0=worst)	Years on the job	Hours worked per week	Base pay
1	39	6	78	44000
2	8	4	45	75600
3	18	7	68	68900
4	52	2	67	13900
5	3	3	76	18700
6	35	5	52	58600
7	45	4	60	50900
8	47	4	29	20300
9	19	6	18	25200
10	95	7	98	42900

livery speed, its website, its sales force, etc.) on a scale from 1 to 7 where 1 corresponds to poor and seven corresponds to excellent. There are also questions about overall satisfaction, the likelihood of continuing to purchase from this manufacturer, as well as other questions about the client firm. The results of the survey are in the file `customersurvey.xls`. What advice can you give to the manufacturer about what's important to its customers?

14. **Customer satisfaction.** Market researchers are trying to determine what determines consumer satisfaction with household cleaning products. A large number of people are asked in a survey to think of a household cleaning product they frequently use and then rate the reputation of the brand, the ease of use, the price, the scent, as well as overall satisfaction the product. The data is in the file `cleaningsurvey.xls` and responses are on a scale from 1 to 5, where 5 corresponds to excellent and 1 corresponds to poor (an excerpt of the data is shown in Table 12.6).

(a) Which of the above seem to have the strongest link with overall satisfaction?

(b) Are the technical assumptions for multiple regression well satisfied here?

15. **Meaning, culture and resources at work.** A management consulting firm advises client companies on how to better engage and retain em-

Table 12.6. From the file `cleaningsurvey.xls`.

Reputation of brand	Ease of use	Price	Scent	Overall satisfaction
4	4	4	3	4
4	4	3	3	4
3	4	2	2	3
2	2	2	2	3
2	3	3	2	3
4	4	3	3	3
4	4	3	3	4
4	4	3	3	4
4	4	3	4	4
3	3	3	3	3

ployees. As part of the consulting firm's research, a large number of working people are surveyed about their jobs. The survey contains questions about how engaged the person feels at work and their intention to stay at their job, and also contains questions about the following three dimensions: how meaningful their job is to them (referred to as "meaning"), the culture of trust and respect at their job (referred to as "culture"), and the resources they have available to do their job (referred to as "resources"). The first worksheet in the file `engagementsurvey.xls` contains responses to the survey (an excerpt is in Table 12.7). Each number in the file is the average of a large number of related questions, and a few representative questions in each category are shown as comments (which can be seen if you move the mouse over each column header). The numbers range from 1 to 5, where 5 corresponds to the best possible and 1 to the worst possible.

(a) What can you say about the relative importance of meaning, culture, and resources for engaging employees?

(b) What can you say about the relative importance of meaning, culture, and resources for retaining employees?

(c) There are three other worksheets in the file that contain responses to questions that may be linked to each of the three dimensions described above. Which seem to be have the strongest links with culture, meaning, and resources?

(d) Briefly discuss the managerial relevance to the consulting firm's

Table 12.7. From the file `engagementsurvey.xls`.

Person #	Meaningful work	Culture of trust	Organizational resources	Engagement	Intention to stay
1	3.5	3.6	3.0	4.4	4.7
2	3.1	2.9	3.1	3.4	3.0
3	3.9	3.7	3.6	4.3	3.7
4	3.8	2.8	3.5	5.0	3.7
5	4.2	3.8	4.1	4.9	4.7
6	4.1	4.1	3.5	4.3	4.0
7	3.4	2.8	1.7	4.6	3.0
8	3.4	3.7	3.6	3.7	2.3
9	1.7	2.8	2.7	2.3	1.0
10	4.4	4.3	2.4	5.0	3.3

Table 12.8. From the file `delivery.xls`.

Trip #	Minutes	Miles	Delivery Stops	Containers Delivered	Containers Picked Up	Driver
1	80	6.3	2	2	2	GC
2	73	5.8	2	4	2	RH
3	78	6.2	3	3	6	KL
4	115	6.7	4	9	2	RS
5	116	8.6	4	2	1	RR
6	89	6.1	5	3	4	EU
7	115	7.7	5	6	8	GC
8	127	7.2	3	7	1	EU
9	137	8.1	5	9	0	RH
10	107	7.9	4	3	9	RS

clients.

16. **Evaluating drivers.** A delivery company would like to know which of its drivers are the fastest and which are the slowest. Also, it would like an easy way to forecast delivery trip times from the trip mileage, the number of delivery stops, the number of containers delivered and the number of containers picked up. Historical data from two months, including the initials of the drivers, are in the file `delivery.xls` (an excerpt of the data is shown in Table 12.8).

Table 12.9. From the file `flow.xls`.

Layer 1 Batch	Layer 1 Color	Layer 2 Batch	Layer 2 Color	Temp (C)	Fold (in)	Time (sec)	Perf
3	Black	4	Black	20	1.0	40	60
4	Black	6	Red	30	0.5	30	63
3	Red	6	Red	30	0.5	30	43
2	Black	5	Red	30	1.0	40	43
3	Red	3	Black	20	0.5	30	78
1	Black	5	Red	20	1.0	40	63
3	Black	2	Black	20	1.0	40	58
3	Red	6	Red	30	0.5	30	55
4	Red	6	Red	20	1.0	40	20
3	Red	6	Red	30	0.5	30	43

(a) Give a method to forecast trip times and use it to give a forecast for an 8 mile trip with two stops, four containers delivered and four containers picked up. How accurate do you expect this forecast to be?

(b) Which of the drivers seem to be the fastest? The slowest?

17. **Filter quality.** A company manufacturing filter cartridges would like to know which manufacturing conditions (temperature, fold size, drying time, and color) yield a product with the best performance. The cartridges are comprised of two layers of filtering material that is produced in batches. The manufacturer is also concerned that there were significant mysterious performance differences across the batches that can't be explained by the different manufacturing conditions. Data from a number of different production runs is given in the file `flow.xls` (an excerpt is in Table 12.9); higher numbers in the last column indicate better performance. What advice do you have for the company on the best manufacturing conditions?

18. **Salary structure.** Investors considering the purchase of a large corporation would like to better understand the current salary structure of its managerial employees. The file `salarystructure.xls` (an excerpt is in Table 12.10) contains the job title, education level, years of experience, and location for its employees. How can you summarize the salary structure?

Table 12.10. From the file `salarystructure.xls`.

Title	Education	Experience	State	Salary
Director	BS	7	NY	$ 117,000
VP	BS	20	NJ	$ 128,100
VP	PhD	2	FL	$ 105,900
First VP	PhD	21	NY	$ 147,600
Senior VP	BS	9	NY	$ 137,300
VP	BS	11	CA	$ 114,500
VP	BS	32	NY	$ 123,900
First VP	BS	4	NY	$ 119,900
VP	BS	4	NJ	$ 104,100
VP	BS	2	CA	$ 101,300

Table 12.11. From the file `performance.xls`.

Performance (mm/sec)	Temperature (degrees F)	Weight (grams)
2236	184	485
2265	178	445
2322	144	358
2114	100	103
2448	182	65
2313	152	342
2275	166	393
2341	139	389
2290	151	409
2272	116	236

19. **Research and development.** A company engaged in research and development manufactures a product under different conditions to learn how to improve its performance. The file `performance.xls` (an excerpt is in Table 12.11) shows the product performance under different product weight and manufacturing temperature conditions. How well can the product performance be predicted using its weight and manufacturing temperature? Which conditions improve performance?

20. The file `xandy.xls` (an excerpt is in Table 12.12) contains data on some dependent variable Y along with three independent variables $X_1, X_2,$

Table 12.12. From the file `xandy.xls`.

Y	X_1	X_2	X_3
308	91	205	16
303	88	198	-51
378	127	287	170
226	54	122	88
281	96	216	88

and X_3. Create a multiple regression equation to predict Y using only X_1 and X_2 and another equation to predict Y using X_1, X_2, and X_3. What can explain why the coefficients are so different between the two models?

21. **Business school rankings.** The file `businessweek.xls` contains data on the top ranked undergraduate business programs by *Business Week* in 2010. The last column contains an index computed from the other data columns, and then the schools are ranked according to this index.

    (a) What factors seem to impact this index?

    (b) Are student surveys weighted more heavily compared with recruiter surveys?

    (c) Approximately how many positions in the ranking would Boston University expect to move if it were to increase its average salary by $10,000 and its student survey results by 10 points?

22. **Fighting employee flight.** A large bank would like to know how to reduce employee turnover. The file `retention.xls` contains data on a large number of its employees as well as the market conditions during the employee's tenure along with information about the bank branch. Whether or not the employee was retained for the full year is indicated in the last column. What factors appear to impact employee turnover the most? What managerial implications do these have?[3]

23. **Well-being survey.** The city of Somerville, MA became the first city in the United States to conduct a well-being survey of its residents.[4] The

---

[3]Based on "How Fleet Bank Fought Employee Flight," by H. R. Nalbantian and A. Szostak, *Harvard Business Review*, April 2004, pages 116–126.

[4]How Happy Are You? A Census Wants to Know, by John Tierney, *The New York Times*, April 30, 2011. http://www.nytimes.com/2011/05/01/us/01happiness.html?pagewanted=all

survey included general happiness questions such as "On a scale of 1 to 10, how happy do you feel right now?" as well as questions about the city such as "Taking everything into account, how satisfied are you with Somerville as a place to live?" The file `Somerville.xlsx` contains the responses of residents who completed the full survey as well as a brief explanation of the variables gathered. What seems to impact satisfaction with the city of Somerville? What other stories do you see in the data?

24. The file `vshape.xlsx` contains data on three variables X_1, X_2, and Y. Use multiple regression to build a model to predict Y from X_1 and X_2. (Hint: consider using an interaction term)

6. Case Study: Commercial Real Estate Leasing[5]

The vice president of a rapidly growing information systems company has given you the task to analyze the commercial rental market for relocation to a Midwestern city in the US. At your first meeting, the problem is described as follows:

> We're planning to relocate to the Midwest, and we're committed to one particular city. We've gotten information on office rental costs for 225 new rentals in that city during the last two years. The market there is pretty flat, so the numbers should reflect the present situation. We're at the wish list stage right now, and we are not yet in a hurry. We can wait for a bargain, but how do we know what is a bargain in that market? We know lots about computer networks, but almost nothing about commercial real estate. Obviously, square footage is going to be the primary driver of rental cost, but we don't know how to price other things. For example, how much can we save by going to a longer lease? How does location and age of the building factor into the cost? Clearly we cannot pinpoint costs down to the last dollar, but there is information in the data. Your task is to provide some cost ranges on the various features of the space that we may want to lease. Our rental needs are 50,000 square feet

---

[5]This case was created by Professor Abba Krieger at the University of Pennsylvania and reprinted here with his permission.

of office space in good repair and with modern wiring. We tend towards a short term lease as a precaution because we may not want to lock ourselves into an unknown facility and area from the start. Of course we would prefer a desirable location, but we don't know what the financial trade-offs are. We hope your analysis will allow us to play with scenarios and to judge the asking rent of new properties that come on the market. We would like to know a bargain when we see one.

Using the data contained in the file `leasing.xls`, build a regression model and convey what the vice-president needs to know in concise language and provide pertinent advice. Some additional questions to cover are below.

Additional Questions

- Are there big differences in rent per square foot between city center, new and old suburb?
- How does the length of the lease affect the rent? Is the relationship linear, or are there "diminishing returns?"
- What is the cost of an executive parking space?

7. Case Study: Locating New Pam and Susan's Stores[7]

Pam and Susan's is a chain of discount department stores. There are currently 250 stores, mostly located throughout the South. Expansion has been incremental, growing from its Southern base into the Border States and increasingly into the Southwest. Identification of the most appropriate sites for new stores is becoming an issue of increasing strategic importance.

Store location decisions are based upon estimates of sales potential. The traditional process leading to estimates of sales potential starts with demographic analyses, site visits and studies by the company's real estate experts (augmented by input from local experts). The demographic data judged relevant for a given store location is for people within a store's estimated

---

[7]This case was created by Professors Paul Berger and Michael Shwartz of Boston University School of Management.

"trading zone," usually defined as consisting of those census tracts within a 15 minute drive of the store. Planners in the real estate department consider current and expected future competition, ease of highway access, costs of the site, planned square footage of the store and estimates of average sales per square foot based on data from all existing stores. They Use their judgment to combine demographic information, site information and overall sales rates to come up with an estimate of sales for a new store. Pam and Susan stores have primarily targeted lower middle class to poorer neighborhoods/trading zones.

Increasingly, actual store sales at new locations have deviated from estimates provided by the real estate department. There is therefore interest in developing better methods for estimating sales potential. A group within the planning department had previously developed a subjective approach in which potential sites are classified according to an assessment of the "competitive type" of the trading zone. Below in Table A, the seven "competitive types" are defined. Each of the existing stores was assigned to one of the competitive type categories based on visits by members of the planning department and an assessment of store sales. There is concern about the subjectivity of the "competitive type" classifications and the difficulties that might be faced using this approach to predict sales at new sites. You have therefore been hired as a consultant to explore the possibility of using the wealth of census data in stores' trading zones, along with data on individual stores, to predict sales potential.

To explore this option, variables derived from the most recent census were compiled for the trading zone of each of the 250 stores (there is no overlap in the trading zones of the 250 stores). For each store there is data gathered on demographics and economics of the trading zones, as well as size, composition and sales of the store. This data is in the file `pamsue.xls` (an excerpt is in Table 12.13).

Questions

1. As a possible alternative to the subjective "competitive type" classifications, how well can you forecast sales using the demographic variables (along with the store size and the percentage of hard goods)? What does your model reveal about the nature of location sites that are likely to have higher sales?

2. How good is the "competitive type" classification method (along with using the store size and the percentage of hard goods) at predicting sales?

Table 12.13. An excerpt from the file `pamsue.xls`.

store #	%black	%spanishsp	%inc0-10	%inc10-14
1	9.6	0.9	7.1	6.6
2	0.7	0.6	8.7	5.2
3	1.8	1.1	6.5	4
4	1.7	1.1	11.2	9.3
5	18.5	7.3	25.3	8.8
6	4.5	0.2	6.1	3.8
7	10.9	0.4	8.4	7.9
8	17.7	2.2	18.9	8.7
9	4.6	1.5	12.3	6.9
10	5.8	0.4	6.7	4.2

What recommendations do you have for simplifying the competitive type categories?

3. Two sites, A and B, are currently under consideration for the next new store opening. Characteristics of the two sites are provided below in Table B. Which site would you recommend? What sales forecasting approach would you recommend?

4. Two of the variables in the data base are under managerial control: the size of the store (square feet of selling area) and the percentage hard goods stocked in the store. Margins on hard goods (house wares, appliances, stationery, drugs) are different from margins on soft goods (clothing, for example). What impact do these variables have on sales?

5. TECHNICAL: For your recommended model, check to make sure the technical assumptions are satisfied. Comment on any points that would concern you based on the diagnostics.

Table A: Competitive Types

Type 1: Densely populated areas with relatively little direct competition

Type 2: High income areas with little direct competition

Type 3: Locations near major shopping centers

Type 4: Stores in downtown areas of suburbs

Type 5: Stores with competition from discounters, but not from department stores

Type 6: Stores in shopping centers

Type 7: Stores located along the sides of major roads

Table B: Proposed New Site Locations

	Site A	**Site B**
Store size:		
gross square feet	170,000	160,000
selling square feet	125,000	120,000
Competitive group	1	5
Population		
black	40.0%	13.8%
spanish speaking	10.8%	6.6%
Family income (in $1,000s)		
0–10	26.6%	19.2%
10$^+$–14	14.0%	13.0%
14$^+$–20	19.9%	22.2%
20$^+$–30	23.9%	27.1%
30$^+$–50	13.3%	15.7%
50$^+$–100	2.0%	2.5%
>100	0.3%	0.3%
Median yearly income	$16,838	$18,802
Median rent per month	$160	$166
Median home value	$46,790	$36,058
% homeowners	10.1	10.7
% with no cars	57.0	44.0
% with 1 car	36.6	45.7
% TV	90.0	93.6
% washer	41.8	53.6
% with dryer	9.0	12.2
% with dishwasher	6.0	4.6
% air conditioner	17.9	39.3
% freezer	6.1	5.0
% second home	1.6	4.6
Education (in years):		
0–8	37.4%	40.1%
9–11	24.1%	23.5%
12	29.0%	25.5%
12 plus	3.9%	5.7%

Total population	955,000	431,285
Average family size	3.7	3.5

84612443R00267